Smart Innovation, Systems and Technologies

Volume 94

Series editors

Robert James Howlett, Bournemouth University and KES International,
Shoreham-by-sea, UK
e-mail: rjhowlett@kesinternational.org

Lakhmi C. Jain, University of Canberra, Canberra, Australia;
Bournemouth University, UK;
KES International, UK
e-mails: jainlc2002@yahoo.co.uk; Lakhmi.Jain@canberra.edu.au

The Smart Innovation, Systems and Technologies book series encompasses the topics of knowledge, intelligence, innovation and sustainability. The aim of the series is to make available a platform for the publication of books on all aspects of single and multi-disciplinary research on these themes in order to make the latest results available in a readily-accessible form. Volumes on interdisciplinary research combining two or more of these areas is particularly sought.

The series covers systems and paradigms that employ knowledge and intelligence in a broad sense. Its scope is systems having embedded knowledge and intelligence, which may be applied to the solution of world problems in industry, the environment and the community. It also focusses on the knowledge-transfer methodologies and innovation strategies employed to make this happen effectively. The combination of intelligent systems tools and a broad range of applications introduces a need for a synergy of disciplines from science, technology, business and the humanities. The series will include conference proceedings, edited collections, monographs, handbooks, reference books, and other relevant types of book in areas of science and technology where smart systems and technologies can offer innovative solutions.

High quality content is an essential feature for all book proposals accepted for the series. It is expected that editors of all accepted volumes will ensure that contributions are subjected to an appropriate level of reviewing process and adhere to KES quality principles.

More information about this series at http://www.springer.com/series/8767

Álvaro Rocha · Teresa Guarda
Editors

Developments and Advances in Defense and Security

Proceedings of the Multidisciplinary International Conference of Research Applied to Defense and Security (MICRADS 2018)

 Springer

Editors
Álvaro Rocha
University of Coimbra, DEI/FCT
Coimbra
Portugal

Teresa Guarda
Escuela Politécnica del Ejército
La Libertad
Ecuador

ISSN 2190-3018   ISSN 2190-3026 (electronic)
Smart Innovation, Systems and Technologies
ISBN 978-3-030-08742-5   ISBN 978-3-319-78605-6 (eBook)
https://doi.org/10.1007/978-3-319-78605-6

Printed on acid-free paper

This Springer imprint is published by the registered company Springer International Publishing AG part of Springer Nature
The registered company address is: Gewerbestrasse 11, 6330 Cham, Switzerland

Preface

This book contains a selection of papers accepted for presentation and discussion at The 2018 Multidisciplinary International Conference of Research Applied to Defense and Security (MICRADS'18). This Conference had the support of the Ecuadorian Naval Academy "Cmdte Rafael Morán Valverde," University of Armed Forces of the Ecuador, and AISTI (Iberian Association for Information Systems and Technologies). It took place at Salinas, Península de Santa Elena, Ecuador, on April 18–20, 2018.

The 2018 Multidisciplinary International Conference of Research Applied to Defense and Security (MICRADS'18) is an international forum for researchers and practitioners to present and discuss the most recent innovations, trends, results, experiences, and concerns in the several perspectives of Defense and Security.

The Program Committee of MICRADS'18 was composed of a multidisciplinary group of more than 100 experts from 35 countries around the world and those who are intimately concerned with Research Applied to Defense and Security. They have had the responsibility for evaluating, in a "double-blind review" process, the papers received for each of the main themes proposed for the Conference: (A) Information and Communication Technology in Education; (B) Computer vision in military applications; (C) Engineering Analysis and Signal Processing; (D) Cybersecurity and Cyberdefense; (E) Maritime Security and Safety; (F) Strategy, geopolitics, and Oceanopolitics; (G) Defense planning; (H) Leadership (e-leadership); (I) Defense Economics; (J) Defense Logistics; (K) Health informatics in military applications; (L) Simulation in Military Application; (M) Computer Networks, Mobility and Pervasive Systems; (N) Military Marketing; (O) Military Physical Training; (P) Assistive Devices and Wearable Technology; (Q) Naval and Military Engineering; (R) Weapons and Combat Systems; and (S) Operational Oceanography.

MICRADS'18 received about 100 contributions from 11 countries around the world. The papers accepted for presentation and discussion at the Conference are published by Springer (this book) and by AISTI and will be submitted for indexing by ISI, Ei Compendex, SCOPUS, and/or Google Scholar, among others.

We acknowledge all of those that contributed to the staging of MICRADS'18 (authors, committees, workshop organizers, and sponsors). We deeply appreciate their involvement and support that was crucial for the success of MICRADS'18.

April 2018   Álvaro Rocha
                                                                                     Teresa Guarda

Organization

Honorary Chair

Álvaro Rocha   University of Coimbra, Portugal

Honorary Co-chair

Lakhmi C. Jain   University of Canberra, Australia

General Chairs

Oscar Barrionuevo Vaca   Ecuadorian Naval Academy, Ecuador
José Avelino Moreira Victor   University Institute of Maia, Portugal

Local Organizing Chair

Darwin Manolo Paredes   ESPE - University of the Armed Forces, Ecuador
  Calderon

Scientific Committee

Teresa Guarda (Chair)   ESPE - University of the Armed Forces, Ecuador
Abderrazak Sebaa   University of Bejaia, Algeria
Abdoulaye Sidibe   Wuhan University of Technology, China
Alessandra Pieroni   Università degli Studi Guglielmo Marconi, Italy
Alex Fernando Jimenez Velez   Fuerza Aérea Ecuatoriana (FAE), Ecuador
Amita Nandal   UIST - University for Information Science
                              and Technology "St. Paul The Apostle,"
                              Ohrid, Macedonia
Ana-Maria Cretu   Carleton University, Canada

André Koscianski   UTFPR, Federal University of Technology - Paraná, Brazil

Andrea D'Ambrogio   University of Rome Tor Vergata, Italy

Andrea Visconti   Università di Milano, Italy

Angelo Borzino   Brazilian Army Technological Center, Brazil

Angelo Brayner   Federal University of Ceará, Brazil

António Abreu   ISCAP/IPP, Portugal

Arjun Singh Saud   Tribhuvan University, Nepal

Arlindo Veiga   Universidade de Cabo Verde, Cape Verde

Asanka Pallewatta   University of Kelaniya, Sri Lanka

Bilan Zhu   Tokyo University of Agriculture and Technology, Japan

Bruno de Pinho Silveira   ESPE - University of the Armed Forces, Ecuador

Calogero Vetro   University of Palermo, Italy

Chandra Kishore Somasundaram   University of Information Science and Technology "St. Paul the Apostle," Macedonia

Chiara Braghin   Università degli Studi di Milano, Italy

Chih-Hsien Huang   Imec, Belgium

Chinenye Ajibo   University of Nigeria Nsukka, Nigeria

Claudia Jacy Barenco Abbas   Universidade de Brasília, Brazil

Damir Blazevic   Faculty of Electrical Engineering, Computer Science and Information Technology Osijek, Croatia

Daniela Suzuki   Institute of Biomedical Engineering (IEB-UFSC), Brazil

David Fernandes Cruz Moura   Centro Tecnológico del Ejército de Brasil, Brazil

Derya Yiltas-Kaplan   Istanbul University, Turkey

Diana Patricia Arias Henao   Universidad Militar Nueva Granada, Colombia

Diego Paes de Andrade Peña   Universidade Federal do Maranhão, Brazil

Dimitrios Dalaklis   World Maritime University (WMU), Sweden

Domenica Costantino   Politecnico di Bari, Italy

Edison Gonzalo Espinosa Gallardo   ESPE - University of the Armed Forces, Ecuador

Elisa Francomano   University of Palermo, Italy

Emanuele Bellini   University of Florence, Italy

Enrique Carrera   ESPE - University of the Armed Forces, Ecuador

Eugen Rusu   Dunarea de Jos University of Galati, Romania

Eugénio Almeida   Instituto Universitário de Lisboa (ISCTE-IUL), Center for International Studies, Portugal

Fabian Ramirez Cabrales   Colombian Naval Academy "Almirante Padilla," Colombia

Fahimina Taranum   Muffakham Jah College of Engineering and Technology, India

Felipe Torres Leite   Universidade do Estado do Rio Grande do Norte - UERN, Brazil

Juan Jorge — Universitat Politècnica de Catalunya (UPC), Spain

Jurij Mihelič — University of Ljubljana, Slovenia

Kalinka Kaloyanova — Sofia University, Bulgaria

Kátia Regina de Souza — Military Institute of Engineering, Brazil

Kleber Cavalcanti Cabral — Universidade Federal Rural do Semi-Árido, Brazil

Luigi Palizzolo — University of Palermo, Italy

Luis Alvarez Sabucedo — University of Vigo, Spain

Luis Carral Couce — Universidad de La Coruña, Spain

Luis Dieulefait — Universitat de Barcelona, Spain

Luis Eduardo Palacios Aguirre — Fuerza Aerea Ecuatoriana, Ecuador

Luis Anido Rifón — Universidade de Vigo, Spain

Luiz Goncalves Junior — Sao Paulo State University - Unesp, Brazil

Manuel Francisco González Penedo — CITIC, Centro de Investigación CITIC, UDC, Spain

Manuel Tupia — Pontificia Universidad Catolica del Perú, Perú

Manuel Vilares Ferro — University of Vigo, Spain

Marcello Vanali — Università di Parma, Italy

Marcelo Henrique Prado da Silva — Instituto Militar de Engenharia - IME, Brazil

Marcelo Carneiro dos Santos — Instituto Militar de Engenharia - IME, Brazil

Marcos Barreto — Universidade Federal da Bahia, Brazil

Marcos Alexandruk — Universidade Nove de Julho - UNINOVE, Brazil

Maria Augusta Fernández Moreno — Universidad Católica de Ibarra, Ecuador

María Carolina Romero Lares — World Maritime University, Sweden

Maria João Marques Martins — Academia Militar de Portugal, Portugal

Maria Manuela Martins Saraiva Sarmento Coelho — Military University Institute, Portugal

Marino Belloni — Dept. Physics, Mathematics and Informatics, Italy

Mario Di Raimondo — University of Catania, Italy

Martín López Nores — University of Vigo, Spain

Mauricio Loachamín Valencia — Universidad de las Fuerzas Armadas - ESPE, Ecuador

Maurizio Migliaccio — Università degli Studi di Napoli Parthenope, Italy

Maximo Jr. Q. Mejia — World Maritime University, Sweden

Milton Itzhak Littuma — Ecuadorian Air Force, Ecuador

Mohamed Abouelela — King Saud University, Saudi Arabia

Mohammad AlRousan — Jordan Univ. of Science & Technology, Jordan

Mohammad Pasha — Muffakham Jah College of Engineering and Technology, India

Mohammed Mahmood Ali — Muffakham Jah College of Engineering and Technology, (Osmania University), India

Muhammed Ali Aydin	Istanbul University, Turkey
Nina Figueira	Brazilian Army, Brazil
Ninoslav Marina	UIST - University for Information Science and Technology "St. Paul The Apostle," Macedonia
Pastor David Chávez Muñoz	Pontificia Universidad Catolica de Peru, Peru
Paulo Afonso Silva	Military Institute of Engineering, Brazil
Paulo Nunes	Portuguese Hydrographic Office, Portugal
Pedro Ferreira	Centre for Marine Technology and Ocean Engineering (CENTEC), Portugal
Rafael Timoteo de Sousa Junior	University of Brasilia, Brazil
Rashed Mustafa	University of Chittagong, Bangladesh
Raul Villa Caro	EXPONAV/Universidad de la Coruña, Spain
Renato Jose Sassi	Universidade Nove de Julho, Brazil
Reza Malekian	University of Pretoria, South Africa
Ricardo Choren	Military Institute of Engineering, Brazil
Robert Beeres	Netherlands Defence Academy, Netherlands
Robson Pacheco Pereira	Instituto Militar de Engenharia - IME/RJ, Brazil
Ronaldo Salles	Military Institute of Engineering - IME, Brazil
Rosalba Rodríguez Reyes	ESPE - University of the Armed Forces, Ecuador
Rui Jorge Palhoto Lucena	Military University Institute - Military Academy, Portugal
Sabrina Silveira	Universidade Federal de Viçosa, Brazil
Salif Silva	University of Cabo Verde, Cape Verde
Santiago Gonzales Sánchez	Universidad Inca Garcilaso de la Vega, Perú
Sarat Mohapatra	Centre for Marine Technology and Ocean Engineering (CENTEC), Instituto Superior Tecnico, The University of Lisbon, Portugal
Sebastião Alves Filho	Universidade do Estado do Rio Grande do Norte - UERN, Brazil
Sidnei Alves de Araújo	Universidade Nove de Julho, Brazil
Silvio Berardi	University Niccolò Cusano - Rome, Italy
Silvio Melo	Universidade Federal de Pernambuco, Brazil
Simona Safarikova	Palacky University, Dpt. of Development and Environmental Studies, Czech Republic
Simone Battistini	Universidade de Brasília, Brazil
Sonia Cárdenas Delgado	Universidad Politécnica de las Fuerzas Armadas - ESPE, Ecuador
Suzana Paula Gomes Fernando da Silva Lampreia	Portuguese Navy, Portugal
Teresa Guarda	ESPE - University of the Armed Forces, Ecuador
Thanaphong Thanasaksiri	Chiang Mai University, Thailand
Thanasis Loukopoulos	University of Thessaly, Greece

Contents

Information and Communication Technology in Education

Cybersecurity and Cyberdefense

Empirical Study of the Application of Business Intelligence (BI) in Cybersecurity Within Ecuador: A Trend Away from Reality

H. Andrea Vaca[1], Rolando P. Reyes Ch.[1,2(✉)], Hugo Pérez Vaca[1,2],
and Manolo Paredes[1]

[1] Universidad de las Fuerzas Armadas ESPE, Sangolquí, Ecuador
andrefa6@hotmail.com,
{rpreyes1,hlperez,dmparedes}@espe.edu.ec
[2] Departamento de Seguridad y Defensa, Universidad de las Fuerzas Armadas
ESPE, Sangolquí, Ecuador

Abstract. Business Intelligence (BI) over the years, has achieved a strong impact within the business world, especially when it comes to gaining knowledge for making strategic decisions. This impact has been successfully in other contexts. However, only exist some examples in Latinamerica (Costa Rica), have started to use BI for the Cybersecurity context, being a strange when several authors have considered this relationship as a global trend. Our objective is to investigate if this trend is happening in Ecuador. For which, we propose to carry out a survey to professionals in the industry and professors in the academy. The results obtained show that the majority of Ecuadorian companies, use BI only to improve their competitiveness, but without considering cybersecurity context, despite being aware of the high risk. The industry, academia and country must enter a stage of reflection of the application of BI in Cybersecurity.

Keywords: Business intelligence · BI · Cybersecurity · Survey
Trend

1 Introduction

Business Intelligence (BI) is an approach that is not new, within the global trend. Since its inception, it has been used to transform the company's data into important information to get later into knowledge [1]. However, in recent years, this focusing has managed to maintain an interesting trend where currently BI tools allow performing systematic and controlled processes for generation of knowledge in aggressive business strategies to the companies [2]. That is why; it is not uncommon to find BI in other fields of application such as: knowledge management, analytics, and cybersecurity, among others. For the particular case of BI in cybersecurity, it is still new, at least in Latin America, except for a specific case in Costa Rica [3]. In Ecuador, BI focus on cybersecurity is practically unknown, although it seems that BI's application is oriented only business part. For this reason, our objective in this empirical study is to establish the current state of the applied BI approach in cybersecurity in Ecuador, and to establish if its application is appropriate within companies and academia. We apply a

Á. Rocha and T. Guarda (Eds.): MICRADS 2018, SIST 94, pp. 3–11, 2018.
https://doi.org/10.1007/978-3-319-78605-6_1

methodology of quantitative research based on a survey, which we were done to professionals and academics. The results indicate that BI is not applied in cybersecurity and that only is being used for reports and analytics in business transactions. In Ecuador, a period of reflection is needed to improve the application of BI in cybersecurity in order to increase prevention.

The present article consists of the following: Sect. 2 describes the background that refers to the research topic. The methodology used and research questions are detailed in Sect. 3. In Sect. 4, execution and results are mentioned. The discussion and conclusions are set out in Sect. 5. Finally, Sect. 6 indicates future work.

2   Background

According to Journal & Conscience [4], Business Intelligence (BI) was used to obtain and collect business transactional information in order to create knowledge or predictions that allow executives or directors to make decisions. Obtaining this type of information required considerable time and specialized personnel. However, at present, Business Intelligence (BI) has advanced a lot in this aspect, to such an extent that now it has software tools that have allowed companies to reduce the time of processing their information to make decisions instantly. The companies have been able to generate more quickly, useful and necessary knowledge of their business, creating competitive advantages in the short and long term [5]. To explain about what BI represents, we will make a brief description of the current situation of BI in the world and its incursion in cybersecurity.

2.1   Business Intelligence Systems and Their Field of Application

According to Barrento et al. [2], a Business Intelligence (BI) system is an integral solution that includes several software tools; allow collecting the largest amount of information, any type of data such as: empty or duplicate data, in order to transform them into information readable that the user can analyze and observe through graphics, among others. The author tells us that the trend of the field of application of BI is strongly linked to:

- **Knowledge Management:** with which has allowed creating useful knowledge for the administration, learning and regulatory compliance of a company.
- **Collaboration platform:** which has allowed the facilities to obtain information from various areas and exchange them with each other based on their data.
- **Reports or reports:** This has improved the visualization of information at a dynamic level or understandable to a normal user, in particular the users who represent the top managers of an organization or Company.
- **Analytics:** to establish quantitative processes in the aid of optimal decision making. At this point is where we consider the terms of: process mining, the generation of patterns from large volumes of data [6], process mining, process management techniques through events [7] and the well-known statistical analysis, which allows investigating individual or collective data samples.

- **Measurement:** This has allowed the creation of hierarchies in performance measures. The measures are commonly used by employers as indicators to determine the status of a company.

2.2 Management of Large Volumes of Data in Cybersecurity

Cybersecurity according to ISACA [8] it is defined as a set of instruments, strategies, knowledge, norms, risk techniques, alignments, experiences and technologies that are strongly related to safeguarding the assets of the organization and users with respect to external intrusions.

In this regard, we can say that thanks to this concept by ISACA, it is reasonable to see that there is currently a large deployment of tools, policies for cybersecurity. Its reason seems to be obvious, because it is a discipline that operationally generates large volumes of information, which could even be said that there are cases where obtaining information is such, that it may have exceeded the capacity of the usual software (e.g., malware-attack). This is the main reason, for management of large volumes in cybersecurity is necessary and paramount when processing information in a reasonable time [9].

In a study conducted by Mondrag [3], tells us that the information of the companies about of cybersecurity refers mostly to cyber-attacks; among the most common we can say: denial of service, port scanning, and malware propagation, among others. What the companies have done in this regard, is to use BI tools as a strategy for the visualization of interpretive graphs of the attacks (histograms, cakes, bars) in real time. The author considers that in this aspect, BI tools are useful to generate patterns that they can be visualized instantly as a means of preventing new cyber-attacks or compare with other attacks or establish regulations and organizations' rules that endorse the proper protection.

2.3 Cybersecurity and Visual Representation of Patterns Based on Large Volumes of Data

According to Arias [10], the construction of patterns from information of the computer attacks provided by a BI tool, have helped companies and academies to create knowledge to prevent and take care of their assets, as well as to create procedures for the protection of its information. A case of this conclusion, the author refers to BI's tendency using in cybersecurity within several Costa Rica companies. The reason is simple, BI techniques have allowed them to transform the information obtained from a cyber-attack, or from a simple data set, to knowledge and prediction of future cyber-attacks [11].

2.4 Cybersecurity and Its Trend in Ecuador

According to Ekos Journal [12], Ecuador companies, in recent years, have begun to evolve their business models thanks BI use, obtaining competitive advantages, improving their added value and creating new business scenarios in real time, fully predictive, based on appropriate decisions to grow their business. This magazine predicts the possibility that companies, over the years, improve BI's concepts within their businesses to match the global trend. However, while commercial experts find themselves boasting about BI use in their businesses, Vargas et al. [13], warn that in recent years, Ecuadorian companies have had cyber-attacks on their assets, being especially alarming in the banking sector due to the few and limited measures that the Ecuadorian government is taking in this regard.

According to what has been found in the literature, we can deduce that BI is a trend which is growing up in the world, especially in business issues, when obtaining competitive strategies. In Ecuador, BI trend has not been the exception. However, the world has started to use the BI approach in several aspects of cybersecurity, a situation that is possibly a distant and unknown issue in Ecuador. So, we ask ourselves, if the BI approach applied in cybersecurity in Ecuador, is the right one within the companies and the academy, considering the current trend. For which, we have raised the following research questions:

RQ1: Is the BI approach and its tools being used as part of the research techniques for cybersecurity within companies and academia in Ecuador?
RQ2: Which is the recent status of BI application, and how are they being associated in cybersecurity in Ecuador?

3   Research Methodology

The research questions previously mentioned raise a survey as a research strategy (quantitative). The survey elaboration and application, we consider the recommendations of Monterrey [14], who establishes 7 stages for the application of the survey, in which we used all the stages of Monterrey [14], because it fits our research. To detail in the Fig. 1, it shows the stages of the research methodology.

Fig. 1. Research methodology

In this regard, the stage of *defining the objective* is oriented to what the researcher wishes to investigate in his research questions. The research questions were defined in Sect. 2. The next stage refers to *selecting the respondents*, where we select our respondents from different companies and academies in Ecuador at random way. The selection of company respondents is aimed at people working in systems departments of public and private companies in Quito city, and with respect to respondents from the academy, it was appropriate to select teachers who know issues related to cybersecurity or BI. For the stage called *determining the instrument* refers to practically survey preparation with 7 questions, where 1, 2 and 3 respond to our RQ1 and questions 4, 5, 6 and 7 respond to our RQ2. The *survey execution* is the application of the survey to the selected respondents. The *processing and analysis of the results* is done with Google tools, and finally the dissemination of the results that is done in this article. The results of the execution of the methodology are detailed in the following section.

4 Execution and Results

4.1 Execution

As mentioned above, the selection of respondents was oriented to public companies that have considerably large computer infrastructure, a situation that also we take into account regarding the academy. The survey was sent by email to a total of 50 respondents, of which 30 are distributed to professionals who belong to private companies, 16 to professionals from public companies and 4 surveys sent to professors of the Academy. The survey was applied during 2 months at the end of the winter of 2016. In total, 47 respondents answered, representing a 94% response rate, considered acceptable in this type of studies.

4.2 Results

RQ1: Is the BI approach and its tools being used as part of the research techniques for cybersecurity within companies and academia in Ecuador?

To answer this research question, it is necessary to initially inquire with information regarding the "use of research techniques and BI tools in the company/academy", where it can be observed (Fig. 2) that 56.5% of respondents affirmatively use research tools and business intelligence solutions, 39.1% of respondents barely know some references and 8.7% have total ignorance of the subject. In addition to the results of this question, we also consider asking the respondents (who answered affirmatively to know BI) if the tools they use are free, proprietary or self-developed (Fig. 3). Results indicate that 50% of respondents use free software tools, 39.1% prefer proprietary software and few respondents develop their own tools (6.5%).

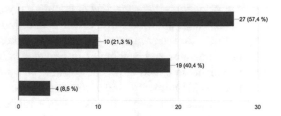

Fig. 2. Use of research techniques and BI tools in companies/academy

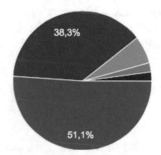

Fig. 3. If you use BI tools, what characteristics are they?

We inquired by asking respondents if they had any difficulty acquiring BI tools. Figure 4 shows that 58.7% consider difficulty in acquiring BI tools due is in its high cost, 43.5% consider not knowing the advantages of the tools to acquire those, 34.8% think that the lack of training can influence, and 15.2% consider the difficulty of finding support.

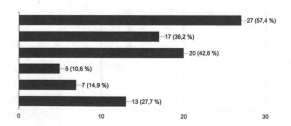

Fig. 4. What do you think would be the main difficulty in using BI tools in your business?

RQ2: What is the status of BI application, and how are they being associated in cybersecurity in Ecuador?

To answer this research question, we considered asking the respondents if they use their Business Intelligence (BI) tools for any particular reason (e.g., fraud, statistics, etc.). Figure 5 shows that 39.1% of respondents mention that they use BI tools to find business characteristics that improve their competitive strategy, 32.6% of respondents

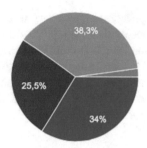

Fig. 5. You think that BI tools or solutions are a great help for?

used for the detection and prevention of computer frauds, 26.1% very closely appears of respondents that used in statistical and economic analysis and only 2.2% for other purposes.

We are very interested in who responded about using BI tools to detect fraud. We wanted to ask, if the company or academy create internal policies for the use of BI tools. Figure 6 shows that 52.2% of companies/academia have not elaborated policies for the use of BI tools, 32.6% have elaborated policies but only related aspects of the business turnaround and only 26.1% have tried to perform internal policies but related to cybersecurity.

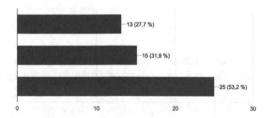

Fig. 6. Where do you work, have you developed internal policies for the use of BI or cybersecurity tools?

With this information we can complement our inquiry, when we ask the respondents, if there is an effective application of their policies. Figure 7 shows that 63% believe that effective training and dissemination is necessary for use, 52.2% think that effective application believes that a national cybersecurity policy is necessary, while

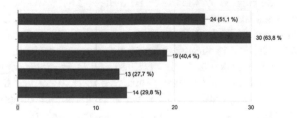

Fig. 7. What do you think is necessary for a proper application of BI tools policies in cybersecurity?

41.3% think that effective application is necessary a national program of diffusion in the use of the tools or solutions of BI within the context of cybersecurity.

Finally, we asked the respondents if authorities, to which they belong, are aware of information security risks and if they know how to mitigate them (Fig. 8). The 34.8% are very aware of the risks of computer attacks on companies, 26.1% mention that they are aware, but do not know how to mitigate the risks, 21.7% are also aware, but are not interested in their safety of information, and finally 17.4% are not aware nor are they interested.

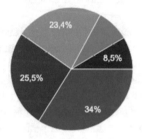

Fig. 8. The institution to which they belong, are aware of the risks in information security and how to mitigate them?

5 Discussion and Conclusions

The results obtained are really interesting as well as worrisome. Ecuadorian companies are using BI tools with the main objective of giving greater added value to their business with the perspective of improving their income. The discussion is clear, in Ecuador, research techniques and BI tools are only oriented to improve the economic companies' environment. Although, these tools, for Ecuadorian companies are extremely useful for their decision making, companies do not invest in them, but opt for the use of free software tools, where they do not pay maintenance subscriptions and updates. Regarding the perspective of using their BI tools in the context of cybersecurity, companies are not interested in investing in this aspect, despite of the fact knowing the risks and opening up possibilities of becoming potential targets of cyber-attacks. We believed, if companies did not use their BI tools in their cybersecurity they could at least have policies in that regard. But to our surprise, internal policies have not been created for the use of BI tools and even worse for cybersecurity, even though managers are aware of the risks. We believe that initiatives should be taken regarding the use of BI tools in cybersecurity since they allow in some way to prevent computer attacks. It is strange to know that the companies are waiting to earn more money while are exposed to being attacked and losing their biggest investment. For this reason, companies must take the initiative to take care of their investments establish internal policies, until the Ecuador's Government defines and presents national cybersecurity policies, Ecuador's companies and academia must enter a period of reflection in this regard.

6 Future Works

Our research leads us to several future works, among which we can mention the creation of a proposal for internal business policies in the BI environment in cybersecurity context. Additionally, the creation of national cybersecurity policies with the use of BI tools possibly created as a proposal by the Ecuadorian State. To this, we can add research concerning the creation of algorithms that obtain patterns of computer attacks using BI tools, among others.

References

1. Perspectiva, R.: Las nuevas tendencias de la inteligencia de negocios. [online] IDE Perspectiva (2018)
2. Barrento, M., Neto, M., Maria, M., Dias, S.: Sistemas de Business Intelligence Aplicados à Saúde (2010)
3. Mondrag, G., Guzm, C.A., Mart, L.: Revisión Sistemática de Literatura: Visualización de Seguridad (n.d.)
4. Calzada, L., Abreu, J.L.: El impacto de las herramientas de inteligencia de negocios en la toma de decisiones de los ejecutivos. Int. J. Good Consience 4(2), 16–52 (2009)
5. Cano, J.L.: Business Intelligence: Competir Con Información, p. 397. Banesto, Fundación Cultural (2007)
6. Riquelme, J.C., Ruiz, R., Gilbert, K.: Minería de datos: Conceptos y tendencias. Inteligencia Artificial 10(29), 11–18 (2006)
7. Pérez Jiménez, S.: Minería de procesos (2015)
8. ISACA: Cybersecurity Fundamentals Glossary (2014)
9. Jiménez, L., Hernandez, S., Mendez, J.: Enfoque Sociotécnico Aplicado a un Sistema de Gestión Business Intelligence (n.d.)
10. Arias, R., Leiva, C.: Representación visual de patrones de ataque en ciberseguridad (n.d.)
11. Laudon, S.: Sistemas Informacion (1999)
12. Ekos Revista, La inteligencia en el negocio (2016). http://www.ekosnegocios.com/negocios/. Accessed 18 Jan 2018
13. Vargas, R., Recalde, L., Reyes, R.: Ciberseguridad. Ciberdefensa Y Ciberseguridad, Más Allá Del Mundo Virtual: Modelo Ecuatoriano de Gobernanza En Ciberdefensa 20 (2017)
14. Universida Virtual del Tecnológico de Monterrey: Metodología para llevar a cabo una encuesta (2005)

Situational Status of Global Cybersecurity and Cyber Defense According to Global Indicators. Adaptation of a Model for Ecuador

Fabián Bustamante, Walter Fuertes(✉), Theofilos Tulkeredis,
and Mario Ron

Department of Computer Science, Universidad de las Fuerzas Armadas ESPE,
Av. General Rumiñahui S/N, Sangolquí, Ecuador
{lfbustamante, wmfuertes, ttoulkeridis,
mbron}@espe.edu.ec

Abstract. The aim of this study has been to establish a Cybersecurity and Cyber defense model in Ecuador, based on the comparison of the worldwide situation status. Therefore, we performed a descriptive research, which has been related to the metrics and indicators of the countries with greater relevance in such areas, which have been considered as important by the International Telecommunications Union (ITU). Subsequently, we applied an incremental methodology in order to perform the comparative analysis. The main findings demonstrate that G8 countries appear or be the most prepared and mature in Cybersecurity/Cyber defense and Cyberwar. This has been also evidenced about countries with high technological level and which historically have maintained warlike conflicts. The final exposed results according to the Cybersecurity pillars of the ITU, yielded details of the relevant points and evidenced that on a global scale the United States and France are the most outstanding countries, which has led to propose a model to follow for Ecuador, which represents a country that still allows great opportunities for growth and improvement in this topic.

Keywords: GCI · ITU · Cybersecurity · Cyber defense · ISMS
ICS

1 Introduction

In the last decade more than ever before, countries have had the need to protect their national assets, critical infrastructures and the information sovereignty. This has been the development based on the severe growth and the complexity of Cyber-attacks [1]. News about computer theft, extortion, new viruses, cyber-attacks and digital cyber wars increases daily, which directly or indirectly affect the states. Therefore, there is an apparent and urgent need to implement Cybersecurity and Cyber defense mechanisms in order to prevent threats and vulnerabilities. Such mechanisms have been developed worldwide through various standardized processes. However, in developing countries like Ecuador, there is evidence of a lack of a clear application of indices or metrics, which have been previously successfully used in more advanced countries.

© Springer International Publishing AG, part of Springer Nature 2018
Á. Rocha and T. Guarda (Eds.): MICRADS 2018, SIST 94, pp. 12–26, 2018.
https://doi.org/10.1007/978-3-319-78605-6_2

Based on the described issue, the scientific community has been concerned with identifying and efficiently applying the best strategies, standards, and metrics at a global level. For example, [2, 3] presented comparisons between countries and defined a global index of security and profiles for each country. In [4–6], they demonstrated initiatives and strategies from several countries on cyber threats. In [7–9] the authors performed a voluntary comparison of Cybersecurity frameworks at the level of some industrially developed countries. In [10, 11], the addressed an industrial comparative and a regulatory analysis for a proactive Cybersecurity. However, all these studies, with the exception of the ITU, contained information and results that have been more qualitative than quantitative, which unfortunately lacks to contribute in the solution of the main problem.

The current study focuses on determining the most appropriate Cybersecurity and Cyber defense model to be applied in Ecuador. In order to fulfill such concern, a global comparative analysis has been carried out using the ITU Global Cyber Security Index (GCI) [2]. The GCI is a joint initiative of different interest groups, determining the commitment of countries to Cybersecurity issues, as well as the ITU Cyber wellness profiles [3], which provide an overview of the countries' levels of Cybersecurity development, being aimed at the public, private and military sectors. Hereby, some eight analysis criteria have been considered as proposed by [4, 7, 10], which require the participation of the following stakeholders: (a) the ten top countries in Cybersecurity; (b) the countries that founded the North Atlantic Treaty Organization (NATO); (c) the group of most developed countries in the world, the so-called G8; (d) the ranking of the best equipped armies in the world [12]; (e) the most representative Latin American countries for Ecuador; (f) the most representative countries in America for Ecuador; (g) the Ranking of countries that most attacked in Cybersecurity in 2017 [13, 14], and (h) the ranking of countries that have been most frequently attacked in 2017 [14, 15]. With such evaluation, we were able to identify the best models of countries developed in Cybersecurity, Cyber defense, Cyberwar, and Cybercrime.

The main aim of this study consists of establishing a model to be followed in the context of Cybersecurity, both in information and communication technologies (ICT), as well as in industrial control systems (ICS) [16–18]. In this way, specialist in information security will be able to follow a clear road map and acquire reference to the presented topics, their results as well as conclusions of the present study, which subsequently will propose national policies of cyber security.

The remainder of the current study has been organized as follows: Sect. 2 presents the background of the proposed topics, while Sect. 3 explains the methodology that has been used for the comparative analysis. Section 4 describes the evaluation results and the established model, whereas Sect. 5 presents conclusions and future work.

2 Initiatives to Measure Countries' Commitment to Cybersecurity Issues

2.1 Global Cybersecurity Index (GCI)

The GCI has been launched in 2014 and its objective has been to help to foster a global culture of Cybersecurity and its integration in the core of ICTs. A second objective has been to measure the commitment of the member countries towards Cybersecurity in order to manage future efforts in the adoption of the integration of Cybersecurity on a global scale.

2.2 North Atlantic Treaty Organization (NATO)

The residence of NATO is in Brussels, Belgium. The NATO has been founded in 1949 by twelve developed countries, being the United States, Norway, France, Canada, Italy, United Kingdom, Holland, Portugal, Belgium, Iceland, Denmark and Luxembourg. At present day, the NATO is conceived as a collective military power of about 3.5 million personnel in service and annual expenditures of about $ 900 million USD. This organization constitutes a system of collective defense, in which the affiliated states agree to defend any of its members in case of an attack by an external faction. In March 2011, the defense ministers of NATO approved the concept of Cyber defense. This concept defines the protection of NATO networks as a fundamental responsibility of the allies. Among the measures adopted has been the protection of cyber-attacks towards information and communication systems of critical importance for the alliance.

2.3 Group of Eight (G8)

The G8 is the group of the eight countries with the globally most industrialized economies. It is composed of the United States, Russia, Canada, France, Italy, Germany, the United Kingdom and Japan. Their representatives of Justice have agreed to join efforts to eradicate cybercrime and, especially, child pornography.

2.4 Cybersecurity, Cyber Defense, Cyberwar, and Cybercrime

According to the ITU, Cybersecurity is a set of tools, policies, security concepts, security safeguards, guidelines, risk management approaches, actions, training, best practices, assurance and technologies that are able to be used in order to protect the cyber environment and organization as well as user's assets. Organization and user's assets include computing devices, personnel, infrastructure, applications, services, telecommunications systems, and the totality of transmitted and/or stored information in the cyber environment [2, 19].

At other side, NATO contemplates about Cyber defense as a proactive measure for detecting or obtaining information of any kind of cyber intrusion, cyber-attack, or impending cyber operation. Additionally, the same organization considers fundamental to determine the origin of an operation that involves launching a preemptive, preventive, or cyber counter-operation against the source [20].

Cyberwar or Cyber warfare has been defined by the Army of the United States and Russia as an escalated state of cyber conflict among states. Hereby, cyber-attacks have been performed by state actors against cyber infrastructure and cybercrime has been the use of cyberspace for criminal purposes, as established by national as well as international laws [21, 22].

3 Methodology to Perform the Comparative Analysis

Figure 1 illustrates the phases of the methodology used to perform the comparative analysis. It starts with the gathering of information and the application of comparison criteria based on the needs of national Cybersecurity and Cyber defense.

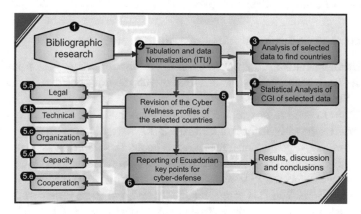

Fig. 1. Schematic illustration of the incremental methodology. (Source: author's contribution)

3.1 Bibliographic Research and Determination of Analysis Criteria

After the bibliographic and descriptive research in scarce sources of benchmarking of Cybersecurity and Cyber defense worldwide, we determined that the ITU through its reports of GCI [2] has been the most reliable resource to perform this comparative analysis. The inputs are distributed in two types of documents, being the Global Cybersecurity Index 2017 report [2] and the Cyber wellness profiles [3]. The most relevant data used for the study have been the "GCI Score", "Country" and "Total Population".

In order to further complete the evaluation, we reviewed some public benchmarking of some governments and ministries of defense [7, 10, 23], as well as Cybersecurity reports from computer security manufacturers. The primary sources of information and the comparison criteria have been Top Ten [2], NATO, G8, Military [7], America, Latin America, Attackers [13, 14] and Attacked [14, 15].

The Top Ten analysis criterion has been considered taking into account the Global Cybersecurity Index 2017 report [2] that indicates the countries that stand out and how they have reached their maturity level. Therefore, Ecuador may be able to follow the steps of these countries as a potential roadmap. The NATO criterion has been chosen, due to their participation in many international war conflicts and since they have been continuously mentioned as a remarkable reference in a variety of documents in the bibliographic survey, confirming one of the main objectives for the creation of this organization. The G8 criterion for security investments has been taken into consideration as they are observing the industrialized countries that belong to this group. The Military criterion has been based on the ranking of the most efficient and well equipped armies, as indicated in international reports [12]. This has been proposed to be kept in mind within the perspectives of Cyber defense and Cyberwar, as searched by the Ecuadorian Army. America and Latin America have been two criteria for analysis taking into account the interdependence and commercial exchange that Ecuador keeps with all these countries and the impact they may have in the field of national information security. Attackers and Attacked have been also an adequate benchmark for comparison in Ecuador's Cybersecurity.

3.2   Tabulation and Normalization of Data Extracted from ITU

After determining fundamental information in the previous step, data from sources in 45 countries have been tabulated with a total of 405 records considered mandatory and critical for the current analysis.

In addition to such information, further data that have been revealed in the Cyber wellness profiles have been normalized, using its five pillars, being legal measures, technical measures, organizational measures, capacity development and cooperation. In turn, these pillars defined fields through which the governments filled them in the form of an interview. Later, the ITU quantifies them in order to reach the Global Cybersecurity Index (GCI). The ITU is responsible for keeping these Cyber wellness profiles up-to-date and governments notify the ITU as soon as there is any related change. Hereby, it is ensured that the information is current in the development of this study.

The attributes of these five pillars that have been normalized and tabulated is the following: LEGAL: "Criminal legislation" and "Regulation and compliance". TECHNICAL: "CSIRT", "Standards" and "Certification". ORGANIZATION: "Policy", "Roadmap for governance", "Responsible agency" and "National benchmarking". CAPACITY: "Development of Standardization", "Development of Human Resources", "Professional Certification" and "Agency Certification". COOPERATION: "Intra-state cooperation", "Intra-agency cooperation", "Association with the public sector" and "International cooperation". Thus, from the 45 countries, a total of 765 data have been obtained from the GCI pillars, considered to be secondary for the current study, but significant in the discussion, results and conclusions.

3.3   Analysis of Data

Once the information matrix has been tabulated, several data views have been generated using dynamic tables with the criteria that have been planned in point A of this analysis methodology.

Table 1 documents the TOP TEN of countries with the best GCI according to the ITU (best values are indicated in green, intermediate in yellow, poorest in red), highlighting the GCI score, the fields Legal, Technical, Organization, Capacity Building, and Cooperation called pillars according to the ITU [2, 3]. For a better understanding they have been briefly described below: (1) Legal: Measure on which the Global Cybersecurity Index (GCI score) has been based to indicate all the institutions, reference frameworks and legal aspects of Cybersecurity, Cyber defense and cybercrime of the countries analyzed by the ITU; (2) Technical: Measure on which the GCI score has been based to indicate all the technical institutions and frameworks of

Table 1. Top ten of countries with better GCI.

Global Rank	GCI SCORE	Country	Total Population	Legal	Technical	Organizational	Capacity Building	Cooperation
1	0.925	Singapore	5,256,000	0.95	0.96	0.88	0.97	0.87
2	0.919	U. States	315,791,000	1	0.96	0.92	1	0.73
3	0.893	Malaysia	29,820,000	0.87	0.96	0.77	1	0.87
4	0.871	Oman	2,904,000	0.98	0.82	0.85	0.95	0.75
5	0.846	Estonia	1,340,000	0.99	0.82	0.85	0.94	0.64
6	0.83	Mauritius	1,314,000	0.85	0.96	0.74	0.91	0.7
7	0.824	Australia	22,919,000	0.94	0.96	0.86	0.94	0.44
8	0.819	Georgia	4,304,000	0.91	0.77	0.82	0.9	0.7
8	0.819	France	63,458,000	0.94	0.96	0.6	1	0.61
9	0.818	Canada	36,675,000	0.94	0.93	0.71	0.82	0.7
10	0.788	Russia	146,300,000	0.82	0.67	0.85	0.91	0.7

Cybersecurity and Cyber defense; (3) Organizational: Measure that indicates in the analyzed country, the existence of coordinating institutions of policies and strategies in the development of Cybersecurity and Cyber defense at the national level; (4) Capacity Building: GCI score measure based on the existence of research and development, education, training programs and professional and public agencies with the capacity to build Cybersecurity and Cyber defense; (5) Cooperation: Measure on the existence of associations, cooperative frames of reference and information sharing networks on the subject of Cybersecurity and Cyber defense.

Table 2 lists Ecuador and the twelve countries that founded NATO, in which it has been ordered by the population number. Based on the amount of citizens, the green zone would be comparable with Ecuador, discarding countries such as Denmark, Norway, Luxembourg and Iceland.

Table 2. NATO member countries, compared with Ecuador.

Global Rank	GCI SCORE	Country	Total Population
2	0.919	U. States	315,791,000
8	0.819	France	63,458,000
12	0.783	UK	62,798,000
31	0.626	Italy	60,964,000
9	0.818	Canada	36,675,000
15	0.76	Netherlands	16,714,000
65	0.466	Ecuador	14,865,000
27	0.671	Belgium	10,788,000
55	0.508	Portugal	10,699,000
34	0.617	Denmark	5,593,000
11	0.786	Norway	4,960,000
36	0.602	Luxembourg	523,000
77	0.384	Iceland	328,000

Table 3 lists Ecuador in comparison with the eight countries of the G8. In this case, the green zone corresponds to all eight countries, due to the fact that these countries are more populated compared to Ecuador.

Table 3. G8 countries compared with Ecuador.

Global Rank	GCI SCORE	Country	Total Population
2	0.919	United States	315,791,000
10	0.788	Russia	146,300,000
11	0.786	Japan	126,435,000
24	0.679	Germany	81,991,000
8	0.819	France	63,458,000
12	0.783	UK	62,798,000
31	0.626	Italy	60,964,000
9	0.818	Canada	36,675,000
65	0.466	Ecuador	14,865,000

Table 4 lists Ecuador and the ranking of the ten countries considered to possess the most efficient and best equipped armies, ordered by the amount of population. The green comparison zone covers all these ten countries, being more populated than Ecuador.

Table 4. Countries of the best equipped armies compared to Ecuador.

Global Rank	GCI SCORE	Country	MILITARY RANKING	Total Population
32	0.624	China	3	1,353,601,000
23	0.683	India	4	1,258,351,000
2	0.919	United States	1	315,791,000
10	0.788	Russia	2	146,300,000
11	0.786	Japan	7	126,435,000
14	0.772	Egypt	10	83,958,000
24	0.679	Germany	9	81,991,000
43	0.581	Turkey	8	74,509,000
8	0.819	France	5	63,458,000
12	0.783	UK	6	62,798,000
65	0.466	Ecuador	0	14,865,000

Table 5 lists Ecuador with the main countries that have been the strongest cyber Attackers in 2017.

Table 5. List of countries that most attacked in 2017.

Global Rank	GCI SCORE	Country	ATTACKERS RANKING	Total Population
65	0.466	Ecuador	0	14,865,000
32	0.624	China	1	1,353,601,000
2	0.919	United States	2	315,791,000
10	0.788	Russia	3	146,300,000
24	0.679	Germany	5	81,991,000
15	0.76	Netherlands	6	16,714,000
8	0.819	France	7	63,458,000
11	0.786	Japan	8	126,435,000
12	0.783	UK	9	62,798,000

Tables 5 and 6 documents has been ordered by the total population and lists Ecuador with the countries that most attacked and the most frequently attacked in 2017. In this case, the green zone corresponds to the countries that have a greater number of citizens compared to Ecuador and that are not in the first five positions in the ranking of countries that have been most frequently attacked.

Table 6. List of most frequently attacked countries in 2017 compared with Ecuador.

Global Rank	GCI SCORE	Country	ATTACKED RANKING	Total Population
65	0.466	Ecuador	0	14,865,000
2	0.919	United States	1	315,791,000
15	0.76	Netherlands	2	16,714,000
24	0.679	Germany	3	81,991,000
32	0.624	China	4	1,353,601,000
12	0.783	UK	5	62,798,000
11	0.786	Japan	6	126,435,000
8	0.819	France	7	63,458,000
10	0.788	Russia	8	146,300,000
58	0.501	Ukraine	9	44,940,000
43	0.581	Turkey	10	74,509,000

After obtaining these lists (Tables 1, 2, 3, 4 and 5), the countries that have been left of the green selection areas have been filtered, while we applied the average of each group. Thus, all countries that are above the average (i.e. without considering Ecuador for the calculation of their position, except for the America and Latin America criteria), will be considered as selected and valid to take them into account as the best examples to follow in Cybersecurity and Cyber defense for Ecuador. We explained the results of this process in Sect. 4.

3.4 Statistical Analysis of Data of the Selected Countries

The determination in this study about the highest efficiency in Cybersecurity and Cyber defense, has been also based on data that has been collected from the "Country" and "GCI SCORE" fields of all countries being selected in each category. Afterwards, they

Table 7. Frequency of use of the country selected in each category.

COUNTRY	TOP TEN	NATO	G8	MILITARY RANKING	LATIN AMERICA	AMERICA	ATTAKERS	ATTACKED	FREQUENCY
Argentina					1				1
Brazil					1	1			2
Canada		1	1			1			3
Colombia					1	1			2
Ecuador					1				1
Egypt				1					1
France		1	1	1			1	1	5
Japan			1	1				1	3
Malaysia	1								1
Mexico						1			1
Netherlands		1							1
Russia			1	1				1	3
UK		1	1	1					3
United States	1	1	1	1		1			5

have been assigned with a weighting point, determining their use of frequency (Table 7). The results have been demonstrated in Sect. 4 of the current article.

Once the most efficient countries in Cybersecurity and Cyber defense have been discovered, we searched the ITU Cyber wellness profiles for quantify them in a table and radar chart and then detailed them one by one according to the profiles based on the GCIs being Legal, Technical, Organizational, Capacity and Cooperation, leading to concrete results and benefits based on the current study.

4 Evaluation Results and Discussion

The results obtained from the methodological process of comparative analysis proposed has been presented below, according to the criteria indicated in Sect. 3.

4.1 Evaluation with the Established Criteria

In the TOP TEN criterion the countries selected with GCI SCORE on the average of 0.84 have been the United States and Malaysia with GCI SCORES of 0.919 and 0.893, respectively (Fig. 2). In the radar diagram, these two countries have been the only ones that stand out in the analysis of this criterion (Fig. 2), because their GCIs are over the average of 0.84. During the review of the pillars on which the GCI index is based on, it

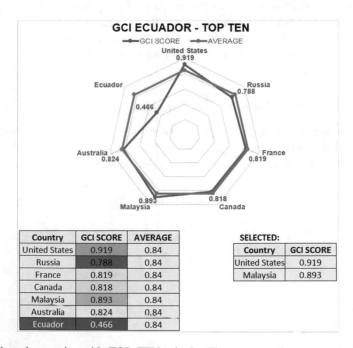

Country	GCI SCORE	AVERAGE
United States	0.919	0.84
Russia	0.788	0.84
France	0.819	0.84
Canada	0.818	0.84
Malaysia	0.893	0.84
Australia	0.824	0.84
Ecuador	0.466	0.84

SELECTED:

Country	GCI SCORE
United States	0.919
Malaysia	0.893

Fig. 2. Selected countries with TOP TEN criteria. The orange color represents the average values, while the GCI SCORE measurements are highlighted in blue color (author's contribution).

has been noted, that the United States and Malaysia are very strong in "Capacity", "Technical", while the United States are also very strong in "Legal". For the remainder of the criteria, the same type of radar diagram has been applied. However, only the currently obtained results will be described.

Within the NATO criteria, the countries selected with a GCI SCORE with an average above 0.738 have been the United States, France, UK, Canada and the Netherlands with GCI SCORES of 0.919, 0.819, 0.783, 0.818 and 0.760, respectively. Therefore, it appears, that not all founding countries of the NATO are strong in Cybersecurity and Cyber defense, as observed with Portugal and Italy, who range below the mentioned average.

In the G8 criterion, the selected countries with a GCI SCORE above the average of 0.777 have been the United States, Russia, Japan, France, United Kingdom and Canada with GCI SCORES of 0.919, 0.788, 0.786, 0.819, 0.783 and 0.818, respectively. Germany and Italy are again below the average, which suggests that they do not have the same progress in terms of Cybersecurity, despite being industrialized countries.

In the Militarily criterion, the selected countries with a GCI SCORE above the average of 0.743 have been the United States, Russia, Japan, Egypt, France, United Kingdom, with GCI SCORES of 0.919, 0.788, 0.786, 0.772, 0.819 and 0.783, respectively. As China and India have a good score in the military ranking but a low GCI, being far below the average, therefore they have not been selected.

In the Latin American analysis criteria, the countries selected with a GCI SCORE above the average of 0.418 have been Brazil, Colombia, Argentina and Ecuador with GCI SCORES of 0.593, 0.569, 0.482 and 0.466, respectively. Brazil and Colombia are very strong in Cybersecurity, being far above average. Bolivia has a very low ranking, while Ecuador has been at average, since it belongs to this region.

In the Americas criterion, the countries selected with a GCI SCORE above the average of 0.522 have been the United States, Brazil, Mexico, Colombia and Canada with GCI SCORES of 0.919, 0.593, 0.660, 0.569 and 0.818, respectively. The United States, Mexico and Canada stand out, being much stronger than those countries selected in the Latin America criterion.

In the Attackers criterion, the country selected with a GCI SCORE above the average of 0.787 has been France with a GCI SCORE of 0.819. All the countries of this criterion have been very close with their GCI, so that their similar measurements may appear confusing around the average values, which leads to the need of a complementary study for a more detailed discrimination. Ecuador may review how these countries handle Cyber defense and cyber war, as these are countries ranking far from the first places of attackers, having simultaneously a good GCI score.

In the Attacked criterion, the countries selected with a GCI SCORE above the average of 0.695 have been Japan, France and Russia with GCI SCORES of 0.786, 0.819 and 0.788, respectively. Unlike within the Attackers radar diagram, we observed (Fig. 3), which France coincides also as a country against cyber war attacks. In such analysis, it stands out, that Japan and Russia, despite being countries with a large population, are well developed in self-protection from cyber war attacks, as indicated by their GCI.

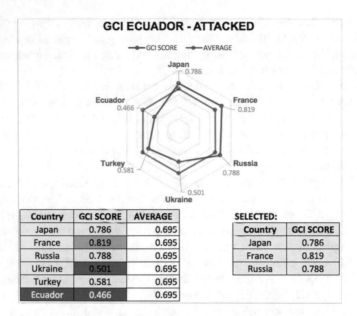

Country	GCI SCORE	AVERAGE
Japan	0.786	0.695
France	0.819	0.695
Russia	0.788	0.695
Ukraine	0.501	0.695
Turkey	0.581	0.695
Ecuador	0.466	0.695

SELECTED:

Country	GCI SCORE
Japan	0.786
France	0.819
Russia	0.788

Fig. 3. Selected countries with ATTACKED criteria. The purple color represents the average values, while the GCI SCORE measurements are highlighted in blue color (author's contribution).

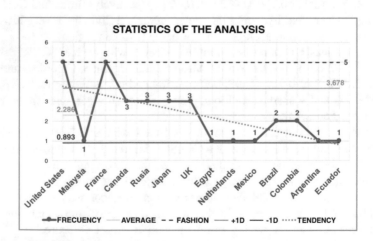

Fig. 4. Cumulative results of the statistical analysis (author's contribution).

Finally, when statistically processing the data (Table 7), the United States and France excel in the most criteria of comparison analysis, as clearly visualized with the highest frequencies, it is 5. The standard deviation also confirms that the data that stand out from the norm, have been the United States and France (Fig. 4). This implies to Ecuador to follow their successful path and how these two countries handle public and private Cybersecurity.

4.2 Model Resulting from the Performed Analysis

In this step, we handled the results obtained from the two best models of the countries in the comparison methodology, evaluating them with the current situation of Ecuador. Hereby, Ecuador would be able to obtain its own model, which includes the strategies and the metrics to be followed by the professionals, who are responsible for Cybersecurity in Ecuador (Fig. 5).

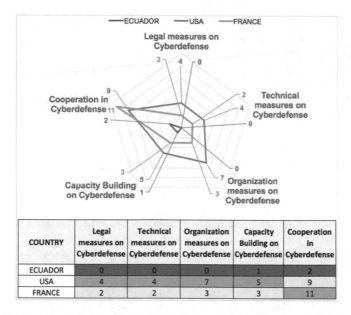

COUNTRY	Legal measures on Cyberdefense	Technical measures on Cyberdefense	Organization measures on Cyberdefense	Capacity Building on Cyberdefense	Cooperation in Cyberdefense
ECUADOR	0	0	0	1	2
USA	4	4	7	5	9
FRANCE	2	2	3	3	11

Fig. 5. Results of the analysis of ITU pillars related to Cybersecurity.

Due to the evaluation of the data and results (Tables 8 and 9; Fig. 5), we may deduce that Ecuador presents serious limitations in Cybersecurity with respect to the USA and France. Moreover, it has been yielded, that the only points obtained by Ecuador have been in "Capacity building" and in "Cooperation". This leads to the conclusion, that Ecuador's cyber wellness indicate that there are training and project generation programs in Cybersecurity to follow, in addition to joint work with Interpol and with the FIRST Security Incidents and Safety Response Forum. When analyzing the Cyber security points in the ITU Cyber wellness of the United States and France, we encountered the priorities and the strategic aspects that guide the case of Ecuador, whose contents have been visualized in the Tables 8 and 9.

Table 8. List of relevant points in cyber defense of the ITU cyber wellness profiles, based on the United States model.

LEGAL:
Legislation in Cybersecurity for electronic communication and transactions.
TECHNICAL:
Use of frameworks that cover traditional IT. Industrial Control Systems and Critical Infrastructure.
ORGANIZATION:
Route maps for industrial control systems and electric power delivery for Cybersecurity.
CAPACITY:
Awareness of Cybersecurity, plans to strengthen cooperation.
COOPERATION:
National plan for infrastructure protection. Agreements for protection of civil and military computer systems. Analysis centres to help protect critical networks and infrastructures. Treaties of cooperation with Europe in cybercrime, among others.

Table 9. List of relevant points in cyber defense of the ITU cyber wellness profiles, based on the French model.

LEGAL:
Treaties to fight against cybercrime, creation of a Cybersecurity agency.
ORGANIZATION:
Policies for the classification of military and national defense information. National Benchmarking to measure the development of Cybersecurity.
CAPACITY:
Best practices and guides to apply them in the public and private sector. Definition of security policies and defense systems.
COOPERATION:
Cooperation in Cyber defense and Cybersecurity with Germany, the Netherlands, Estonia, the United Kingdom, the United States, among others.

5   Conclusions

The main objective of this study has been to present a comparative analysis, based on the strategic approaches of Cybersecurity and Cyber defense of global relevance. Considering that the comparison criteria have been addressed to the needs of improving information security in Ecuador at the public, private and military levels, it has been evident that the countries that Ecuador should follow as a model are the United States (GCI = 0.919) and France (GCI = 0.819). Hence, the pillars of Legal, Technical,

Organization, Capacity and Cooperation, together with the Cyber wellness profiles that the ITU offers as a reference, have been vital to achieve this goal, since they clearly highlight the roadmap to be followed mainly in Cyber defense.

As future research, we have planned the application of the model used by the United States and France for the development of national policies on information security, as well as the implementation of an Information Security Management System (ISMS) for both traditional IT as for the Industrial Control Systems for public and private institutions. In this way, the national strategic areas that could be threatened from the field of information technology would be evident. In addition, this would allow to design strategies to deny or minimize the effects of the occurrence. Finally, the effects that the country would suffer with a potential impact on the economy as well as the welfare of its population could be measured.

References

1. Watkins, B.: The Impact of Cyber-Attacks on the Private Sector, p. 12. Association for International Affair, Briefing Paper (2014)
2. Brahima, S.: Global Cybersecurity Index 2017. International Telecommunication Union (ITU), pp. 1–77 (2017)
3. International Telecommunications Union. Cyber wellness Profiles. http://www.itu.int/pub/D-STR-SECU-2015
4. Davara, F.: Documentos de seguridad y defensa 60 Estrategia de la información y seguridad en el ciberespacio, pp. 1–120 (2014)
5. Parraguez, L.: The State of Cybersecurity in Mexico: An Overview, pp. 1–23. Wlson Center Mexico Institute, Mexico, January 2017
6. Australian Cyber Security Centre: Strategies to Mitigate Cyber Security Incidents – Migration Details, Australian Government, pp. 1–42, February 2017
7. Shackelford, S., Russell, S., Haut, J.: Bottoms Up: A comparison of voluntary Cybersecurity Frameworks, pp. 1–39 (2016)
8. Cioaca, C., Bratu, A., Stefanescu, D.: The Analysis of Benchmarking Application in Cybersecurity, Romania, pp. 1–6 (2017)
9. National Institute of Standars and Technology: Framework for Improving Critical Infrastructure Cybersecurity, United States Government, pp. 1–61, January 2017
10. Craig, A., Shackelford, S., Hiller, J.: Proactive cybersecurity: a comparative industry and regulatory analysis, pp. 1–63 (2016)
11. Department of Commerce United States: Meeting of the Commission on Enhancing National Cybersecurity. New York University – School of Law, pp. 1–26, May 2016
12. Military Strength Ranking (2017). www.globalfirepower.com/countries-listing.asp
13. Symantec Corporation: Internet Security Threat Report, United States, vol. 22, pp. 1–77 (2017)
14. Medina, M., Molist, M.: Ciberseguridad: Tendencias 2017, Universidad Internacional de Valencia, pp. 1–35, February 2017
15. Pilkey, A., Ahmad, A., Patel, A., et al.: 2017 State of cyber security, F-SECURE, pp. 1–77 (2017)
16. Bustamante, F., Fuertes, W., Díaz, P., Toulkeridis, T.: Integration of IT frameworks for the management of information security within industrial control systems providing metrics and indicators. IEEE Xplore, pp. 1–4, August 2017

17. Bustamante, F., Fuertes, W., Díaz, P., Toulkeridis, T.: Methodology for management of information security in industrial control systems: a proof of concept aligned with enterprise objectives. Adv. Sci. Technol. Eng. Syst. J. **2**(3), 88–99 (2017)
18. Bustamante, F., Fuertes, W., Díaz, P., Toulkeridis, T.: A methodological proposal concerning to the management of information security in Industrial Control Systems. In: Ecuador Technical Chapters Meeting (ETCM), vol. 1, pp. 1–6. IEEE, October 2016
19. ICS-CERT: "Year in Review", Industrial Control Systems Cyber Emergency Response Team U.S. Department of Homeland Security, pp. 1–24 (2014)
20. Vishik, C., Matsubara, M., Plonk, A.: Key Concepts in Cyber Security: Towards a Common Policy and Technology Context for Cyber Security Norms, pp. 221–242. NATO CCD COE Publications, Tallinn (2016)
21. Tabansky, L.: Basic concepts in cyber warfare. Mil. Strateg. Aff. **3**, 75–92 (2011)
22. Boopathi, K., Sreejith, S., Bithin, A.: Learning cyber security through gamification. Indian J. Sci. Technol. **8**(7), 642–649 (2015)
23. Abrahams, L., Mbanaso, U.: State of Internet Security and Policy in Africa, pp. 1–7, October 2017

Applicability of Cybersecurity Standards in Ecuador - A Field Exploration

Mario Ron[1], Marco Bonilla[1(✉)], Walter Fuertes[1], Javier Díaz[2],
and Theofilos Toulkeridis[1]

[1] Universidad de las Fuerzas Armadas ESPE,
Av. General Ruminahui, Sangolquí, Ecuador
{mbron, wmfuertes, ttoulkeridis}@espe.edu.ec,
marcounidad4@gmail.com
[2] Universidad Nacional de la Plata,
Avenida 7 877, 1900 La Plata, Buenos Aires, Argentina
jdiaz@unlp.edu.ar

Abstract. The evolution of information and communications technology (ICT) as well as cyberspace, walks along with collateral hazards. These need to be managed, in order to minimize the affecting impacts on information, which often appear to be vital for the operability of nations. The fulfillment of such risk reduction has been achieved with the development of worldwide, well known and widespread regulations, rules, manuals, guides and procedures for good Cybersecurity practices. Its impact has been significant in the construction of national policies within each country, due to its influence beyond the borders and the needed international cooperation for its elaboration. However, the current study presents a description, scope and coverage of both the standardization organizations as well as standards and specifications related to Cybersecurity, in order to establish a guide for researchers and information security specialist in the protection of assets and information of companies. Furthermore, it also demonstrates a systematic non-experimental field research that considered Ecuadorian institutions and companies to identify the current status of their adoption. Finally, the obtained results allow the applicability of standards, in order to protect enterprise information, which in turn will be able to serve in the future as input for the definition of national policies in such context.

Keywords: Cybersecurity · National policy · System · Hazard
Risk reduction

1 Introduction

Historically, public and private companies have been a permanent target of cyber-attacks, electronic fraud, and identity theft, extortion, among other hazards, which have affected the institutional image, critical mission infrastructures, and even state sovereignty. The need to ensure cyberspace has accelerated the evolution of the current approach of Cybersecurity, which should be based on appropriately accepted international standards, in order to protect the assets of the organization and users in cyberspace.

© Springer International Publishing AG, part of Springer Nature 2018
Á. Rocha and T. Guarda (Eds.): MICRADS 2018, SIST 94, pp. 27–40, 2018.
https://doi.org/10.1007/978-3-319-78605-6_3

The development of standards has been encouraged by international organizations with the participation of national groups responsible for standardization, which adhere to such regulations. In some cases there are national agencies of great importance and power, which influence the international environment. For example, the International Organization for Standardization (ISO), the National Institute of Standards for Technology (NIST), the Institute of Electrical and Electronics Engineers (IEEE), among others, which have been analyzed in the current study.

Many states have officially adopted international standards as part of their national regulations, like Ecuador did for its public sector, while its private sector neglected such initiative. In addition, despite the emergence of globally established standards and norms, even public and private institutions have not been able to perform the necessary precautions, in order to protect the assets and information of companies.

Based on the aforementioned scenario, some fundamental issues arose, which need to be addressed in this study: What are the standardization organizations and the regulations associated to Cybersecurity? What is the scope and coverage of the standards related to Cybersecurity? What are the recommended best practices for establishing security within national policies? How would it be possible to achieve the applicability of Cybersecurity standards in Ecuador? What is the current status of the adoption of these standards in the private and public sector in Ecuador? Considering that the standards present a spectrum of the state of knowledge of a given issue and contribute to the solution of needs in a structured manner, therefore, the hypothesis of this research focuses that standards and good practices need to be adapted to the national and enterprises realities, which may differ in their social, economic, and political conditions.

Consequently, the main aim of this study has been to conduct a field research for the application of Cybersecurity standards in Ecuador, which allows to establish a useful implementation guide. In order to achieve such goal, a set of planned actions has been performed, which included (1) a review of standardization frames, standards and norms related to Cybersecurity; and (2) a field study based on the data collection of Ecuadorian institutions and companies. One of the main contributions will be the presentation of the coverage of such standards and the current status of their adoption in Ecuador. This will ultimately serve to select appropriate standards within the framework of development of Ecuadorian cybersecurity policies.

The rest of the study has been subdivided in four further segments. Section 2 reviews the standardization frames and the standards as well as norms related to Cybersecurity. Section 3 demonstrates the field research for the applicability of these standards in Ecuador. Section 4 presents the obtained results and their analysis. Finally, we present in Sect. 5 the conclusions and future work.

2 Organizations of Standardization and Standards Related to Cyber Security

This section describes the function and characteristics of the organisms that generate standards. Additionally, we will present some well-known and currently used standards. Furthermore, we conducted an analysis of their scope and coverage in relation to Cybersecurity. This will allow to orientate interested parties in national policies, in the collaboration obtained from international organizations for their development, as well as the adoption of standards by public and private institutions.

2.1 Standardization Organizations

European Committee for Standardization (CEN) [1]. This committee develops and publishes standards and technical specifications that respond to the requirements of companies and other European organizations, in order to improve the safety, quality and reliability of its products, services or processes, as well as innovation and the diffusion of new technologies. The Vienna Agreement establishes a framework for technical cooperation with the International Organization for Standardization (ISO). Beneath the CEN, there is the CEN-CENELEC specialized Group on Cybersecurity (CSCG-), which aims to analyze technological evolution, its impact for the digital market and evaluate how standards will be able to support cybersecurity policies and regulations, protection of data, as well as establishing recommendations for the international standardization environment.

International Organization for Standardization (ISO) [2]. This is an independent international non-governmental organization, with the membership of 165 national standardization associations. The ISO has 309 Technical Committees, among them is the ISO/IEC JTC 1 Committee on Information Technology, which has published 2956 standards and has 523 work programs. This group also has an advisory subgroup, four working subgroups and 20 subcommittees, including the ISO/IEC subcommittee JTC 1/SC 27 on IT Security Techniques, which in turn has 7 subcommittees/sub-working groups.

International Telecommunication Union (ITU) [3]. This is the specialized agency of the United Nations for Information and Communication Technologies (ICT), which allocates radio electric spectrum as well as satellite orbits worldwide and develops technical standards for the interconnection of networks and information technologies. The members of the ITU officially represent the ICT sector in the 193 member states. They are ICT regulatory bodies, academic institutions and technology companies. The Study Group 17 (SG17) of the ITU coordinates the issue related to trust and security in the use of information and communication technologies. It is currently working on cyber security, security management, security architectures and frames, combat spam, identity management, personal data protection as well as the security of applications and services for the Internet of objects (IO), smart networks, smartphones, web services, social networks, cloud computing, mobile financial systems, IPTV and tele biometrics.

Institute of Electrical and Electronic Engineers (IEEE) [4]. This is the worldwide largest technical professional organization dedicated to advancing technology for the benefit of humanity, through publications, conferences, technology standards and professional and educational activities. The IEEE has more than 426,000 members in more than 160 countries, with Societies and Communities or Technical Councils. Such institute publishes a third of the world's technical literature in electrical, computer and electronic engineering and develops international standards that underpin many of the telecommunications, information technology and energy generation products and services. It has the IEEE Xplore digital library with about four million documents. Among the 39 technical societies, is the Computer Society, the largest company within the IEEE, with free access to courses and online technical and business books, electronic bulletins, magazines, discounts on conferences and software development certifications. Its field of interest is in computer science and information technology. It also has 18 communities among which it allows to find the Cybersecurity Community, with attention to cybersecurity, the Center for Safe Design (CDS) and several more initiatives in this area.

Internet Engineering Task Force (IETF) [5]. This is an international non-profit standardization organization, which regulates Internet proposals and standards, known as RFCs that are available on the Internet and are able to be reproduced. The organization covers several areas and is grouped into a variety of zones. The Internet Assigned Numbers Authority (IANA) assigns and coordinates the use of the parameters of Internet protocols, commissioned by the Internet Society (ISOC). The Security Area has 18 groups on various topics related to Internet security.

International Electro Technical Commission (IEC) [6]. This is a leading organization in the preparation and publication of international standards for electrical, electronic and related technologies, based on consensus.

Specialized Technical Standards Organizations. There are several non-profit organizations formed by companies, technicians and professionals interested in technology development, which propose standards declared as de facto standards in the industry. Among them, there are:

The World Wide Web Consortium (W3C), for Web standards; International Society of Automation (ISA), for automation and control systems used in industry and critical infrastructures; Organization for the Advancement of Structured Information Standards (OASIS), of open standards for the global information society, related to security, Internet of Things, the cloud, computing, energy, content technologies, emergency management and other areas; Global Platform, which are specifications for applications in secure chip technology; Open Mobil Alliance (OMA), which are open specifications for mobile services; Committee for Standardization (CS), responsible for the standardization policy of the North Atlantic Council (NAC) with the NATO Standardization Office (NSO) assists the NATO Military Committee in the development of military operational standards; Near Field Communication Forum (NFC), use of near-field communication; Groupe Speciale Mobile Association (GSMA) GSM mobile phone system; Telecommunications Industry Association (TIA); European Association for the standardization of information and communication systems (ECMA International); European Conference of Postal and Telecommunications Administrations (CEPT).

Regional Standardization Organizations. There are regional standardization bodies in Europe, Asia, America and other intercontinental. Europeans have generally become global, the best known in South America and the Caribbean are:

Mercosur Normalization Association (AMN); Pan American Standards Commission (COPANT); CARICOM Regional Organization for Standards and Quality (CROSQ); Andean Standardization Network (RAN), charged with harmonizing technical standards in the Andean Community of Nations (CAN) with international practices, to facilitate trade, technology transfer and improve the competitiveness of products and services of the Andean Community of Nations (CAN). 25 member countries, following the recommendations of the World Trade Organization (WTO).

National standardization organizations. There are national organizations with international transcendence whose publications have become international standards. Generally they are part of the international organizations previously named as the ISO, IEC or ITU. Among the most important are the UK National Standards Body (NSB) in the United Kingdom with the British Standards Institution (BSI), Association Française de Normalization (AFNOR) in France, Deutsches Institut für Normung (DIN) in Germany, Japanese Industrial Standards Committee (JISC) in Japan, Spanish Association for Standardization (AENOR) in Spain, China Communications Standards Association (CCSA) in the Republic of China, Federal Agency on Technical Regulation and Metrology (GOST) in Russia, National Institute of Standards and Technology (NIST) and the American National Standards Institute (ANSI) of the United States of America, Brazilian Association of Technical Standards (ABNT) in Brazil, General Directorate of Standards of the Ministry of Economy (DGN) in Mexico, among others.

2.2   Standards and Related Specifications

The aforementioned organizations have generated a multiplicity of standards that are a great technical reference for the development and application of technology, among which are those that have caused the greatest impact and influence in Cybersecurity and that have been established as necessary references for consultation in the elaboration of organizational policies.

ISO/IEC 27000 [7]. Framework which contains published standards and others in development, to establish an information security management system (ISMS) and certify it on the basis of the ISO 27001 standard, which is the main standard, being part of the work of the subcommittee JTC1/SC27. The standards range from 27000 to 27019 and from 27030 to 27044 and the standard 27799 that ends the series (they are not freely disseminated).

The aspects addressed by the ISO/IEC 27000 reference framework, in the corresponding subseries refer to: scope and purpose, definitions, requirements for certification, good practices to apply security controls, specification, design, conception and start-up of an ISMS, metrics and techniques to determine the effectiveness of an ISMS, risk management approach, requirements for the accreditation of audit entities, audit guide for an ISMS use of the standard, in the industrial sector, exchange and dissemination of sensitive information, information security for telecommunications sector, integrated implementation of 27001 and 20000-1, corporate governance of information security, Security for financial and insurance sector, assessment of the financial aspects of information security, security for Cloud Computing, control systems related to the sector of the energy industry, support the transition of versions, continuity of the business, cyber security, Internet security and critical infrastructure protection information (CIIP), network security, security in computer applications, information security incident management, security in relations with suppliers, collection of digital Evidence, security in the digital newsroom, intrusion detection and prevention systems, security in storage media, suitability and adequacy of research methods, analysis and interpretation of digital evidence, principles and research processes for digital evidence, information security events (SIEM) and security for patient health data.

NIST - Framework for improving the cybersecurity of critical infrastructures [8]. Created by the government of the US of America and the private sector of that country that allows managing the risk of Cybersecurity, considering business requirements without establishing additional regulatory requirements. This framework uses business drivers to determine risks and establish cybersecurity activities as part of the organization's risk management processes. It consists of three parts, being the basic framework, the environment or profile and the levels or layers for its implementation.

The first part is a set of cybersecurity activities, effects and common information references in critical infrastructure sectors that provide a detailed guide for the development of individual organizational profiles. The second part empowers the organization to align its cybersecurity activities with its business requirements, risk tolerances and resources. The third part allows that through layers, organizations are able to see and understand the characteristics of their approach to Cybersecurity risk management.

The framework allows to apply the best practices, in order to improve the security and resilience of critical infrastructure [9, 10]. It provides a reference of standards, guidelines and updated good practices, being worldwide recognized. Based on this framework, the NIST has developed the Bald bridge Performance Program and a tool to determine the level of maturity in safety and develop an action plan for the improvement of safety practices and management.

The NIST 800 Series is a set of documents that describe the information security policies, procedures and guidelines of the federal government of the USA. The NIST uses Special Publications (SP) to define security and privacy guidelines, recommendations and reference materials. It includes the SP 800 subseries, the SP 1800 subseries (NIST Cybersecurity Practice Guidelines) [11] and selected publications of the SP 500 series (information technology), which are directly related to computer security, cybernetics, information and privacy.

The publications of the NIST 800 series evolved as a result of exhaustive research on viable and cost-effective methods to optimize the security of information technology (IT) systems and networks in a proactive manner. The publications cover all procedures and criteria recommended by NIST to evaluate and document threats and vulnerabilities and to implement security measures to minimize the risk of adverse events [12]. Publications have been useful as guidelines for compliance with safety standards and as legal references in case of litigation involving safety issues.

The SP 800 series contain several standards in which 256 safeguards have been organized into 18 categories, some of which are very useful [13] and the aspects addressed there are: contingency planning and safety and emergency plans, development of policies and administration of security servers, security for the operation of public access web servers and network infrastructure, security for email systems, risks associated with the technologies, awareness and training program, security incidents, security key responsibilities in the development of information systems, prevention of attacks and malware incidents, sanitation program for the means, techniques and controls for the disinfection and elimination of confidential information, risk management practices for support, authentication mechanisms to manage physical access, protection of the confidentiality of personnel identification information and risk management in the supply chain.

2.3   Analysis of the Scope and Coverage of Related Standards

According to a comparative study conducted by Rodrigo Carvajal, between the ISO and NIST standards [13], a series of indicators have been presented that significantly differ between these two normative frameworks and serve as a guide for professionals and stakeholders in the management of the security of the information. However, it is fundamental to consider that regulatory frameworks need to be adapted to the national realities of each country, in which their social, economic, political and cultural realities are taken into account.

Therefore, NIST proposes a legal framework for public entities, being limited for domestic use within the United States of America. The opposite occurs with the ISO/IEC that proposes a commercial regulatory framework of an international nature and as such has spread throughout the world. Although technically the NIST standard has been more extensive and detailed, the ISO/IEC standard has been easier to read and to understand. Nonetheless, both regulatory frameworks are still under development, in order to comply with the requirements of technological evolution and their inherent risks. In Table 1, a practical summary of this study is presented with indicators and significant differences in regulatory frameworks.

Table 1. Comparative analysis between NIST and ISO/IEC

Norm/Indicator	NIST	ISO/IEC
Applicability	Public entities, limited to USA	Commercial use, international standard
Scope	Focused on information systems	Centered on the organization
Risk management	Defined context, detailed risk treatment, specific monitoring	Particular context of each organization, high-level risk assessment
Baselines	Three levels for controls according to the potential impact	Does not use the concept, controls are defined in the risk management process
Certification and accreditation	Measurement of the implemented controls	Independent certification body
Readability	More complex and difficult to read	Easier to understand, available also in Spanish
Availability	Free and easy to access	It has a cost and a purchase process
Security coverage	Greater number of domains and controls	Some domains still in development

3   Applicability of Cybersecurity Standards in Ecuador, Field Exploration

In order to determine the current status of the application of the Cybersecurity Standards or related in the Republic of Ecuador, a field research has been conducted, in order to collect the relevant information [13]. This shall serve as a guide to identify the most optimal methodology, the classification described by [14], while a Holmes matrix has been used for its subsequent selection.

The developed labor has been a descriptive study in which data has been collected from Ecuadorian institutions and companies in a single systematic non-experimental research process, without deliberately manipulating the independent variable that refers to the Cybersecurity standards. The survey elaborated refers to data of the companies as they are in their natural context, which will be analyzed later. The research has been of a Tran-sectional or Transversal type, as there has been a single data collection at a certain time by means of a survey. For this purpose, we conducted an applied research, which aims to solve the problems of Cybersecurity.

The studied population has been determined by Ecuadorian institutions and companies, being identified on the page of www.ekosnegocios.com [15], which includes a group of approximately 1,000 companies. As such research has been of transection design by survey, it has a probabilistic sample, where all the elements of the population have the same possibility of being chosen. The sample has been determined by a random selection, using probability formulas, in order to obtain a representative sample of the population, allowing an association between variables and results that become generalized with respect to the population.

An analysis of geographical areas has been performed at administrative and planning levels (Fig. 1) and industrial sectors. Statistically, 277 randomly selected companies in each sector have been obtained as sample size, as described in Table 2.

Fig. 1. Geographical Zones in Ecuador Zones at administrative and planning levels. Source: http://www.planificacion.gob.ec/folleto-popular-que-son-las-zonas-distritos-y-circuitos/

Table 2. Division of companies by sector

Ord	Sector	Assigned company	Total sample
1	Fertilizers and chemical products for agricultural use, aquaculture and fishing	82	23
2	Wholesale, retail and diverse	124	34
3	Construction of roads, streets and others	38	11
4	Cooperatives, Mutualists and Banks	37	10
5	Manufacture of various foods	120	33
6	Automotive manufacturing	16	4
7	Other specialized activities	63	17
8	Energy production	47	13
9	Insurance	28	8
10	Information technology	74	20
11	Various shops	371	103
	Total	1000	277

An independent variable has the Cybersecurity Standards, which are represented by technical-legal documents with rules to fulfill an activity. The definitions of Cybersecurity standards, standardization bodies, types of standards and related legislation have been used as indicators.

As a Dependent variable, Cybersecurity has been considered in Ecuadorian Institutions and companies, as the set of protocols established for the protection of critical business information. Its classification indicators have been the geographical areas (Political-Administrative Division of Ecuador) and the productive sectors defined in the Ecuadorian Chamber of Commerce.

In order to fulfill the objective of the current research, a survey has been drawn, being aimed at Ecuadorian institutions and companies as well as computer executives from several public companies in the country. In order to prepare this survey, important issues related to the variables of the research have been determined. In order to obtain a viable survey, a reduction analysis has been performed, where finally 10 closed multiple-choice questions have been defined for two groups of respondents. The first consisted of 277 companies, as determined in the statistical process and the second group consisted of some 140 executives of public companies.

The survey aimed to acquire a general description of the surveyed organization and its relationship with Cybersecurity, through data, which allow to classify the organization according to the size and importance of IT in the business, its relationship with Cybersecurity, the use of standards and the way in which the company may be able to participate in the proposal of National Policies of Cybersecurity.

4 Evaluation Results and Discussion

4.1 Description of the Results of the Field Exploration

The obtained results of the surveyed enquiry are presented in Table 3, being grouped in two types of questions. There, some consulted aspects have been related to planning and government, while others have been to infrastructure and services. Each aspect conducts to categories for response, in order to yield a percentage of the given answers.

Table 3. Results of the field exploration.

Planning and government			Infrastructure and services		
Consulted aspect	Components	%	Consulted aspect	Components	%
Strategic or business planning	Developed by an own team	34.4	Size of the information security team	1–5 people	62.7
	Developed by an individual consultant	21.5		6–10 people	20.5
	Developed by a mixed team	17.2		>10 people	6.0
	Approved and disseminated	9.7		No personnel	10.8
	Periodically updated	6.5	IT infrastructure ownership	Own infrastructure	47.8
Importance of IT in the company	Very important	26.5		Leasing	15.6
	Important	60.2		Outsourced	8.9
	Unimportant	13.3	IT infrastructure to be used in BYOD	Owned by the company and remains inside	47.8
Participation in the creation of national cybersecurity policies	Online forums	21.3		Owned by each employee (smart phones)	23.9
	Face-to-face forums	19.9		Owned by each employee (laptops)	1.1
	Tables and discussion groups	10.3		Owned by the company and enabled to be used outside	20.7
	Receiving training	14.0		Owned by the employee for work at home	2.2
	Other forms of participation	34.6		Owned by the company for work at home	4.4

(*continued*)

Table 3. (*continued*)

Planning and government			Infrastructure and services		
Consulted aspect	Components	%	Consulted aspect	Components	%
Preferences in cybersecurity standards	ISO/IEC 27001	34.4	IT services used by the company	Service Desk (outsourced 13.75%)	36.3
	NIST Cybersecurity Framework	22.2		On-site support (outsourced 30%)	18.5
	COBIT for Security	22.2		Support in the field (outsourced 20%)	7.4
Application of cybersecurity policies in the company	Formally approved policies	45.8		Support applications (outsourced 17.5%)	21.5
	Not been formally approved	19.3		Infrastructure management (outsourced 11.25%)	14.8
	Elaborated and implemented	20.5		Other kind of services (outsourced 7.5%)	1.5
	Have not been elaborated	14.6			

4.2 Discussion

The obtained results of the survey demonstrate that there has been a relationship between the need to apply the Cybersecurity norms and the standards.

Based on the results of the first part of the questionnaire, we may conclude that companies in Ecuador have not yet become aware of the importance of security standards and their use in information security. Therefore there is a pending work to define an appropriate action in the National Cybersecurity Policies.

From the second result, we may deduce that a large percentage of companies (73.10%) have been concerned with preparing their strategic planning. However, its dissemination and updating has been less than 10%. We encountered an urgent need to motivate companies to encourage active participation, dissemination and updating of their planning.

Although the business turn has been mandatory, in order to consider the importance of IT in the company, it may be inferred from the results that a large majority of them assign importance (76.24%) to IT. This result should be used to obtain the necessary support in the application of Cybersecurity measures in business processes.

The response regarding participation in the creation of national Cybersecurity policies has been very important. All companies desire to be part of the process, although in different ways, which should be taken into account in the development project of these policies, especially in the participation methodology.

Regarding the application of Cybersecurity policies in the company especially in developing countries, there has been a high percentage (85.53%) that have been concerned with preparing such. However, its implementation appeared to be very low with some 20.48%. In a future study, we may need to overview the specific reason for such result, which ultimately causes the initiated effort of policy making to be useless. In addition, this reality need to be considered and contrasted with the lack of the National Cybersecurity Policy in Ecuador.

The majority of companies maintain their own IT infrastructure (47.77%) and only a minority (8.89%) has been outsourced. This reality determines a risk in the evolution and technological update regarding IT investments and the definition of a scalable methodology in terms of the application of Cybersecurity measures in the future, being aspects which should be reflected in order to define the National Cybersecurity Policy.

The size of information security equipment in most companies is small (62.65%). It also depends on the size of the company and its business. It is a manageable number for the purposes of training and education. However, there is a 10.84% that does not have personnel, which demonstrates their lack of commitment and interest in safety. It would be necessary to determine if such result obeys to a process of outsourcing.

The use of Service Desk (36.3%) is important to consider for the centralized application of protection actions or to guide them to the support providers, either on site, in the field, application support or infrastructure management. However, in a concatenated way with this response it might be seen that the outsourced services do not have much interference. Therefore, the attention of the National Policy would be to the companies themselves, rather than to the service providers.

For the application of the modality of work outside the company and the trend of BYOD (Bring Your Own Device), it should be considered that a percentage of them (52.17%) have infrastructure that have been able to be used outside the facilities. This leads to define a different and complex scenario of information security for business management, an aspect that should be considered in the National Cybersecurity Policy.

Finally, in relation to the content of the current study, what has been described will certainly serve as a reference guide for interested researchers and professionals responsible for the security of information, in the help of the protection of IT infrastructures, assets and information of companies and the states. Furthermore, it also allows to guide participants in the development of national cybersecurity policies in Ecuador.

4.3   Applicability, A Cross Analysis

In order to determine the applicability, it has been necessary to perform an analysis of the best conditions, which combined may be successful when applying a standard. This is the sum of the best planning conditions, importance of IT in the company, using the adopted standard for Ecuador in those companies in which the policies have been drawn up and applied (as described with the route of the arrows in Fig. 2).

With the analysis of the obtained results in the survey, we most certainly consider that the conditions of applicability of this route would be: 17.2% + 26.5% + 10.3% + 34.4% + 20.5%. Thus, might not be the best, therefore, it would need to improve those indicators, in order to achieve optimal conditions.

The equivalent analysis may be performed considering the conditions of the infrastructure and services as illustrated in Fig. 3, where the most optimal combined

Strategic or business planning	%	Importance of IT in the company	%	Participation in the creation of national	%	Use of Cybersecurity standards	%	Application of cybersecurity policies	%
Developed by an own team	34.4	Very important	26.5	Online forums	21.32	ISO / IEC 27001	34.44	Formally approved policies	45.78
Developed by an individual consultant	21.5	Important.	60.24	Face-to-face forums	15.85	NIST Cybersecurity Framework	22.22	Not been formally approved	19.27
Developed by a mixed team	17.2	Unimportant.	13.25	Tables and discussion groups	10.29	COBIT for Security	22.22	Elaborated and implemented	20.48
Approved and disseminated	9.67			Receiving training	13.97			Have not been elaborated	14.57
Periodically updated.	6.45			other forms of participation	34.57				

Fig. 2. Cross analysis considering aspects of TI planning and government

Size of the information security team	%	IT infrastructure	%	IT infrastructure ownership	%	IT services used by the company	%
1-5 people	62,7	Own infrastructure	47,8	Owned by the company and remains inside	47,8	Service Desk	36,3
6-10 people	20,5	Leasing	15,6	Owned by each employee (smart	23,9	On-site support (outsourced 13.75%)	18,5
> 10 people	6,0	Outsourced.	8,9	Owned by each employee (laptops)	1,1	Support in the field (outsourced 30 %)	7,4
no personnel	10,8			Owned by the company and can be used outside	20,7	Support applications (outsourced 20%)	21,5
				Owned by the employee for work at	2,2	Infrastructure management	14,8
				Owned by the company for work at home.	4,4	Other kind of services (outsourced 11.25%)	1,5

Fig. 3. Cross analysis considering aspects of TI infrastructure and services.

conditions would be in large companies with security equipment of more than 10 people that have outsourcing contracts, whose own infrastructure enables to be used outside of their facilities and have a service desk. According to the available data, there would be a sum of: 6% + 8.9% + 20.7% + 36.3%. These would not be the optimal condition of applicability, allowing the option to improve the indicators to ensure adequate applicability.

5 Conclusions and Future Work

A vision of the standardization organizations has been presented, both globally and regionally and its importance. Such afford and result may guide the agencies responsible for the National Cybersecurity in the participation of them and obtaining support in this effort, as well as national companies and citizens concerned about information security. In the same way, we presented the main standards currently in use within this topic, its scope and coverage, in order to help in the decision making that organizations and companies should result regarding its application. Such advice should be considered to be always convenient, as the application of obtaining knowledge and good existing practices in accordance with the local reality, will result to be beneficial. Additionally, the obtained information in the current study contributes significantly in knowing the national reality that serves as the basis for the development of the National

Cybersecurity Policy. This will guide specific actions to protect the critical information of the organizations and therefore provide protection of business operations that affect the national economy.

As future work we have planned to develop a methodology for the construction of National Cybersecurity Policies in developing countries.

Acknowledgment. The authors would like to express special recognition to all companies and professionals who participated in the survey and who have the firm intention of collaborating in the development of Ecuador's National Cybersecurity Policy.

References

1. CEN: European Committee for Standardization. http://www.cen.eu/. Accessed 15 Oct 2016
2. ISO: International Organization for Standardization. http://www.iso.org/. Accessed 15 Oct 2016
3. ITU: International Telecommunication Union. https://www.itu.int. Accessed 15 Oct 2016
4. IEEE: Institute of Electrical and Electronics Engineers IEEE. https://www.ieee.org/index.html. Accessed 12 Oct 2017
5. IETF: Internet Engineering Task Force. https://www.ietf.org/. Accessed 15 Oct 2016
6. IEC: International Electro technical Commission. http://www.iec.ch/. Accessed 15 Oct 2016
7. ISO: ISO 27000.es. http://www.iso27000.es/iso27000.html. Accessed 18 Oct 2016
8. NIST: Framework for Improving Critical Infrastructure Cybersecurity, USA, p. 39 (2014)
9. Bustamante, F., Fuertes, W., Díaz, P., Toulkeridis, T.: A methodological proposal concerning to the management of information security in industrial control systems. In: IEEE Ecuador Technical Chapters Meeting (ETCM). IEEE (2016)
10. Bustamante, F., Fuertes, W., Diaz, P., Toulqueridis, T.: Methodology for management of information security in industrial control systems: a proof of concept aligned with enterprise objectives. Adv. Sci. Technol. Eng. Syst. J. **2**(3), 88–99 (2017). https://doi.org/10.25046/aj020313
11. Shen, Lei: The NIST cybersecurity framework: overview and potential impacts. SciTech Lawyer **10**(4), 16 (2014)
12. Bustamante, F., Fuertes, W., Díaz, P., Toulkeridis, T.: Integration of IT frameworks for the management of information security within industrial control systems providing metrics and indicators. In: Electronics, Electrical Engineering and Computing (INTERCON) (2017)
13. Carvajal, R.M.: Estudio de las normas españolas y estadounidenses de seguridad de la información. Universidad de Valladolid (2015)
14. Hernández Sampieri, R., Fernández Collado, C., Baptista Lucio, M.P.: Metodología de la investigación, 5ta. McGraw-Hill, Mexico (2010)
15. EKOS (2010). http://www.ekosnegocios.com. Accessed 15 Oct 2017

Computer Vision in Military Applications

Mathematical Models Applied in the Design of a Flight Simulator for Military Training

César Villacís[1], Fabián Romero[1], Mario Navarrete[1],
Walter Fuertes[1(✉)], Santiago Chamorro[1], Iván Rodríguez[1],
Manolo Paredes[1], Vladimir Bastidas[2], Camilo Lozano[2], Juan Solano[2],
Christian Aguirre[2], and Juan Betancourt[2]

[1] Computer Science Department, Electrical and Electronics Department,
Universidad de las Fuerzas Armadas ESPE, Sangolquí, Ecuador
{cjvillacis, faromero, mfnavarrete3, wmfuertes,
smchamorro, iwrodriguez, mdparedes}@espe.edu.ec
[2] Escuela de Aviación Naval "TNFG Rómulo Donoso Ramirez",
Manta, Ecuador
{vbastidas, clozano, jsolanol, caguirre,
jbetancourt}@armada.mil.ec

Abstract. The current study presents mathematical models of an aircraft, which has been considered as a mass subjected to different forces, performed in classical physics. The main objective is to design algorithms, used to simulate the flight of an aircraft and to create an interactive simulator based on mathematical models. For this purpose, the different forces that the mass of the aircraft has been subjected to in the air have been analyzed and interpreted numerically, in order to generate a mathematical model that makes it possible to reproduce the flight of an airplane within a simulation software developed with Unity. Finally, the performance of the algorithms within Unity's game engine has been evaluated, before and after using threads in order to be able to conduct communication and evaluate the data transmission analysis.

Keywords: Flight simulator · Mathematical models of aircraft
Aerodynamic forces · Kinematic control

1 Introduction

The training of pilots of commercial and military aircrafts should contemplate the simulation of real situations and those similar, including normal flight cases, abnormal cases, and failure cases, so that there is an overview of possible scenarios and how to face critical and difficult to solve situations [1]. In this sense, it is necessary that the training of pilots be carried out safely and also at low cost. Therefore, flight simulators and space disorientation should be used to reproduce real situations in both favorable and unfavorable conditions, allowing pilot students develop the necessary skills and abilities that allow them to face a real flight, such as flexibility, resilience, decision making, among others [2].

© Springer International Publishing AG, part of Springer Nature 2018
Á. Rocha and T. Guarda (Eds.): MICRADS 2018, SIST 94, pp. 43–57, 2018.
https://doi.org/10.1007/978-3-319-78605-6_4

The current flight simulators use mathematical models to simulate the behavior of the airplane in the air [3, 4], and Stewart platforms from 3 degrees of freedom, allowing the pilot to feel the same sensations as when a real plane is flying [5, 6]. Flight cabins are even used to emulate real ones, which exist in the industrial market, as if they were made with the same instruments. This platform is responsible for transmitting to the pilot the accelerations and gravities, as well as a visual and sound system. Supported by mathematical models, this platform allows the pilot to have a real flight sensation.

The main objective of this study was to design a flight simulator based on mathematical models and programming algorithms that allowed us to recreate a Cessna type airplane in a virtual world and to test the proper functioning of it, for which it has been used by Unity 3D Game Engine as a development tool [7–9].

The main contributions of this study included: Mathematical model of: (1) the frontal velocity of the airplane; (2) the airplane's vertical speed; (3) the plane's lateral velocity; (4) airplane turns.

The remainder of this study is organized as follows: Section 2 discusses related work, while Sect. 3 explains the background of this study. Later, Sects. 4 and 5 explain the experiment configuration and present the experimental results, respectively. Finally, Sect. 6 presents the conclusions and future work lines.

2 Related Works

Merk [10], proposes training combat aircraft pilots using simulations of computer-controlled attackers or opponents to improve the tactical training of such pilots. Boril [11], focuses on the description of a simulation center used in the training of the Czech Air Force personnel. Cen [12] describes the free flight test facility, with emphasis on the design, development and verification of the simulation platform. Louali [13], proposes real-time flight simulators for training military pilots to complement their skills. Therefore, it allows to pilots to be immersed in a virtual environment, thus contributing to the development of their abilities for flight decision making.

In regards to the development of flight simulation platforms, Mauro [14] proposes an architecture to develop a dynamic flight simulator based on a model of 3DOF actuators with a parallel platform. Cheon [15], presents a simulation platform to verify the image-based object tracking method adopted in the small unmanned aerial vehicle (UAVs). Wu [16] developed a low-cost, high fidelity flight simulation-training device for research and educational purposes. Munzinger et al. [17], present a simulation for a remotely controlled aircraft of much smaller size, which needed to be developed to test a modern controller. Chen [18], explains the real-time simulation environment that was developed for helicopter aviation training. Liu [19], proposed a flight simulation platform for dynamic soaring. Setiawan et al. [20], intended the development of a real-time flight simulator and verified if it worked in real time.

To improve flight simulator analysis and evaluation, Hays et al. [21] performed a meta-analysis on flight simulation to identify the important features associated with efficacy in traditional training. Lofaro [22] explained the role of pilot training in assessing performance in the identification and management of hazards, especially in the air and under changing conditions through an automated collaborative system for aviation safety. Gervais et al. [23] and Chaudron [24], addressed the problem of achieving optimum real-time performance with the High-Level Architecture (HLA). Lorains et al. [25] performed the evaluation and description of flight training simulator Microsoft Flight Simulator 2004 for an integration with HIL Architecture that has been used to develop, test, and validate embedded systems. Scamps [26], developed a flight simulator Evaluation Course at QANTAS Airways. Khan et al. [27], presented a modular flight control strategy to demonstrate the improved command tracking performance with fault tolerance and reconfiguration capability.

In comparison with the analyzed works, our study proposes several mathematical models to calculate the frontal, lateral and rotation speed. In addition, these models have been tested on a scale Stewart platform with three degrees of freedom.

3   Theoretical Framework

3.1   Mathematical Model of the Basic Forces of an Airplane

A series of forces act on an airplane in flight, some of which are favorable and others unfavorable, so it is the pilot's primary task to exercise control over them in order to maintain a safe and efficient flight. According to [28–30], the four basic forces that affect the maneuvers of the pilot on an airplane in flight are: (1) Thrust (T); (2) Weight (W); (3) Lift (L); (4) Drag (D), as illustrated in Figs. 1 and 2.

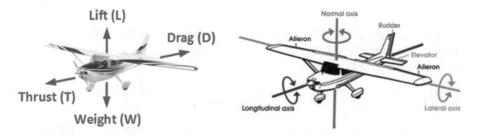

Fig. 1. Basic forces of an airplane.   **Fig. 2.** Rotational movements of the airplane.

Thrust: It is the pushing force that allows the plane to move through the air mass which is provided by the aircraft engine, and is opposite to the strength of the resistance, which equation is expressed as:

$$F_t = V_m \cdot k_h \tag{1}$$

Where: F_t is the thrust force; V_m is the motor speed of the propeller; k_h is the thrust constant depending on the type of propeller.

Weight: It is the result of the force of attraction exerted by gravity on all bodies located on the surface of the earth, attracting them to its center. The force of gravity opposes lifting on the plane, both on the ground and during the flight, whose equation may be expressed as:

$$W_a = m_a \cdot g \tag{2}$$

Where: W_a is the weight of the airplane; m_a is the mass of the plane; g is the constant of the acceleration of gravity.

Lift: It is the ascension force that allows the plane to stay in the air. The lift is created mainly in the wings, tail and to a lesser extent, in the fusclage or structure. In order to the aircraft to be able to fly, the lift force must equal its weight $(L = W)$, thus counteracting the force of gravity. Its equation is:

$$L = \frac{1}{2}\rho \cdot V^2 \cdot S \cdot C_L \tag{3}$$

Where: L is the sustaining force; ρ is the density of the fluid where the body moves, for this case in the air; V is the speed of the airplane; S is the wing surface; C_L is the coefficient of lift.

Drag: It is the resistance that exerts the air as the airplane moves and it tries to minimize to make more efficient the consumption of fuel and increase the speed. The corresponding equation may be expressed as:

$$D = \frac{1}{2}\rho \cdot V^2 \cdot S \cdot C_D \tag{4}$$

Where: D is the strength of the drag; ρ is the density of the fluid where the body moves, being the air; V is the speed of the airplane; S is the wing surface; C_D is the coefficient of the resistance.

3.2   Rotational Movements of the Airplane

An airplane can rotate around three axes perpendicular to each other, whose point of intersection coincides with the airplane's center of gravity. According to [9, 29, 30], the three axes of rotation are: (a) Transverse or lateral axis (y-axis), which is an imaginary axis that extends from one end of the wing of the plane to the other. The pitch is a rotation with respect to the transverse axis of the airplane that crosses the wings, controlled primarily by the elevators; (b) Longitudinal axis (x-axis), which is an imaginary axis that extends from the nose to the tail of the plane. The roll is a rotation with respect to the longitudinal axis, controlled primarily by the ailerons and secondarily by the rudder to generate the effect of banking; (3) Vertical axis (z-axis), which is an imaginary axis that passes through the plane's center of gravity, is perpendicular to the transverse and longitudinal axes. The yaw is a rotation with respect to the vertical axis of the plane, controlled primarily by the rudder (see Fig. 2).

4 Research Design and Implementation

For the design and development of the flight simulator it has been necessary to define four mathematical models, considering that the aircraft is a mass subjected to different forces. In this way, an analysis has been made of each of these forces and the causes that produce them. This has allowed the flight simulator to function correctly and for the pilot to be able to manipulate the airplane's controls, as if he were in the real world.

4.1 Mathematical Model of the Front Velocity

Front velocity: The thrust force generates the frontal displacement of the airplane. This displacement can be explained according to the classic physical laws taking to F_t like a punctual force applied to the mass of the airplane. Considering the following equations:

$$F_t = m_a \cdot a_a \tag{5}$$

$$W_a = m_a \cdot g \tag{6}$$

$$V_x = V_{ix} + a_a \cdot t \tag{7}$$

The following equation is obtained:

$$V_x = V_{ix} + \left(\frac{V_e \cdot k_h}{m_a} \right) \cdot t \tag{8}$$

Where: F_t is the thrust force; V_e is the engine speed or helix; k_h is the thrust constant; ma is the mass of the airplane; a_a is the front acceleration of the airplane; V_x is the front speed; V_{ix} is the initial velocity of the airplane; t is the time.

4.2 Mathematical Model of the Vertical Velocity

In the case of lift, it must be considered that the wing surface corresponds to two identical wings, so the following equations are obtained:

$$L_T = L_l + L_r \tag{9}$$

$$L_T = \frac{1}{2} \rho \cdot V^2 \cdot (S_l + S_r) \cdot C_L \tag{10}$$

Where: L_l is the left lift; L_r is the right lift; S_l is the left wing surface; S_r is the right wing surface;
Thus:

$$S_l = S_r \tag{11}$$

Considering that each of the wings of the airplane has an aileron that helps it change the wing surface in order to vary the lift provided by it, the numerical variation of the surface can be modeled by the following equation:

$$S_w = S_w - (S_w \cdot sin(\beta_t) \cdot k_a) \tag{12}$$

Where: S_w is the wing surface; β_t is the angle of inclination; k_a is the aileron constant depending on the type of material construction.

According to Eq. (12), it is indicated that if the angle of inclination β_t is positive the surface of the wing decreases, and if β_t is negative the surface of the wing increases. The pilot controls the angle of inclination of the ailerons β_t through the angular movement of the rudder. The angular movements of the ailerons are asymmetric due to the design characteristics of the airplane. The control of the rudder on each of the ailerons can be represented by these two equations:

$$\beta_{tr} = -\beta_{trd} \tag{13}$$

$$\beta_{tl} = +\beta_{trd} \tag{14}$$

As it can be seen that when a turn occurs in the rudder, one of the wings gains surface while the other loses, which allows us to model the total lift in the following way. First, we replace Eq. (12) in (10), and get the Eq. (15):

$$L_T = \frac{1}{2}\rho V_x^2 C_L(S_l - (S_l \cdot sin(\beta_t) \cdot k_a) + S_r - (S_r \cdot sin(\beta_t) \cdot k_a)) \tag{15}$$

Then, we replace Eqs. (13) and (14) in (15), and get the Eq. (16):

$$L_T = \frac{1}{2}\rho V_x^2 C_L(S_l - (S_l \cdot sin(\beta_{trd}) \cdot k_a) + S_l - (S_r \cdot sin(\beta_{trd}) \cdot k_a)) \tag{16}$$

Considering the following trigonometric identities, and the Eq. (11):

$$sin(\alpha) = -sin(\alpha) \tag{17}$$

$$-sin(\alpha) = sin(-\alpha) \tag{18}$$

We replace Eqs. (11), (17) and (18) in (16) and obtain Eq. (19):

$$L_T = \frac{1}{2}\rho V_x^2 C_L(S_l - (S_l \cdot sin(\beta_{trd}) \cdot k_a) + (S_l - (S_l \cdot sin(-\beta_{trd}) \cdot k_a))) \tag{19}$$

Obtaining thus the Eq. (20):

$$L_T = \rho V_x^2 C_L \cdot S_l \tag{20}$$

Where: L_T is the total lift; V_x is the front velocity; S_l is the left wing surface; S_r is the right wing surface; β_{tr} is the turn angle to the right; β_{tl} is the turn angle to the left; β_{trd} is the turning angle of the rudder; C_L is the coefficient of lift.

The lift force of the airplane is created mainly in the wings and tail, where the fuselage generates a minimum lift force, which will not be taken into account for this study.

For the plane to fly the L_T must overcome W_a, this is achieved thanks to the speed of the aircraft and this is produced by the thrust force of the engine. Then, the lifting force of the aircraft depends on the frontal velocity, and additionally, the equilibrium point that occurs when the magnitude of the lifting force L_z is equal to the magnitude of the weight of the aircraft W_a can be achieved.

The lift force generates the vertical displacement of the airplane. This displacement can be explained according to the classic physical laws taking to this lift force like a punctual force applied to the mass of the airplane. At this point it is necessary to consider that the aircraft can have lateral inclinations which generate an angular displacement of the L_T. This angular displacement generates a cosine component on the vertical axis, the cosine component will be lower than the L_T, and the airplane will lose the lift force for its flight, as can it be seen in Figs. 3 and 4.

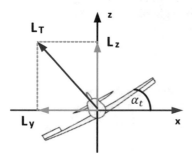

Fig. 3. Tilt of the aircraft.

Fig. 4. Turns of the airplane made through the ailerons

Considering the following equations:

$$L_T = \frac{1}{2}\rho \cdot V^2 \cdot S \cdot C_L \tag{21}$$

$$L_Z = L_T \cdot \cos(\alpha_t) \tag{22}$$

$$F_L = L_z - W_a \tag{23}$$

$$F_L = m_a \cdot a_v \tag{24}$$

$$V_V = V_{vi} + a_a \cdot t \tag{25}$$

The following equation is obtained:

$$V_V = V_{vi} + \left(\frac{L_T \cdot cos(\alpha_t) - W_a}{m_a}\right) \cdot t \tag{26}$$

Where V_v is the vertical velocity as a function of the vertical lift force. As seen, the lift force of the airplane depends on the speed of the aircraft and this, in turn, depends on the thrust force of the engine. By replacing (8) and (10), the Eq. (27) can be reached where the vertical speed of the aircraft becomes directly dependent on the force of the engine thrust and the time of flight:

$$V_v = V_{vi} \frac{t \cdot \left(\left(\rho\left(V_{ix} + \left(\frac{V_e + k_h}{m_a}\right) \cdot t\right)^2\right) \cdot S_W \cdot C_L \cdot cos\left(\alpha_t\right) - W_a\right)}{m_a} \tag{27}$$

Where: V_v is the vertical velocity of an airplane; V_{vi} is the initial vertical velocity of an airplane; V_{ix} is the initial velocity of the airplane; V_e is the engine velocity of an airplane; k_h is the thrust constant; m_a is mass of an airplane; t is the flight time; S_w is the wing surface; C_L is the coefficient of lift; α_t is the angle of rotation; W_a is the weight of an airplane.

4.3   Mathematical Model of Lateral Velocity

The velocity in the y-axis or V_y is the speed of lateral displacement of the airplane, being a function dependent on the resultant force L_y of the sinusoidal component of L_T. This can be seen in Fig. 4, and whose equation is as follows:

$$V_L = V_{iy} + \left(\frac{L_y}{m_a}\right) \cdot t \tag{28}$$

The lateral velocity is to be dependent on the total lift force produced by the wings, and this, in turn, makes it directly dependent on the speed of the aircraft is proceeded to make an analysis similar to the vertical velocity, where it has considered the horizontal movement does not affect the weight of the plane. The equation obtained is (29):

$$V_L = V_{iy} \frac{t \cdot \left(\left(\rho\left(V_{ix} + \left(\frac{V_e + k_h}{m_a}\right) \cdot t\right)^2\right) \cdot S_W \cdot C_L \cdot sin(\alpha_t)\right)}{m_a} \tag{29}$$

Where: V_L is the lateral velocity of an airplane; V_{iy} is the initial velocity on the y-axis of an airplane; V_{ix} is the initial velocity on the x-axis of the airplane; V_e is the engine velocity of an airplane; k_h is the thrust constant; m_a is mass of an airplane; t is the flight time; S_w is the wing surface; C_L is the coefficient of lift; α_t is the angle of rotation.

4.4   Mathematical Model of the Turns of the Plane

The turn of the plane is the maneuver by which the direction of flight of the airplane is changed, and basically consists in the displacement of the aircraft around its longitudinal axis. In the turn, the lift force L remains perpendicular to the wings while the W_a remains directly at the ground. Therefore, the lift force L is broken down into Ly, having to be equal to the lift. For this reason, every turn power will be applied on the controls so that the aircraft does not descend and lose control.

The turns are made through the ailerons, which have an asymmetric movement. As the rudder is turned to one side, the aileron of the wing on that side rises while the opposite wing lowers, both at a deflection angle proportional to the amount of rotation given by the rudder. The aileron is arranged upward in the wing where the rudder moves and involves less curvature in that part of the wing and therefore less lift, which causes that wing to go down while the aileron below the wing has greater curvature and lift, making that wing rise. This combination of opposite effects is what produces the movement of roll towards the wing that descends, as shown in Fig. 4:

If the plane is in a straight and level flight, then: $L_{left} = L_{right}$; if the plane is in a turn to the left, then: $L_{left} < L_{right}$; if the plane is in a turn to the right, then: $L_{left} > L_{right}$.

According to this analysis and using the Eq. (3) of Lift, it can be deduced that:

$$L_w = \frac{1}{2}\rho \cdot V^2 \cdot S \cdot C_L \tag{30}$$

$$L_{wl} = \frac{1}{2}\rho \cdot V^2 \cdot (S_l - S_{al} \cdot sin\,(\beta_t) \cdot k_a) \cdot C_L \tag{31}$$

$$L_{wr} = \frac{1}{2}\rho \cdot V^2 \cdot (S_r - S_{ar} \cdot sin\,(\beta_t) \cdot k_a) \cdot C_L \tag{32}$$

Where: L_w is the alar lift; L_{wl} is the left alar lift; L_{wr} is the right alar lift; S_l is the left surface; S_r is the right surface; S_{al} is the left surface of the left aileron; S_{ar} is the right surface of the left aileron; β_t is the angle of inclination of the aileron; k_a is the aileron constant depending on the type of material construction.

The movement of the elevator generates a force that allows the plane to rotate on the y-axis. Thus, the plane redirects the force of the thrust Fe vertically, allowing the plane to rise or fall faster, as it is deduced in the following equations:

$$W_y = W_{iy} \cdot r + \left(\frac{F_{ele} \cdot r}{m_a}\right) \cdot t \tag{33}$$

$$F_{ele} = f(\gamma_{rud}) \tag{34}$$

$$F_{ele} = \gamma_{rud} \cdot k_{ele} \tag{35}$$

Replacing Eqs. (34) and (35) in (33) gives the following equation:

$$W_y = W_{iy} \cdot r + \left(\frac{\gamma_{rud} \cdot k_{ele} \cdot r}{m_a}\right) \cdot t \tag{36}$$

The rudder produces a turn in the z-axis allowing the aircraft to redirect the lift force F_e in the horizontal direction, thus helping the aircraft to make a tangential shift in the curve during a turn, as is deduced in the following equations:

$$W_z = W_{iz} \cdot r + \left(\frac{F_{rud} \cdot r}{m_a}\right) \cdot t \tag{37}$$

$$F_{rud} = f\left(\theta_{ped}\right) \tag{38}$$

$$F_{rud} = \theta_{ped} \cdot k_{rud} \tag{39}$$

Replacing Eqs. (38) and (39) in (37) gives the following equation:

$$W_z = W_{iz} \cdot r + \left(\frac{\theta_{ped} \cdot k_{rud} \cdot r}{m_a}\right) \cdot t \tag{40}$$

The mathematical models discussed above have been implemented in the Unity 3D game engine for simulation software, which will be explained below.

4.5 Explanation of the Main Equations and Functions of the Flight Simulator

The main scalar and vector equations used were: (1) Kinematic equations, such as the relationship between position and linear velocity [29, 30]; the relationship between attitude and angular velocity [29, 30]; the relationship between the frontal velocity, the lateral speed and the speed of rotation with the lift force that were the focus of this study; (2) Dynamic equations, such as the relationship between linear and angular velocities and applied forces and moments [29, 30].

The main functions developed in the programming language C# with Unity that allow control the entire mathematical model of the flight of the plane in a virtual environment, for the Cessna 172 plane model, were the following:

The Awake() function, prevents Unity from having to search for the object to which we want to access using threads.

The FixedUpdate() function, controls the cross-entry platform to obtain the values of the axes and buttons of the joystick using threads.

The Start() function obtains a rigid body which is an airplane, and from it the vector resistance force (Drag) is generated by considering the angular forces to which the airplane is subjected and the motor torque is generated to move the airplane.

The seven performance control values of the Cessna 172 airplane recommended by [31], are implemented through variables with Unity, as indicated below: (1) Cruise speed: 226 km/h; (2) Stall speed: 87 km/h; (3) Never exceed speed: 302 km/h: (4) Range: 1,289 km; (5) Service ceiling: 4,100 m; (6) Rate of climb: 3.66 m/s; (7) Wing loading: 68.6 kg/m².

The Move() function is executed within the Unity Update () function so that the rest of the functions are executed, which will simulate the flight of the plane, such as the following:

- ClampInputs() function, which controls the inputs to assign a value between −1 and 1, to the roll, pitch, yaw and the plane's accelerator.
- CalculateRollAndPitchAngles() function, which calculates the angles of rotation using the mathematical model of plane turns.
- AutoLevel() function, which auto-levels the flight of the plane, controlling the roll and pitch values.
- CalculateForwardSpeed() function, which calculates the speed in advance and maintain it between 0 and 120 knots, for practical purposes of the simulation.
- ControlThrottle() function, which controls the accelerator, which is linked to the z axis of the joystick.
- CalculateDrag() function, which controls the resistance generated by part of the air during the flight, taking into account the factors of friction, temperature and, if necessary, the brakes.
- CalculateAerodynamicEffect() function, which calculates the aerodynamic effect of the movement of the plane, in which the drag effect is applied.
- CalculateLinearForces() function, which calculates the linear forces that are applied to the surfaces of the wings of the airplane such as Lift (Lift), among others.
- CalculateTorque() function, which calculates the torque of the aircraft engine.
- CalculateAltitude() function, which calculates the altitude at which the plane is located to handle the virtual instruments of the plane.

5   Experimental Results and Discussion

For the design and development of the flight simulator, the basic forces, speeds, and turns to which the airplane is subjected to during flight time has been controlled. For this, the mathematical models presented in this study that generate the movements of the airplane has been used. Three prototypes of the flight simulator has been built to test the functionality of the algorithms: (a) Prototype 1, to control the movement of the airplane with a joystick, without the mathematical models; (b) Prototype 2, to control the movement of the airplane with keyboard and applying mathematical models; (c) Prototype 3, to control the movement of the airplane with a joystick applying the mathematical models.

In addition, three experiments have been performed in which the algorithms of the mathematical models are integrated into algorithms for the handling and communication of the simulator, which functions within a possible Stewart platform of 3 degrees of freedom. This has been operated by a control system that might develop in LabView. This algorithm receives the variables of the transformation of the rotation in the axes x, y, z, so that Stewart's platform, which is connected to the simulator, obtains the roll, pitch and yaw values. The evolution of the communication algorithm for controlling the movement of a Stewart platform has been as follows:

Version 1 of the WriteCoordinates() function, in which the movement control variables are written separately and sent to be written in a text file, obtaining a latency time of 0.63 s, and a sending of 22.3 frames per second.

Version 2 of the WriteCoordinates() function, in which the movement control variables are written at the same time that the Update() function is executed within a floating-type one-dimensional array. This should be written in a text file, obtaining a latency time of 0.37 s, and a delivery of 43.28 frames per second.

Version 3 of the WriteCoordinates() function, in which the movement control variables are sent during each Update() to the necessary variables. This is done by means of threads that are activated at the time of the call to the Update() function in the simulation, obtaining a latency time of 0.02 s, and a sending of 59.8 frames per second.

Figures 5 and 6 presents the result obtained in the different versions of the simulation software developed:

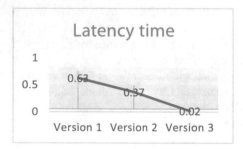

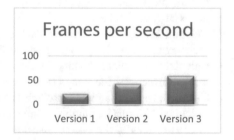

Fig. 5. Mean scores for latency time.   **Fig. 6.** Mean scores for frames per second.

Based on these results, the latency time was reduced from 0.63 s to 0.02 s, thus achieving a transmission of almost 60 frames per second. This would simulate in real time the movement of the airplane in the virtual world with the reproduction of the movements used by the selected Stewart platform, as illustrated in Fig. 7:

Fig. 7. Software interface of the version 3.

6   Conclusions and Future Work

We design mathematical models of the physical phenomenon that occurs during the flight of an airplane based on its frontal velocity and the lifting force. Hereby, vertical and lateral forces have been obtained in relation to the lift forces of the aircraft, which corresponded to equations, which have been generated for the speeds and the movements of the rudder and the elevator that allow the control of the flight of the airplane within the simulation software. The use of threads in the flight simulator allowed the response times to be short as evident by the Update() function of Unity 3D. These are always running and only send values the moment the state of movements within the simulator changes or has been altered. The numerical verifications and the calibration process of the flight simulator were based on the existing manuals for the Cessna 172 aircraft model [31], where both the general characteristics and the performance features of this model were considered for the parameterization of the simulation system.

As future work, we have planned the construction of a real-scale simulator for the verification of mathematical models and interconnected to a Stewart platform of 3 to 6 degrees of freedom, which reproduces the movements of an airplane.

References

1. Allerton, D.: Principles of Flight Simulation, 1st edn. Wiley (2009). ISBN: 0470754362
2. Stevens, B., Lewis, F., Johnson, E.: Aircraft Control and Simulation: Dynamics, Controls Design, and Autonomous Systems, 3rd edn. Wiley-Blackwell (2015). ISBN: 1118870980
3. Alexandrov, V.V., Bolotin, Y., Lemak, S.S., Parusnikov, N.A., Zlochevsky, S.I., Guerrero, S.W.F.: Introduction to Control of Dynamic Systems, p. 239. Ediciones BUAP (2009). ISBN: 978-607-487-095-4
4. Matlalcuatzi, E., Alexandrov, V.V., Matlalcuatzi, F., Altamirano, L., Reyes, M., Moctezuma, J., Hernandez, O.: Diseño del simulador dinámico para pilotos como un sistema biomecatrónico. In: Memorias del XVI Congreso Latinoamericano de Control Automático, CLCA 2014, 14–17 Octubre 2014, Cancún (2014)
5. Rekdalsbakken, W.: Design and application of a motion platform in three degrees of freedom. In: Proceedings of the 46th Conference on Simulation and Modelling (SIMS 2005), pp. 269–279, Tapir Academic Press (2005). NO-7005 TRONDHEIM
6. Rekdalsbakken, W.: Design and application of a motion platform for a high-speed craft simulator. In: IEEE 3rd International Conference on Mechatronics (ICM 2006) (2006). https://doi.org/10.1109/icmech.2006.252493
7. Henson, R.: Unity 4.x Game Development by Example Beginner's Guide. Packt Publishing (2013). ISBN: 1849695261
8. Patrick, F.: Unity from Proficiency to Mastery (C# Programming): Master C# with Unity. Amazon Digital Services LLC (2017). ASIN: B076ZQHKQT
9. Villacís, C., Fuertes, W., Bustamante, A., Almachi, D., Prócel, C., Fuertes, S., Toulkeridis, T.: Multi-player educational video game over cloud to stimulate logical reasoning of children. In: IEEE/ACM 18th International Symposium on Distributed Simulation and Real Time Applications. https://doi.org/10.1109/ds-rt.2014.24
10. Merk, R.J., Roessingh, J.J.M.: Assessing behaviour of cognitive agents in a flight simulator with fighter pilots. In: IEEE International Conference on Systems, Man, and Cybernetics (SMC). IEEE (2016)

11. Boril, J., Leuchter, J., Smrz, V., Blasch, E.: Aviation simulation training in the Czech air force. In: IEEE/AIAA 34th Digital Avionics Systems Conference (DASC), p. 9A2-1. IEEE, September 2015
12. Cen, F., Li, Q., Fan, L., Liu, Z., Sun, H.: Development of a pilot-in-loop real-time simulation platform for wind tunnel free-flight test. In: IEEE International Conference on Information and Automation, pp. 2433–2438. IEEE, August 2015
13. Louali, R., Belloula, A., Djouadi, M.S., Bouaziz, S.: Real-time characterization of Microsoft flight simulator 2004 for integration into hardware in the loop architecture. In: 19th Mediterranean Conference on Control & Automation (MED), pp. 1241–1246. IEEE, June 2011
14. Mauro, S., Gastaldi, L., Pastorelli, S., Sorli, M.: Dynamic flight simulation with a 3DOF parallel platform. Int. J. Appl. Eng. Res. **11**(18), 9436–9442 (2016)
15. Cheon, S.-II., Ha, S.-W., Moon, Y.-H.: Hardware-in-the-loop simulation platform for image-based object tracking method using small UAV. In: IEEE/AIAA 35th Digital Avionics Systems Conference (DASC). IEEE (2016)
16. Wu, L.-N., Sun, Y.-P.: Development of a low-cost flight simulation training device for research and education. In: Proceedings of the 2nd International Conference on Intelligent Technologies and Engineering Systems (ICITES2013). Springer (2014)
17. Munzinger, C., Anthony, J.C., Prasad, J.V.R.: Development of a real-time flight simulator for an experimental model helicopter (1998)
18. Chen, J.: Helicopter real-time flight simulation environment development and digital signal processor applications (2006)
19. Liu, D.-N., Hou, Z.-X., Gao, X.-Z.: Flight modeling and simulation for dynamic soaring with small unmanned air vehicles. Proc. Inst. Mech. Eng. Part G J. Aerosp. Eng. **231**(4), 589–605 (2017)
20. Setiawan, J.D., Setiawan, Y.D., Ariyanto, M., Mukhtar, A., Budiyono, A.: Development of real-time flight simulator for quadrotor. In: International Conference on Advanced Computer Science and Information Systems (ICACSIS), pp. 59–64. IEEE, December 2012
21. Hays, R.T., et al.: Flight simulator training effectiveness: a meta-analysis, p. 63 (1992)
22. Lofaro, R.J., Smith, K.M.: The aviation operational environment: integrating a decision-making paradigm, flight simulator training and an automated cockpit display for aviation safety. In: Technology Engineering and Management in Aviation: Advancements and Discoveries, pp. 241–282. IGI Global (2012)
23. Gervais, C., Chaudron, J.B., Siron, P., Leconte, R., Saussié, D.: Real-time distributed aircraft simulation through HLA. In: IEEE/ACM 16th International Symposium on Distributed Simulation and Real Time Applications (DS-RT), pp. 251–254. IEEE, October 2012
24. Chaudron, J.B., Saussié, D., Siron, P., Adelantado, M.: Real-time distributed simulations in an HLA framework: application to aircraft simulation. Simulation **90**(6), 627–643 (2014)
25. Lorains, M., MacMahon, C., Ball, K., Mahoney, J.: Above real time training for team invasion sport skills. Int. J. Sports Sci. Coach. **6**(4), 537–544 (2011)
26. Scamps, A., Gibbens, P.: Development of a flight simulator evaluation course at QANTAS. In: AIAA Modeling and Simulation Technologies Conference and Exhibit (2005)
27. Khan, A.H., Khan, Z.H., Khan, S.H.: Optimized reconfigurable autopilot design for an aerospace CPS. In: Computational Intelligence for Decision Support in Cyber-Physical Systems, pp. 381–420. Springer Singapore (2014)
28. Motes, A.: Physics of Flight: An Introduction. AM Photonics, 3rd edn. (2016). ASIN: B01AIPUIU8

29. Vepa, R.: Flight Dynamics, Simulation, and Control: For Rigid and Flexible Aircraft, 1st edn. CRC Press (2014). ASIN: B00MOU4S66
30. Raol, J.R.: Flight Mechanics Modeling and Analysis, 1st edn. CRC Press (2008). ASIN: B005H6YDUK
31. Roud, O., Bruckert, D.: Cessna 172SP Training Manual. CreateSpace Independent Publishing Platform (2017). ISBN-10: 1519617070

Autonomous Video Surveillance Application Using Artificial Vision to Track People in Restricted Areas

Yordi Figueroa, Luis Arias, Dario Mendoza$^{(\boxtimes)}$ ⑩, Nancy Velasco,
Sylvia Rea, and Vicente Hallo

Universidad de las Fuerzas Armadas ESPE, Sangolquí, Ecuador
{ywfigueroa,lnarias,djmendoza,ndvelasco,
snrea,vdhallo}@espe.edu.ec

Abstract. The present project implements an application for the search, recognition and monitoring of people based on artificial vision algorithms. The OpenCV libraries are used to process the images, which were obtained from a conventional IP video surveillance camera. This type of cameras can be used in different environmental conditions (high, medium and low lighting) and up to an effective distance of 70 m. In the detection and search phase, cascade classifiers are used with local binary patterns LBP (Local Binary Patterns). Subsequently, in the follow-up phase, a tracking algorithm is implemented, addressed only to the person detected through kernelized correlation filters KCF (Kernelized Correlation Filters), so that the objective is not lost. A graphical interface was developed in the Qt Software which allows an easy use of the application. The average effectiveness of the algorithm is 90% in different environments and places by mitigating the different luminosity changes.

Keywords: Video surveillance · Artificial vision · OpenCV · Tracking
ANN · Autonomous surveillance · Recognition

1 Introduction

The detection of people by artificial vision is a topic that has been taking more and more importance during decades but the detection is not a trivial task. This detection systems are fundamental for autonomous systems such as vehicles that through artificial vision avoid people involving accidents [1]. Artificial vision also allows to improve video surveillance systems that are activated by non-human forms such as animals or environmental conditions such as dust and wind.

Background subtraction is one of the key techniques for automatic video analysis, especially in the domain of video surveillance [2].

The detection of people involves several techniques. In [3], people are tracked through mutual occlusions by an adaptive background subtraction method that combines color and gradient information. The algorithm of detection [4] employs a combination of shape analyses to identify and to locate a person and its body parts (head,

© Springer International Publishing AG, part of Springer Nature 2018
Á. Rocha and T. Guarda (Eds.): MICRADS 2018, SIST 94, pp. 58–68, 2018.
https://doi.org/10.1007/978-3-319-78605-6_5

hands, feet, torso) and to create a reference so that people can be tracked through interactions such as occlusions.

In [5] authors present a 3D people detection and tracking approach using RGB-D data. By enriching the algorithm with information allows its goals self-learning to improve the detection and association of information. In [6] an algorithm that track people with twists and exponential maps is implemented.

This research project develops an algorithm to search, detect and track people based on artificial vision libraries, which works in varying light conditions and up to a distance of 70 m varying the zoom of the camera.

The system implemented uses a phase to recognize only people based on neural networks. This recognition phase is designed to detect a standing person from their front, back or side.

On regards of the processing needs, as the recognition phase needs a bigger "computing" processing capacity than the tracking one, there is an intermediate phase between those two.

This innovative intermediate phase gives control to the system to recognise and to commute automatically from the recognizing phase to the tracking phase. By the implementation of the intermediate phase, the system becomes more robust and efficient in its processing.

2   Development

2.1   Training Phase

The recognition of a human form needs a learning method that discriminates whether it is a person (positive image) or not (negative image). This discrimination requires a large amount of information and training as mentioned in the work of Viola-Jones [7], which uses the cascade method to decide whether a training image is positive or negative.

There are several training files of people presented in OpenCV, but the one made by Vision-Ary [8] contains a large database and stages. This detection algorithm is more efficient and less prone to detect false positives, reaching an effectiveness of 93%, however the training done by OpenCV (haarcascade_fullbody) has a 78% effectiveness in tests and similar conditions Fig. 1.

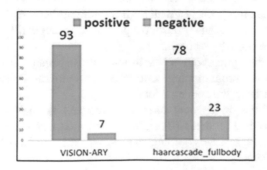

Fig. 1. Testing of training records

2.2   Training Phase

The tracking is executed through kernelized correlation filters KCF (Kernelized Correlation Filters). This algorithm is a follower based on the Kernel Ridge Regression function as a mathematical function to predict the position of the object. The benefit is reduced storage and computational cost and is a faster tracking method than a traditional detector.

The linear regression equations can calculate the following value of movement by knowing the value of certain variable such as: the position [9]. Figure 2(b) while the non linear have the capacity to predict the movement in different directions Fig. 2(a).

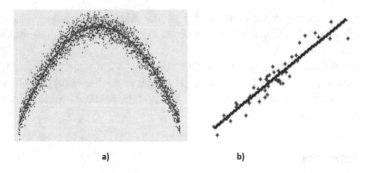

a)   b)

Fig. 2. Linear regression of several positions (a) non lineal, (b) lineal

The sensor that gets the visual information is an IP camera connected to a Local Area Network (LAN Network). This device is used in several professional video surveillance fields due to its stability and fiability. The system uses be controlled through C++ Programming Language and OpenCV Libraries [13, 14].

An advantage of using IP camera is that the images can be seen at the same time by how many users connected to the network. Real Time Streaming Protocol (RTSP) is used process the images in OpenCV because it works in low resolution allowing real time processing. Audio and video are commonly delivered by RSTP communication protocol.

The majority of computerized vision processes require of a big quantity of computational cost. When using images with high resolution, there might be a few seconds delay in the IP camera because transferring images through IP Cameras are much slower than the one done by webcams.

The control of the optic zoom is key in the process because it should be adjusted automatically from the programming code to detect and track individuals in the rage between 3 to 70 m from the camara's lens.

The control of the optic zoom uses CGI protocol through Hypertext Transfer Protocol (http), Physical Address (IP address) and a data frame. These components control its absolute and relative coordinates to zoom in and zoom out based on the area of the detected person.

The CGI are applications that control different features, for instance alarms, alerts or controls to store pictures and videos. The CGI have a mechanism to store in which each user has an individual data based [10].

Both, the client (Qt program and OpenCV) and the server (HIKVISION IP camera) intervene in the CGI functioning. The client send a petition to the server. The server receives the petition and sends it with the data frame in a CGI extension. The petition is processed and a notification is send back to the client informing them that the operation has been done.

The Qt Program sends information using HTTP protocol through the camera [12] IP 192.168.1.64 to control the absolute position of the zoom 10 m (1x), 20 m (2x), 100 m. When the request has been executed, a notification is delivered to the client informing them that the operation has been done.

Figure 3, an example of the CGI data frame code to get the absolute position of the zoom is: http://admin:ABC12345@192.168.1.64/PTZCtrl/channels/1/status.

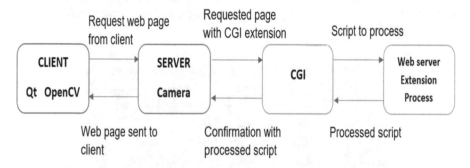

Fig. 3. Operation of CGI commands

To control the depth is needed to consider the detecting area as the biggest rectangle that contains the person detected in the algorithm. The rectangle is calculated using the formula of the area as follows: P (X,Y) the initial point of the rectangle, and P (X + width, Y + height) the final point of the rectangle.

$$Area = (rectangle.width) * (rectangle.height)$$

Considering that the initial point (origin) is the upper left corner in the rectangle: X is the initial point in the abscissa, the coordinate (X + width) is the final point in the abscissa, Y is the initial point in the ordinate, and the coordinate (Y + height) is the ending point in the ordinate.

The area should fulfill certain conditions to zoom in and zoom out the person detected.

$$Area > Amin = \text{ZOOM OUT} \tag{1}$$

$$Area < Amax = \text{ZOOM IN} \tag{2}$$

Amin and Amax are constant values established by test within the program and determine if the person gets closer or away from the camera.

The Region Of Interest (ROI) or detection area contains values that range from 50 to 80 pixels^2, these values are gotten after testing physical characteristics from different people, such as height or wide.

There are interactions between ROI and AMIN-AMAX, for instance, if Amin has a value of 50 and ROI has a value less than 50, the individual is getting away. In that case the zoom has to increase.

In the other hand, if Amax, which has a value of 80, is exceeded the zoom is reduced to guarantee the reference to the ROI.

As can be seen in the Fig. 4, when the ROI contains values from 50 to 80 px the person can be detected because they are within the image (a). When the person is too far, they can not be detected and it is required an increase in the zoom (c). When the individual is too close, a zoom out is required because the silhouette is not completely visualized so the person cannot be detected (b). To avoid ROI data loss (b) (c), it is necessary to use Amin and Amax, which controls the zoom to avoid ROI loss and to fully visualize the silhouette.

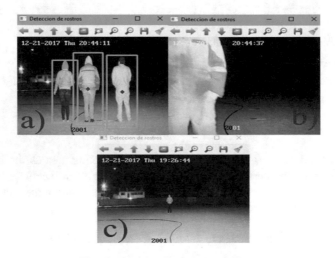

Fig. 4. Process displayed in images

3   Experimental

The algorithm is composed by a series of processes that are summarized in Fig. 5.

3.1   Image Obtaining

Image obtaining is done through an IP camera, using the real-time transmission protocol (RTSP protocol), to communicate with OpenCV [11].

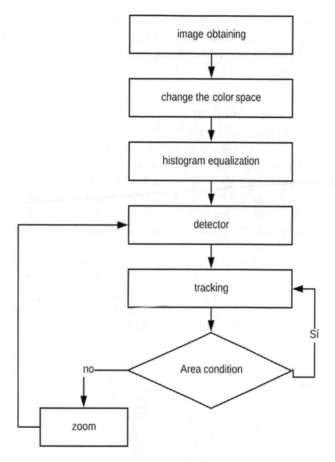

Fig. 5. Flowchart of system processes

3.2   Change the Color Space

Images captured from the environment are in the RGB color space. However it is better to apply the algorithm for object recognition in grayscale space because its main advantage is to work with a single channel image.

3.3   Histogram Equalization

Normalizes the brightness and contrast of the image to improve its properties and ease people detection.

3.4   Detector

OpenCV includes functions to detect objects and people as well as to allow a cascades training, through the algorithm of Viola-Jones [7].

It is necessary to configure the parameters of the multiscale function. If the configuration is correct, the areas of the objects are sent back as vectors. The vector's data is used to the next phase which is the tracking one.

3.5   Tracking

The tracking is done through the KCF Kernelized Correlation Filters algorithm. Using the information from the previous coordinates of the detector, the next movement can be predicted. KCF works in a range of tests of 50 videos, overcoming the last generation like: DCF Dual Correlation Filter, Struck (Structured output tracking with cores), TLD Tracking-Learning-Detection, MOSSE (Monitoring of visual objects) using adaptive correlation filters), MILL, ORIA, and CT Real-time Compressive Tracking. KCF's main advantage is that it is much faster since it uses regression equations of circulating matrices that considerably reduces the computational cost and allows simpler implementation [9].

Tracking is robust because it can track a detected person even though they to move freely (jumping, walking, running, sitting, and in some postures.) The tracker is very important in the project since if only one detector is used. It is possible that the person cannot be detected due that the training of the classifier was realized only with images of people in front, back and side, but not while they were doing actions and movements different to the ones mentioned above (Fig. 6).

Fig. 6. Tracking robust even IF the person move freely

3.6   Area

In order to manage a stable area in the recognition of the person, the limits were established through experimental tests where the correct functioning of the tracking algorithm can be ensured.

The IP camera can automatically adjust the zoom by commands from the programming code ensures to be able to track the person detected efficiently and reliably between 3 and 70 m.

4 Results and Discussions

The tests were carried out in a rural cobblestone road and in an open stadium with organic grass (approximately 90 × 50 m). The tests were performed up to 70 m. Fifty tests were done targeting people with different physical characteristics and after clothes changes.

In the rural cobblestone road, the performance of the algorithm was monitored in two scenarios: in the first scenario, there were animals, trees, poles, and cars. The changes in luminosity minimal impact on the correct identification of the person.

The implemented algorithm does not use the conventional techniques of background subtraction, nor optic flow because those methods are not efficient. These methods activate the system by any element that changes the image of reference, for instance animals or moving vehicles.

For this reason, the algorithm implements a recognizing phase based on neuronal networks to be activated only by people. Ensuring an optimal operation, the system does not activate only with moving objects for this reason without risk for animals or vehicles (Fig. 7).

Fig. 7. Scene with animals and vehicle do not detected by algorithm

The algorithm only detects and tracks a person. Once detected the person is framed within a sky blue rectangle. Other objects entering in the scene are framed in a yellow circle just for visualization. Even though they appear in the scene, they do not activate the system.

In the second scenario, the stadium was considered as a restricted area, since there were no animals, nor circulating cars. The goal of this scenario is to evaluate whether the algorithm of detection and tracking of people works correctly together with the implemented system.

The tests were performed in different hours of the day from 6 am until 6 pm and during the night from 6 to 10 pm to evaluate how the algorithm behaves when faced with different light changes.

Figure 8 shows the results from the tracking tests carried out during day and night up to 70 m.

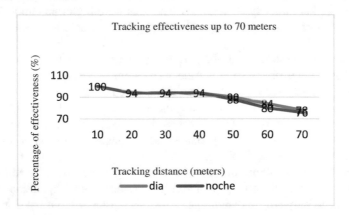

Fig. 8. Effective tracking percentage in both, day and night

Figures 9 and 10, it can be seen that the effective tracking percentage in both, day and night time are equal until 40 m because during the night the IP camera has its own infrared LEDs that compensate the daylight to obtain a sharper image.

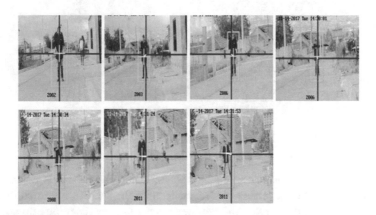

Fig. 9. Tracking in the day

After this, the percentage the tracking decreases to 78% of effectiveness in the day and 76% in the night until the 70 m.

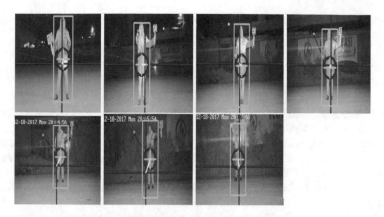

Fig. 10. Tracking in the night

5   Conclusions

The algorithm to search, to detect and to track people by using artificial vision is capable to perform in different light conditions (day, afternoon, night), and surveillance restricted areas in a distance up to 70 m.

The detection and tracking people is robust to outdoor lighting changes but is not able to mitigate the conditions generated by a backlight from sunlight. This issue occurs when the backlight is oriented directly the lens and when a zoom in an over-illuminated person is done.

Optical zoom is necessary for the project because the algorithm is designed to bring the lens closer to detect people up to 70 m. To ensure a big area for surveillance, the use of optical zoom is better than digital zoom because digital zoom's poor resolution can cause the loss of the person's shape due to the noise generated.

There are many advantages of IP camera against of a traditional webcam. For example, with IP Camera the user can watch the video from several computers on the same network. The user can also obtain information and control many parameters of the camera using HTTP protocols, for example: control the light on infrared LEDs, autofocus, brightness control, among others.

The efficiency of the people detection system depends on the number of tests. If more positives and negatives images are used, the results obtained are better, the false positive results diminish, and the recognition improves.

The system requires a robust tracking algorithm because the detector is trained with images of people taken from the front, back, and sides. However, the classifier does not recognize other postures. For this reason, it is necessary to use positions intended to track a person even if they are lying down, sitting or in other positions.

The tracking algorithm works in a complementary way with the stop and zoom algorithm, thus achieving in a dynamic way, a system capable to follow the people identified up to 70 m.

Acknowledgment. The teachers of the career of Mechatronics Engineering from Universidad de las Fuerzas Armadas ESPE – Latacunga, supported this research.

References

1. Yan, J., Zhang, X., Lei, Z., Liao, S., Li, S.Z.: Robust multi-resolution pedestrian detection in traffic scenes. In: CVPR, p. 1 (2013)
2. Sebastian, B., Höferlin, B., Heidemann, G.: Evaluation of background subtraction techniques for video surveillance. In: 2011 IEEE Conference on Computer Vision and Pattern Recognition (CVPR). IEEE (2011)
3. McKenna, S.J., et al.: Tracking groups of people. Comput. Vis. Image Underst. **80**(1), 42–56 (2000)
4. Large, E.W., Jones, M.R.: The dynamics of attending: how people track time-varying events. Psychol. Rev. **106**(1), 119 (1999)
5. Pountain, D.: Track people with active badges. BYTE **18**(13), 57–64 (1993)
6. Bregler, C., Malik, J.: Tracking people with twists and exponential maps. In: Proceedings of 1998 IEEE Computer Society Conference on Computer Vision and Pattern Recognition. IEEE (1998)
7. Viola, P., Jones, M.: Rapid object detection using a boosted cascade of simple features. In: Proceedings of the IEEE Conference on Computer Vision and Pattern Recognition, Hawaii, pp. 1–5 (2001)
8. Vision Ary Team: Boost the World: Pedestrian Detection (2015)
9. Henriques, J., Caseiro, R., Martins, P., Batista, J.: High-speed tracking with kernelized correlation filters, pp. 1–8 (2015)
10. Fenyo, D., Beavis, R.: The GPMDB REST interface, p. 1 (2015)
11. Golondrino, C., Elías, G., Ordoñez, U., Arturo, F., Muñoz, C., Yesid, W.: Stress tests for videostreaming services based on RTSP protocol (2015)
12. D'Amato, J., Dominguez, L., Perez, A., Rubiales, A.: Open platform managing IP cameras and mobile applications for civil security/plataforma abierta de gestion de camaras IP y aplicaciones moviles para la seguridad civil ciudadana, p. 1 (2016)
13. Gan, J., Liang, X., Zhai, Y., Zou, L., Wang, B.: A real-time face recognition system based on IP camera and SRC algorithm, p. 1 (2014)
14. Bapayya, K., Sujitha, K., Basha, A.: RTSP based video surveillance system using IP camera for human detection in OpenCV, pp. 1–3 (2015)

Health Informatics in Military Applications

VR-System ViPoS: VR System with Visual and Postural Stimulation Using HMDs for Assessment Cybersickness in Military Personnel

Sonia Cárdenas-Delgado(✉), Mauricio Loachamín-Valencia,
and Manolo Paredes Calderón

Departamento de Ciencias de la Computación, Centro de Investigación Científica y
Tecnológica del Ejército - CICTE, Universidad de las Fuerzas Armadas - ESPE,
Sangolquí 170501, Ecuador
{secardenas,mrloachamin,dmparedes}@espe.edu.ec

Abstract. Previous studies have suggested that cybersickness could occur among users that interact with simulators, HMDs, or 3D technology. In the military sphere, this also is affecting due to use of different simulators in which the personnel is being trained. This situation could limit the learning and performance during the training. In this work, we propose an approach to developing a VR-system with visual and postural stimulation. The objective is to assess the symptoms associated to cybersickness in military personnel and determine if these symptoms could be mitigated with previous training using two types of HMDs. Assessment methods will include two questionnaires (SSQ and PQ), the head tracking, and time task. Also, we will compare two experimental conditions (Walking-Condition and Sitting-Condition). Our work will contribute to the occupational safety and health of military personnel, and guarantee an effective training in the different simulators according to their missions by mitigation of incidents.

Keywords: Cybersickness · Virtual reality · Health
Occupational safety · Military applications · Multidisciplinary projects
Oculus Rift

1 Introduction

Technological advances have led to a reduction in costs and have increased the manufacturing of 3D screen displays, TVs with 3D technology, and different kinds of head-mounted displays (HMDs) for Virtual Reality (VR). Regarding to the HMDs, they have been used widely for several years in a lot of activities such as training with virtual reality simulations to pilots, sailors, soldiers [1–3], surgeons [4]; also in education as teaching tools [5,6]; and in neuropsychological assessment [7], memory assessment [8], treatments and rehabilitation cognitive

© Springer International Publishing AG, part of Springer Nature 2018
À. Rocha and T. Guarda (Eds.): MICRADS 2018, SIST 94, pp. 71–84, 2018.
https://doi.org/10.1007/978-3-319-78605-6_6

72   S. Cárdenas-Delgado et al.

[9]; and others works as [10] that proposes the use of new tool to reduce the stigma related to schizophrenia, including military applications.

Previous studies performed with adults, children and adolescents confirmed that the video game consoles and VR devices commercially available could provoke motion sickness due a wide range of conditions for active players or passive viewers (e.g. seated or standing) [11–13].

Since the 70s the risk of motion sickness has been studied along with the effects and factors [14]. However, due to the recent popularity of consumer of HMDs, this topic has become relevant over the last couple years. Several studies have considered that human factors could influence in the symptoms of motion sickness and these can vary greatly between individuals (due to the physical and psychological state), the technology used in the design of the virtual environments, and the tasks performed [15–19]. Another type of motion sickness is cybersickness, it is a condition where a person exhibits symptoms similar to motion sickness caused by playing compute, simulation and video games [20]. In this context, there are previous studies that have reported adverse symptoms when training in flight simulators [21]. There are some prominent theories regarding what could cause motion sickness such as: postural instability theory [22], or sensory conflict theory [23–25]. The sensory conflict theory studies have suggested that this theory appears to be the best description of the effect of the conflict between the visual and vestibular systems when individuals are in a virtual experience [14, 24, 26]. The postural instability theory demonstrated that changes in body sway could provoke motion sickness in a wide variety of contexts (e.g. playing the video games and interact with virtual environments [27, 28].

Based on these theories, we propose the development of a VR-system with visual and postural stimulation (*VR-system ViPoS*). We will analyze and compare two experimental conditions of visualization and navigation inside a same virtual environment. The experimental conditions will be: Walking-Condition and Sitting-Condition. Each condition will use two types of HMDs. The purpose is to assess and compare the symptoms of cybersickness between both conditions before and after exposure in the virtual environment while the participants use the HMDs.

Our hypothesis is that the condition that combines the walk and the head motion could produce more symptoms of cybersickness than the condition that combines the postural position sitting and the head motion. Our contribution is to mitigate incidents during training in the different simulators according to their missions in each Force (Army, Navy and Air Force) to which they belong.

We will assess symptoms of motion sickness using the Simulator Sickness Questionnaire (SSQ), and the Total Severity Score will be computed according to [29, 30]. We will administer the SSQ two times, once before ("Pre-test SSQ") exposure and once after ("Post-test SSQ") exposure. For the visualization, we will use two HMDs of low-cost; one that allows tracking the head movements performed by participants and another that does not allow. For the interaction and navigation into the virtual environment, we will use a device with motion control technology. The participants also will evaluate the satisfaction and interaction after using the *VR system-ViPoS*.

This paper is structured as follows. Section 2 a review of related studies with the VR head-mounted displays, motion sickness, airsickness and seasickness in military personnel, and questionnaires standard widely used for this types studies. Section 3 describes research proposal, approach to development involved in this work, hardware, software, measures and a brief description of the procedure that will be carried out. Finally, Sect. 4 presents the conclusions and future work of research.

2 Background

There are many of people who experience motion sickness either because of vestibular stimulation or visually induced stimulation. This incidence is predicted to increase due to development technological, manufacture of new VR devices, or virtual environments that include also methods of locomotion, or with the way, we visualize information, and the tasks that we perform [31]. Several researchers report that up to 60% of the population has some motion intolerance [32]. Some studies show susceptibility to motion sickness begins at about age two, and for most will peak in adolescence and decline gradually. However, many adults remain highly sensitive particularly when there is either an absence of a visual reference or exposed to significant levels of visual stimuli [32,33].

Motion sickness is a problem that affects humans by exposure to different situations (transportation by bus, airplane, submarines, and trains, or simply by using a swing at a playground, or also using different types of simulators as car sickness, sea sickness, air sickness, space sickness or flight simulators) and conditions (sitting, standing, and walking) [1,3,25,34,35]. The people could feel unpleasant symptoms such as nausea, dizziness, headaches, and vomiting. Specifically, motion sickness is common in systems with optical depictions of inertial motion, such as flight and driving simulators, many virtual environments, or VR systems [36–38]. Motion sickness includes cybersickness, which is experienced by users in the virtual scene while they remaining stationary [39,40]. Some works, such as the work by Kennedy et al. [41] suggested that historically people have experienced motion sickness symptoms in response to inertial motion. However, this also could occur due to exposure to some types of visual displays, even with no physical motion, because it may be a result of the effects of visually induced motion sickness (VIMS). VIMS is a condition that occurs when people are exposed to stimuli of real or apparent movement by viewing 3D stereoscopic images. This also includes a number of symptoms such as visual fatigue, headaches, paleness, sweating, dizziness, ataxia, nausea, or vomiting [42,43]. In this article, when we refer to motion sickness, we include cybersickness and visually induced motion sickness because all them show similar symptoms.

Over the last decade the symptoms of motion sickness and its causes by various VR devices have been studied extensively. Some researchers such as Solimini [44] and Naqvi et al. [45] have carried out studies comparing the side effects that can occur as a result of watching 2D and 3D movies. These results showed that viewers who watched 3D movies reported more symptoms of visually induced motion sickness than those who viewed 2D movies. They suggested that 3D movies induce higher levels of motion sickness symptoms compared to the 2D movie. Treleaven et al. (2015) performed a study to assess postural stability using a HMD with a three-dimensional motion tracker built in, and a sampling rate of 30 Hz (Wrap^{TM}1200VR by Vuzix, Rochester, New York). They found that using the HMD provoked motion sickness symptoms in about one-third of the participants, but these seem to be of a lesser severity than previously reported for other device. Kuze and Ukai [46] assessed visual fatigue caused by viewing various types of motion images when playing a game using a HMD, and found that individuals felt symptoms such as eye strain, general discomfort, nausea, focusing difficulty and headache i.e., some motion sickness symptoms. Howarth [15] in his study refers to the use of display screens or head mounted displays, which generate some problems for users, and there are many causal factors; these include the physical aspects such as the display system, the weight and the users physical and psychological state.

Other studies demonstrated the severity of motion sickness symptoms that users might experience when using game controller for a locomotive task in virtual environments with an Inertial Measurement Unit (IMU) included inside the Oculus Rift [47]. Also, Llorach et al. [48] showed severe and high symptoms of motion sickness when using the Oculus Rift DK1 and a game controller. These results also showed that after exposure, the motion sickness symptoms increased.

Other works have compared different versions of the same environment using the Oculus Rift. Davis et al. [39] used the Oculus Rift and compared two different virtual roller coasters, each with different levels of fidelity. They found that the more realistic roller coaster with higher levels of visual flow had a significantly greater chance of inducing cybersickness.

On the other hand, seasickness and airsickness are a common form of motion sickness, is also frequent among military personnel. For example, in US Navy flight officers, 74% of student pilots experienced airsickness [49]. Data from the Royal Air Force revealed 50% of aviators in high performance aircraft had airsickness. Symptoms appear to be more common when military personnel has no control of the aircraft. Symptoms of airsickness increased in more provocative environments with increased turbulence [32]. Regarding naval personnel has been reported even that 60% to 90% of inexperienced sailors can suffer seasickness [50–53].

To assess motion sickness, susceptibility and the symptoms that are experienced by the users in virtual environments with different VR devices, particularly with HMDs, there are some widely used standard questionnaires: Motion Sickness Questionnaire (MSQ) [30], Revised Simulator Sickness Questionnaire (RSSQ) [54], Presence Questionnaire (PQ) [55], a questionnaire to objectively assess symptoms caused with virtual reality systems [56], along with others not yet widely accepted and used.

In this work, we will evaluate cybersickness in the personnel of pilots and student pilots of the armed forces to determine whether they are or not susceptible to feel the cybersickness symptoms. Our contribution is to mitigate incidents during training in the different simulators according to their missions in each Force (Army, Navy and Air Force) to which they belong.

3   Approach to Develop the *VR-System ViPoS*

3.1   Problem and Motivation

The governments have trimmer operational budget, its Armed Forces need to reduce training costs using different types simulators. However, the use of these simulators can produce cybersickness. According to the literature reviewed in Sect. 2, there are related works that have reported motion sickness while the military personnel train on different simulators. The impact of motion intolerance with resultant motion sickness is significant. In fact, for those who experiencing the symptoms, the result is often disabling with associated drowsiness, lack of concentration and disorientation [60]. This situation could limit the learning and performance during the training that performs the military personnel.

3.2   Development Proposal and Research

The limited learning ability during the training in a simulator could be due to the above-mentioned problems and mainly human factors like the physical and psychological state of the people. This could even increase the susceptibility to feeling the symptoms of motion sickness.

Therefore, this work proposes to develop a *VR-system ViPoS* that allows assessing the symptoms of cybersickness in military personnel and to perform previous training to the identified personnel like susceptible to feeling these symptoms before using a simulator. This work will contribute to: (a) to achieve an effective training in the simulators; (b) to mitigate symptoms of cybersickness while train in the simulators for different missions; (c) to guarantee the occupational safety of personnel military of the Army, Navy, and Aviation.

The goal is to diagnose and mitigate the symptoms of motion sickness that the military personnel could feel when they perform training using simulators. This work allows contributing to the occupational safety of military personnel, and to guarantee an effective training.

Based on two theories concerning to motion sickness (postural instability theory [19], and sensory conflict theory [23–25]), we propose developing a VR system with visual and postural stimulation (*VR-system ViPoS*). See the scheme of the VR-system proposed in the Fig. 1. Also, in Fig. 2., can be seen the architecture of the proposed *VR-system ViPoS*.

Also, the VR system is based on the prototyping method. This method consists in that each prototype acts as a model for the next and evolves based on new requirements to obtain the desired system [61]. In the Fig. 3 can be seen the phases of the prototyping.

Furthermore, we will create an algorithm for optimization through a meta-heuristic technique of the Simulated Annealing to found the location optimal of

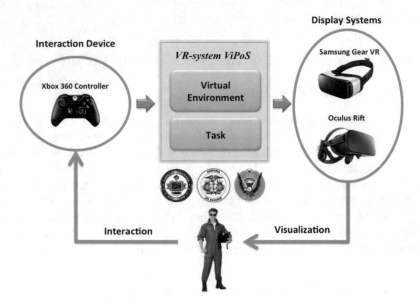

Fig. 1. Scheme of the VR-system

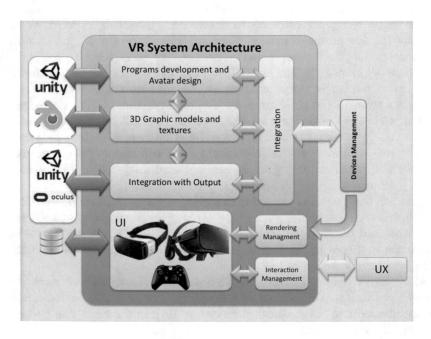

Fig. 2. Architecture of the VR-system

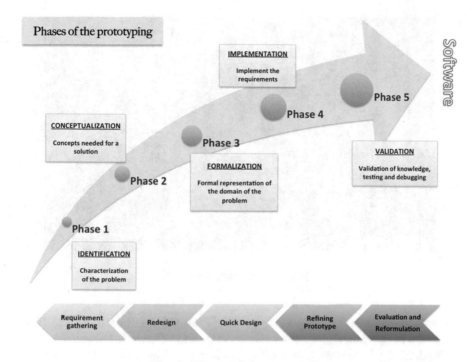

Fig. 3. Phases of the prototyping

the 3D objects into of the virtual environment. For additional details about the simulated annealing technique, refer to [62,63].

We will analyze and compare two experimental conditions of visualization and navigation inside a same virtual environment. The experimental conditions will be: Standing-Condition and Sitting-Condition. Each condition will use two types of HMDs, a Gear VR and an Oculus Rift. These devices will display a virtual environment with 3D objects for visual fixation, while experiencing visual and postural stimulation.

We can see in Fig. 4. The stimulation points involved in motion sickness. Different devices or systems can be used to display a virtual environment. Different taxonomies have been established according to the level of immersion. For example, Muhanna [64] classified the VR systems as:

– **Basic:** hand-based and monitor-based.
– **Partially immersive:** wall projectors, immersive desk, and monocular head-based.
– **Fully immersive:** room-based (vehicle simulation and CAVE) and binocular head-based.

Since the appearance of the first HMD developed by Sutherland [23], many different commercial devices and noncommercial prototypes have been developed. Some of the companies that have shown interest in VR systems are

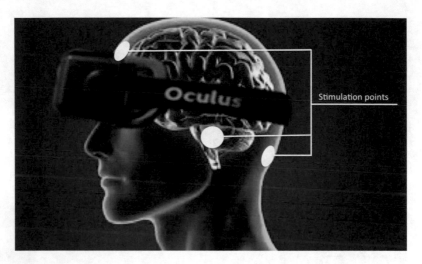

Fig. 4. Stimulation points

Facebook (Oculus Rift), Sony (Playstation VR), HTC (VIVE), Google (Cardboard and Magic Leap), Samsung (Gear VR), and Microsoft (Hololens).

Particularly, HMDs such as the Oculus Rift provides a stereoscopic effect by splitting the screen, this means that a section of the display is dedicated for each eye. The stereoscopic image must constantly update in relation to the users viewpoint. The Oculus Rift tracks the users head in real-time to calculate the viewpoint based on three degrees of freedom for position and three for orientation. This VR device includes sensors (accelerometer, gyroscope, and magnetometer), has higher resolution, higher refresh rate, low persistence to remove motion blur, and positional tracking for low latency [65].

Moreover, the Gear VR is a mobile virtual reality helmet headset developed by Samsung Electronics[1], in collaboration with Oculus. This device is designed to work with Samsungs smartphones (e.g. Galaxy S6 Edge, Galaxy S6 Edge+, Galaxy S7 Edge, Galaxy S8, Galaxy S8+, and Samsung Galaxy Note 8). The Gear VR could support MTP (Motion to Photon) latency less than 20 ms. The lenses field of view is 96′. It also has: Accelerator, Gyroscopic, Magnetic, Proximity and focal adjustment for eyes with myopia or hyperopia. The user interface is through a touch panel, back button and a volume key. And its dimensions are: 198(width) × 116(length) × 90(height) mm.

For the interaction in both conditions we will use a gamepad *Xbox 360* Controller. The participant will be able to move forward, turn right or left by using the controller of the gamepad with her/his hands in order to navigate in the virtual environment. For navigation into the virtual maze, we will create an avatar from a first-person perspective; this will allows us to represent the participants point of view and personify the movements in the virtual environment.

---

[1] http://www.samsung.com.

The *VRsystem ViPoS* will have two scenes. The first scene will have an interface to enter the date of birth, gender, force, and choose the type of condition. The second scene will have a virtual maze to carry out the task; the path will have fifteen intersections. When finishing the task this scene, will show a menu of options that allows returning to the first scene.

The *VR-system ViPoS* is being developed in an Intel Core i7 computer, processor with 32 GB RAM, an NVIDIA GeForce GTX-1080 with a video card of 4 GB, and Windows 10 Operating System. The virtual scenes is being developed using Unity Edition Professional[2] and we will develop the programs with C# and Javascript. To create 3D objects that will be included in the virtual environment we used Blender, and for texturing the images we will use GIMP and Inkscape. For integration of both HMDs we will use the plugins provided by the manufacturer (Oculus SDK).

3.3   Measures and Procedure

Before each session, all participants received written information about the purpose and procedures of the study. They signed an informed consent form. They were free to leave the experimental task at any time, and the study was conducted according to the principles stated in the Declaration of Helsinki.

All the military personnel (pilots and student pilots) will be divided randomly into two groups for each experimental condition. This grouping will be carried out in the experimental validations with the pilots and student pilots of each force respectively. Then, the participants will be able to perform the task.

We will apply three types of questionnaires (PDQ, SSQ, and PQ) to assess motion sickness. The first questionnaire that will be administered is the Personal Data Questionnaire (PDQ) and this has to per each participant will fill out before of exposure.

The second questionnaire is the Simulator Sickness Questionnaire (SSQ) [30] will be use to assess the symptoms of motion sickness. This questionnaire will be applied two times, once before ("Pre-test SSQ") exposure and once after ("Post-test SSQ") exposure. This questionnaire consisted of a checklist of 16 items with four score levels of severity for each symptom (0 - *none*, 1 - *slight*, 2 - *moderate*, 3 - *severe*). These symptoms have three weighted subscales: Nausea, Oculomotor and Disorientation.

The *Total Severity-Score* reflects the general level of the sickness symptoms for each condition. In addition, based on previous studies [66,67] we determined a scale for the Total Severity-Score as *Not significant* (less than 10), *Partially significant* (greater than 10 and less than 20), and *Significant* (grater than 20).

The third questionnaire that will be applied is the Presence Questionnaire (PQ) [55]. This questionnaire allows assessing the perception of the participants satisfaction and interaction with the VR system. The PQ will be administered after exposure.

The proposed experimental validation protocol can be seen in the Fig. 5.

---

[2] http://unity3d.com.

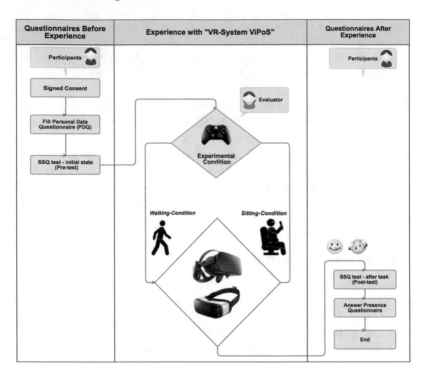

Fig. 5. Experimental validation protocol

4   Conclusions and Future Work

This work will aim to assess motion sickness in military personnel (pilots and student pilots), and identify the symptoms that could provoke the use of two types of HMDs (Oculus Rift and Gear VR) coupled with two experimental conditions (walking and sitting).

To perform the current work, we propose developed *VR-system ViPoS*, it will integrate a virtual maze configured to work with two HMDs. To collect the data we establish a protocol and we will apply questionnaires specialized to assess the motion sickness before and after of exposure. We will analyze the SSQ, the differences found between the ("Pre-test SSQ") and ("Post-test SSQ") scores. With regards to motion sickness symptoms (*Nausea, Disorientation and Oculomotor*) we will assess the level of severity.

As far as we known, VR systems such as the one proposed in this work have not been developed yet. The goal is to diagnose and mitigate the symptoms of cybersickness that the military personnel could feel when they perform training using simulators. This work also will contribute to the occupational safety of military personnel and guarantee an effective training. In addition, this work allows creating a multidisciplinary group composed of researchers, engineers and military personnel of the three armed forces.

Currently, we are developing the virtual environment based on the prototyping method and later we will apply the technique of simulated annealing to found the location optimal of the object into of the virtual environment.

Our future work is oriented to carry out the experimental validation of our *VR-system ViPoS*, to perform the comparison and analyses of the results, and to verify the hypothesis of our research.

Acknowledgment. We wish to express our recognition to the students who participate in the development of this work; to Director of CICTE; to the members of the Armed Forces of Ecuador who collaborate as participants; and to the reviewers of this article.

References

1. Kennedy, R.S., Lilienthal, M.G., Berbaum, K.S., Baltzley, D.R., McCauley, M.E.: Simulator sickness in U.S. Navy flight simulators. Aviat. Space. Environ. Med. **60**, 10–16 (1989)
2. Rogers, S.P., Asbury, C.N., Szoboszlay, Z.P.: Enhanced flight symbology for wide-field-of-view helmet-mounted displays. In: International Society for Optics and Photonics, AeroSense 2003, pp. 321–332 (2003). https://doi.org/10.1117/12.487289
3. Stoffregen, T.A., Hove, P., Schmit, J., Bardy, B.G.: Voluntary and involuntary postural responses to imposed optic flow. Mot. Control - Champaign **10**, 24–33 (2006)
4. Rosser Jr., J.C., Lynch, P.J., Cuddihy, L., Gentile, D.A., Klonsky, J., Merrell, R.: The impact of video games on training surgeons in the 21st century. Arch. Surg. **142**, 181–186 (2007)
5. Barab, S.A., Gresalfi, M., Ingram-Goble, A.: Transformational play: using games to position person, content, and context. Educ. Res. **39**, 525–536 (2010)
6. Mayo, M.J.: Games for science and engineering education. Commun. ACM **50**, 30–35 (2007)
7. Rizzo, A.A., Schultheis, M., Kerns, K.A., Mateer, C.: Analysis of assets for virtual reality applications in neuropsychology. Neuropsychol. Rehabil. **14**, 207–239 (2004)
8. Cárdenas-Delgado, S., Méndez-López, M., Juan, M.-C., Pérez-Hernández, E., Lluch, J., Vivó, R.: Using a virtual maze task to assess spatial shortterm memory in adults. In: Proceedings of the 12th International Joint Conference on Computer Vision, Imaging and Computer Graphics Theory and Applications (VISIGRAPP 2017) - Volume 1: GRAPP, pp. 46–57 (2017)
9. Parsons, T.D., Courtney, C.G., Dawson, M.E., Rizzo, A.A., Arizmendi, B.J.: Visuospatial processing and learning effects in virtual reality based mental rotation and navigational tasks. LNCS, pp. 75–83 (2013)
10. de Silva, R.D.C., Albuquerque, S.G.C., de Muniz, A.V., Filho, P.P.R., Ribeiro, S., Pinheiro, P.R., Albuquerque, V.H.C.: Reducing the schizophrenia stigma: a new approach based on augmented reality. Comput. Intell. Neurosci. **2017** (2017)
11. Dong, X., Yoshida, K., Stoffregen, T.A.: Control of a virtual vehicle influences postural activity and motion sickness. J. Exp. Psychol. Appl. **17**, 128–138 (2011)
12. Merhi, O., Faugloire, E., Flanagan, M., Stoffregen, T.A.: Motion sickness, console video games, and head-mounted displays. J. Hum. Factors Ergon. Soc. **49**, 920–934 (2007)

13. Stoffregen, T.A., Yoshida, K., Villard, S., Scibora, L., Bardy, B.G.: Stance width influences postural stability and motion sickness. Ecol. Psychol. **22**, 169–191 (2010)
14. Reason, J.T., Brand, J.J.: Motion Sickness. Academic Press, Oxford, England (1975)
15. Howarth, P.A.: Potential hazards of viewing 3-D stereoscopic television, cinema and computer games: a review. Ophthalmic Physiol. Opt. **31**, 111–122 (2011)
16. Johnson, D.M.: Introduction to and review of simulator sickness research. Research Report 1832. Soc. Sci. (2005)
17. Maino, D.M., Chase, C., October, C.C.: Asthenopia: a technology induced visual impairment. Rev. Optom. Suppl. **2**, 28–35 (2011)
18. Maino, D.M.: You can help your patients see 3-D!: 3-D is not just hype. It can help you diagnose binocular vision disorders and build your practice. You can even help patient overcome their problems with 3-D viewing. Rev. Optom. **148**, 54–63 (2011)
19. Solimini, A.G., Mannocci, A., Di Thiene, D., La Torre, G.: A survey of visually induced symptoms and associated factors in spectators of three dimensional stereoscopic movies. BMC Public Health **12**, 779 (2012)
20. Ungs, T.J.: Simulator induced syndrome in Coast Guard aviators. Aviat. Space Environ. Med. **59**, 267–272 (1988)
21. Blok, R.: Simulator sickness in the US Army UH-60A Blackhawk flight simulator. Mil. Med. **157**, 109–111 (1992)
22. Riccio, G.E., Stoffregen, T.A.: An ecological theory of motion sickness and postural instability. Ecol. Psychol. **3**, 195–240 (1991)
23. Cobb, S.V.G., Nichols, S., Ramsey, A., Wilson, J.R.: Virtual reality-induced symptoms and effects (VRISE). Presence Teleoperators Virtual Environ. **8**, 169–186 (1999)
24. Kolasinski, E.M.: Simulator sickness in virtual environments (1995)
25. LaViola, J.: A discussion of cybersickness in virtual environments. ACM SIGCHI Bull. **32**, 47–56 (2000)
26. Kolasinski, E.M., Gilson, R.D.: Ataxia following exposure to a virtual environment. Aviat. Space Environ. Med. **70**, 264–269 (1999)
27. Bos, J.E.: Nuancing the relationship between motion sickness and postural stability. Displays **32**, 189–193 (2011)
28. Villard, S.J., Flanagan, M.B., Albanese, G.M., Stoffregen, T.A.: Postural instability and motion sickness in a virtual moving room. Hum. Factors **50**, 332–345 (2008)
29. Drexler, J.M.: Identification of system design features that affect sickness in virtual environments (2006)
30. Kennedy, R.S., Lane, N.E.: Simulator sickness questionnaire: an enhanced method for quantifying simulator sickness. Int. J. Aviat. Psychol. **3**, 203–220 (1993)
31. McCauley, M.E., Royal, J.W., Wylie, C.D., OHanlon, J.F., Mackie, R.R.: Motion sickness incidence: exploratory studies of habituation, pitch and roll, and the refinement of a mathematical model (1976)
32. Benson, A.J.: Motion sickness. In: Encyclopaedia of Occupational Health and Safety, pp. 12–14. International Labour Organization, Geneva (1998)
33. Dobie, T.G., May, J.G.: Cognitive-behavioral management of motion sickness. Aviat. Space Environ. Med. **65**, C1–C2 (1994)
34. Golding, J.F.: Motion sickness susceptibility questionnaire revised and its relationship to other forms of sickness. Brain Res. Bull. **47**, 507–516 (1998)
35. Golding, J.F.: Motion sickness susceptibility. Auton. Neurosci. Basic Clin. **129**, 67–76 (2006)

36. Bos, J.E., Damala, D., Lewis, C., Ganguly, A., Turan, O.: Susceptibility to sea-sickness. Ergonomics **50**, 890–901 (2007)
37. Regan, E.C., Price, K.R.: The frequency of occurrence and severity of side-effects of immersion virtual reality. Aviat. Space Environ. Med. **65**(6), 527–530 (1994)
38. Stanney, K.M., Salvendy, G., Deisinger, J., DiZio, P., Ellis, S., Ellison, J., Fogleman, G., Gallimore, J., Singer, M., Hettinger, L., Kennedy, R., Lackner, J., Lawson, B., Maida, J., Mead, A., Mon-Williams, M., Newman, D., Piantanida, T., Reeves, L., Riedel, O., Stoffregen, T., Wann, J., Welch, R., Wilson, J., Witmer, B.: After-effects and sense of presence in virtual environments: formulation of a research and development agenda. Int. J. Hum. Comput. Interact. **10**, 135–187 (1998)
39. Davis, S., Nesbitt, K., Nalivaiko, E.: Comparing the onset of cybersickness using the Oculus Rift and two virtual roller coasters. In: Proceedings of 11th Australasian Conference on Interactive Entertainment (IE 2015), vol. 167, pp. 3–14 (2015)
40. Van Emmerik, M.L., De Vries, S.C., Bos, J.E.: Internal and external fields of view affect cybersickness. Displays **32**, 169–174 (2011)
41. Kennedy, R.S., Drexler, J., Kennedy, R.C.: Research in visually induced motion sickness. Appl. Ergon. **41**, 494–503 (2010)
42. Diels, C., Howarth, P.A.: Visually induced motion sickness: single-versus dual-axis motion. Displays **32**, 175–180 (2011)
43. Diels, C.: Visually induced motion sickness (2008)
44. Solimini, A.G.: Are there side effects to watching 3D movies? A prospective crossover observational study on visually induced motion sickness. PLoS One **8**, e56160 (2013)
45. Naqvi, S.A.A., Badruddin, N., Malik, A.S., Hazabbah, W., Abdullah, B.: Does 3D produce more symptoms of visually induced motion sickness? In: 2013 35th Annual International Conference of the IEEE Engineering in Medicine and Biology Society (EMBC), pp. 6405–6408. IEEE (2013)
46. Kuze, J., Ukai, K.: Subjective evaluation of visual fatigue caused by motion images. Displays **29**, 159–166 (2008)
47. Llorach, G., Evans, A., Agenjo, J., Blat, J.: Position estimation with a low-cost inertial measurement unit. In: 2014 9th Iberian Conference on Information Systems and Technologies (CISTI), pp. 1–4. IEEE (2014)
48. Llorach, G., Evans, A., Blat, J.: Simulator sickness and presence using HMDs. In: Proceedings of the 20th ACM Symposium on Virtual Reality Software and Technology - VRST 2014, pp. 137–140. ACM Press, New York (2014)
49. Army Medical Department, US: Motion Sickness. Medical Aspects of Harsh Environments (2002)
50. Lawther, A., Griffin, M.J.: A survey of the occurrence of motion sickness amongst passengers at sea. Aviat. Space Environ. Med. **59**, 399–406 (1988)
51. Chan, G., Moochhala, S.M., Zhao, B., Wl, Y., Wong, J.: A comparison of motion sickness prevalence between seafarers and non-seafarers onboard naval platforms. Int. Marit. Health **57**, 56–65 (2006)
52. Turner, M., Griffin, M.J.: Motion sickness in public road transport: passenger behavior and susceptibility. Ergonomics **42**, 444–461 (1999)
53. Stevens, S.C., Parsons, M.G.: Effects of motion at sea on crew performance: a survey. Mar. Technol. **39**, 29–47 (2002)
54. Kim, D.H., Parker, D.E., Park, M.Y.: A New Procedure for Measuring Simulator Sickness the RSSQ. Human Interface Technology Laboratory, University of Washington (2004)
55. Witmer, B.G., Singer, M.J.: Measuring presence in virtual environments: a presence questionnaire. Presence Teleoperators Virtual Environ. **7**, 225–240 (1998)

56. Ames, S.L., Wolffsohn, J.S., McBrien, N.A.: The development of a symptom questionnaire for assessing virtual reality viewing using a head-mounted display. Optom. Vis. Sci. **82**, 168–176 (2005)
57. Gamberini, L.: Virtual reality as a new research tool for the study of human memory. Cyberpsychol. Behav. **3**, 337–342 (2004)
58. Ku, J., Cho, W., Kim, J.-J., Peled, A., Wiederhold, B.K., Wiederhold, M.D., Kim, I.Y., Lee, J.H., Kim, S.I.: A virtual environment for invetigating schizophrenic patients characteristics: assessment of cognitive and navigation ability. Cyberpsychol. Behav. **6**, 397–404 (2003)
59. Elkind, J.S., Rubin, E., Rosenthal, S., Skoff, B., Prather, P.: A simulated reality scenario compared with the computerized Wisconsin card sorting test: an analysis of preliminary results. Cybcrpsychol. Behav. **4**, 489–496 (2004)
60. Krueger, W.W.: Method to mitigate Nystagmus and motion sickness with head worn visual display during vestibular stimulation. J. Otolaryngol. Res. **7** (2017)
61. Onuh, S.O., Yusuf, Y.Y.: Rapid prototyping technology: applications and benefits for rapid product development. J. Intell. Manuf. **10**, 301–311 (1999)
62. Kirkpatrick, S., Gelatt, C.D., Vecchi, M.P.: Optimization by simulated annealing. Science **80**(220), 671–680 (1983)
63. Aarts, E., Korst, J., Michiels, W.: Simulated annealing. In: Search Methodologies, pp. 187–210 (2005)
64. Muhanna, M.A.: Virtual reality and the CAVE: taxonomy, interaction challenges and research directions (2015)
65. Antonov, M., Mitchell, N., Reisse, A., Cooper, L., LaValle, S., Katsev, M.: SDK Overview Version 0.2.5. Oculus VR, pp. 1–48 (2013)
66. Kennedy, R.S., Drexler, J.M., Compton, D.E., Stanney, K.M., Lanham, D.S., Harm, D.L.: Configural scoring of simulator sickness, cybersickness and space adaptation syndrome: similarities and differences. In: Virtual and Adaptive Environments Applications Implications and Human Performance Issues, pp. 247–278 (2003)
67. Stanney, K.M., Hale, K.S., Nahmens, I., Kennedy, R.S.: What to expect from immersive virtual environment exposure: influences of gender, body mass index, and past experience. Hum. Factors **45**, 504–520 (2003)

Computer Networks, Mobility and Pervasive Systems

Security over Smart Home Automation Systems: A Survey

Jhonattan J. Barriga A$^{(\boxtimes)}$ ⓘ and Sang Guun Yoo ⓘ

Facultad de Ingeniería de Sistemas, Escuela Politécnica Nacional, Quito, Ecuador
{jhonattan.barriga, sang.yoo}@epn.edu.ec

Abstract. Internet of Things (IoT) is growing every day since it has allowed the interconnection of different devices existent in smart environments e.g. home environment, and because it has allowed improving or supporting different services such as physical security, energy consumption management, entertainment among others. The smart home environment, also known as Smart Home Automation Systems (SHAS), has properly attacked and solved the domestic needs of families around the world. However, in spite of having a protocol stack architecture to govern the rules of communication and exchange of information, the security and privacy requirements have been displaced to a second place. For this reason, this paper surveys several previous works from the security perspective with the purpose of providing the most important approaches in this field. Additionally, this work provides recommendations and best practices from different security perspectives.

Keywords: IoT · Security · SHAS · Smart home · Mechanisms

1 Introduction

The world of smart things is rapidly growing, and according to Gartner more than 20 billion devices would be connected by 2020 [1]. Nowadays, not only computers and servers are connected to Internet, but also home appliances and small devices; and we call this kind of environment the world of Internet of Things (IoT). Several devices like lightbulbs, house locks, house alarms, smoke detectors, temperature sensors among others, have internet connection capabilities so that users can monitor and manage their home remotely ensuring sustainability and money saving [2].

Smart Home Automation Systems (SHAS) are a piece of IoT since they are composed of a series of smart devices that are connected to the Internet. Although they are focused on making homes smarter, they are based over a layered architecture schema that let them communicate and exchange information. The IoT architecture shown (see Fig. 1) describes the common layers of the IoT world [3].

In SHAS schema, connecting a TV or a refrigerator to the Internet might be considered as a normal scenario, since it would make our life easier. However, the single fact of connecting such node to the IoT world might generate a potential vulnerability since a hardening standard is still not in place to protect such devices [4]. In addition, the risk arises as the SHAS is being used to handle physical security services, such as opening doors or preventing burglars from entering a place. For this

© Springer International Publishing AG, part of Springer Nature 2018
Á. Rocha and T. Guarda (Eds.): MICRADS 2018, SIST 94, pp. 87–96, 2018.
https://doi.org/10.1007/978-3-319-78605-6_7

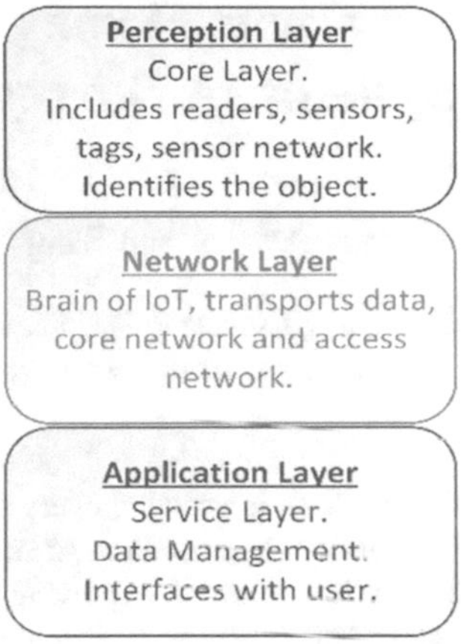

Fig. 1. IoT architecture

reason, appropriate measures to address security and privacy issues have to be in place to guarantee availability, integrity and confidentiality. These needs are supported by industry reports such as [5, 6] that indicates several issues and considerations in regards of different features such as authentication, encryption, and authorization. Moreover, end-users who are the beneficiaries of this technology are mainly exposed, since compromising a device attached to a SHAS might affect their privacy in case of a camera is affected. A worse scenario would be a theft if an attacker is able to vulnerate a secure smart-lock, letting a burglar to enter the house. Therefore, any new vulnerability on any new connected device pose a threat to end-users.

The rest of the paper is structured as follows: SHAS architecture and protocols is briefly described in Sect. 2. Then, Sect. 3, reviews proposed mechanisms to address security issues. Section 4, discusses improvements and security issues that are still to be addressed. Finally, this paper is concluded in Sect. 5.

2 Smart Home Automation Architecture and Protocols

2.1 IoT Protocol Stack Architecture

Network communications are implemented based on a protocol stack (e.g. TCP/IP Stack). Likewise, IoT makes use of a protocol stack, which has been defined with the purpose of designing standards to handle communications between different devices (see Fig. 2). In contrast to TCP/IP stack, we have an extra layer called Adaption Layer, which mainly encapsulates data frames for use in IoT applications [7].

2.2 Smart Home Automation Architecture

A Smart Home Automation Environment is basically composed by three main elements i.e. indoor environment, outdoor environment, and a residential gateway which acts a bridge between the exterior world and the smart home.

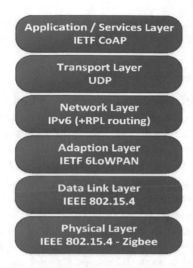

Fig. 2. IoT protocol stack [3, 8, 9, 11]

The gateway plays the role of a concentrator; it collects all information sent by wireless or wired sensors, and make it available through the Internet for consumers. This device acts a as a basic access control element. Although this architecture seems to be simple to implement from a technical perspective, it needs security control points to protect information.

2.3 Communication Protocols and Technologies

Currently, there are several wired and wireless protocols used by different devices, in order to communicate between each other. In regards of wireless technologies, we have WM-Bus, Z-Wave, WiFi (802.11), EnOcean, ZigBee, Bluetooth (IEEE 802.15.1), Wavenis. Insteon, SimpliciTI, 6LoWPAN, IEEE 802.15.4, DECT, BACnet, BidCos, CoAP, MQTT. From the wired perspective, we can enumerate KNX, Home Plug, X10, Insteon, ITU G.hn, M-Bus, MoCA, LonWorks, MQTT [10–12].

3   Existing Secure Mechanisms

Each day, more people are adopting the use of IoT devices. However, their security has not increased in an important manner. It is important to note that the lack of patch management processes and short-time supports are weaknesses of this industry. Vulnerabilities that are present in IoT market are not new at all and they are well known in the field. Lack of authentication and encryption, weak passwords and security misconfigurations are common errors found in several commercial devices. In summary, privacy and security are the main concerns that have not been properly addressed yet [5, 6, 13].

Researchers are proposing new techniques and mechanisms to increase the level of security in SHAS, by attaching to the energy-efficiency directive.

3.1 Privacy Approaches

A novel security algorithm is discussed in [4]. The proposed solution i.e. Triangle Based Security Algorithm (TBSA) pretends to protect data that is being transmitted through an insecure channel. First, the proposed solution assigns keys to all participant sensors. It also uses tracking sequence values to speed authentication, prevent replay attacks, identify source nodes, and guarantee message authenticity. Moreover, a reserved key is present to identify tracking sequence for emergency data. In addition, validating SSID along with password provides an extra level of authentication. Once the node has executed all authentication steps, it could generate a cypher text using TBSA. Moreover, recognizing users and devices as well as access control policies, the system prevents unwanted users accessing the system. Finally, in this system, data would be transmitted over HTTP/1.1 and if there is a match between channel and field ID, destination node should be able to decrypt the message. In regards of TBSA, it is based on the triangular logical function, which is used to produce a key (K) that will be used to cypher the message. To produce such key, the author considers time (t) which is produced by a sensor TMP36, the sensor unique identification (u), the triangular logical parameter (s), and the number of hours per day used as a nonce (n), and uses the Eq. (1). To encrypt and decrypt the text, the solution uses the Eqs. (2) and (3), respectively. This approach, provides three authentication levels, complies with low energy consumption (compared with traditional cryptographic solutions), and provides novel and secure mechanisms to guarantee data integrity and confidentiality [4].

$$K = \sum_{i=1}^{n} (s_i + t_i + u)/2 \tag{1}$$

$$C = (u \oplus t) * M/K \tag{2}$$

$$M = C * K/(u \oplus t) \tag{3}$$

3.2 Security Approaches

SHAS environments are threatened by different techniques such as eavesdropping, masquerading, message modification, replay attacks, DoS, and malwares. Authentication and authorization are mechanisms that would make a smart home network more secure. At least, it would be important the implementation of user credentials to prove user's identity. Additionally, in order to address authorization, it would be recommended to use a firewall combined with an IDS in the network segment where the residential gateway is connected [2].

Event logging is another essential security solution in SHAS environment; and such log mechanism has to guarantee the confidentiality and integrity of logs. Secure Host-based Event Logging (SHEL) [14] is an approach that fulfills such requirements. Forward Secure Sequential Aggregate (FssAgg) authentication system and Bitcoin blockchain are the main features of this novel approach. Six main requirements have been considered within this proposal: (i) log entry integrity, (ii) stream integrity, (iii) data confidentiality, (iv) forward security, (v) processing time and (vi) log access. To address confidentiality, logs are ciphered with Hardware Security Module (HSM) and they are stored in a local storage accessible to home owners only. For integrity, two mechanisms are considered: (i) chain of Bitcoin transactions to ensure log sequence to prevent data loss if an attacker has complete access to the system, and (ii) Forward Stream Authentication (FSSA) with FssAgg signature (SHA256) to allow adding a signature to every log entry. The proposed architecture is a cloud based approach and suggests three layers (Network of things for sensors, Common Service for a hub and storage, and application layer that consists of SHEL, Home control unit and billing cryptocurrency) [14]. This solution is only focused on guaranteeing logs rather than securing communication protocols; however, the use of blockchain is new and might be introduced to address other issues as well. On the other hand, it does not seem to be cost free for customers; they will have to pay for bitcoin transactions in order to preserve log integrity. Finally, in terms of performance, it seems to be costly because it takes around 3 min to generate, sign and verify 100000 log entries.

Protection measures can be applied to any layer of the scope, as long as they are used to increase the level of security. A novel approach to prevent attacks in Routing Protocol for Low-Power and Lossy Networks (RPL) with fake parent nodes is proposed in [15]. Attacking nodes pretend to be real parent nodes by broadcasting a lower rank. In order to prevent such scenario a threshold is defined to discard malicious nodes. If the rank of a neighbor is lower than a pre-defined threshold, such node will be discarded. [15]. This work discards malicious nodes that are not properly attached to the topology; however, it is prone to errors as it may reject authentic nodes. Therefore, it needs to be complemented with a mechanism that authenticates with some sort of key exchange rather than relying on threshold based on number of hops.

Another essential security issue is securing the residential gateway to prevent outside attacks. IOTURVA suggests a monitoring device that identify and blocks anomalous network interactions by using Software Defined Networks (SDN) [16]. The system adds three new components to the Software Defined Controller (Monitoring, Enforcement and Classification) to examine all traffic and apply control policies. In case that there is no match, such information is collected and sent to a service which is in charge of classifying such behavior and generate a signature to identify a pattern [16]. This solution still needs to be improved as the accuracy ratio is low with a bunch of false-positives.

According to [17], a new network-layer architecture might be the solution to protect from external threats. In this approach, the authors propose important features as follows; (i) support for user actions and applications where the inclusion of network access control should be considered, (ii) secure policy creation and distribution, (iii) resource and privacy limitations to consider sending raw data to centralized sources, and (iv) device installation and user interaction to properly set the device up in terms of security configuration [17]. This approach presents considerations for designing a complete new security layer.

Multi-user is another property that SHAS must meet. By defining user profiles, a system could be secured [18]. Most systems only focus on the final customer rather than understanding its real needs. In addition, the smartphone plays the role of user and hence it does not require to be a high-privileged one. In terms of roles, at least a primary user, his backup, a secondary user with restricted privileges and a guest should be defined; each of them with a particular role of access according to environment where he/she will interact. Then, it is important to analyze several input modalities that might come with different devices and the possibility to combine them in order to provide access. Besides, analysis of the impact or possible breach of a specific input device could be carried out. It is important that manufacturers and developers understand the usage environment of a device; indecd, providing fine-grained access-control and authentication would considerably increase the level of security.

3.3   Authentication Approaches

A biometric authentication approach is proposed in [19]. The author uses Fingerprint Recognition Technology (FRT) and Iris Recognition Technology (IRT) using an ARM7TDMI-S processor. Firstly, user's fingerprint has to be enrolled and stored in a database, to be later compared for giving access. Also, user's iris has to follow the same procedure for properly authenticate a user. To accomplish this task, an iris and fingerprint modules are attached to the processor, which has the algorithm running inside it. Only if both IRT and FRT match, the user would have access to the device. This approach seems to be valid; however, it might not comply with energy-saving directive. Yet, this is a hardware dependent solution that might have to be added physically to a device in order to protect it.

Wireless communications are part of SHAS environments. Hence, a proper authentication schema has to be in place to prevent unauthorized access. EAKES6Lo [20] aims to improve 6LoWPAN security by designing a pair of protocols. It setups the system by distributing a unique id and secret key to every node. Then, on the authentication and key establishment phase, all nodes exchange messages to perform mutual authentication, besides registering new users into the network. Although this approach prevents Replay, Man in the Middle (MiTM), impersonation and Sybil attacks, it might present performance issues when mutual authentication is performed, as all nodes involved in the environment will have to exchange credentials [20]. Asymmetric cryptography degrades performance and using symmetric keys might lead attackers to decrypt information if those keys are not safely shared or stored.

Using a Key distribution center (KDC) might help to address authentication issues [21]. The use of digital certificates is helpful when dealing with many devices. Hence, the work reviewed propose a three-protocol based architecture to register a device and authenticate mobile devices against IoT devices. In the first stage, the mobile is registered against the KDC. Then, KDC generates a set of credentials that would be used to access the IoT Device. Finally, at the communication phase, it uses the CA and exchanged keys to send out information using the KDC as an intermediary for all authentication and communication steps [21]. This solution suggests the use CA, which is a strength to guarantee confidentiality and privacy; however, it still requires a secure channel to protect information being sent and to prevent information leakages due to attackers capturing packets in unauthorized way.

Likewise, a Physically Unclonable Function (PUF) is proposed in [22]. This approach proposes a lightweight identity-based cryptosystem for IoT, it uses PUF to generate asymmetric keys for each device. PUFs are physical entities embodied in physical structure which is infeasible to clone. The proposed protocol consists of four phases. First of all, during the enrolment phase, every node is registered to a server node generating a challenge-response pair (CRP) database for the PUF contained in each node. Then, in the Authentication and Key Sharing Phase, if two nodes want to communicate, they are first authenticated by the server node, helping them to generate their key pair and enabling secure key sharing. Later, in the Secure Communication Phase, the message is sent using the keys of sender and receiver. Finally, if nodes move to a new location, it would have to go through the whole authentication process again [22]. It is important that this approach needs nodes to know server address. It seems to be in compliance with energy-efficiency directive; however, there are still improvements to make for in regards of trustiness, securing and guaranteeing CRP database integrity.

3.4 Risk Analysis Approaches

Other works, review SHAS' security from the risk analysis perspective. For example, authors in [23] propose an approach called Information Security Risk Analysis (ISRA) on Smart Home Automation Systems (SHAS). The author compares and analyze 14 different contributions considering two major aspects. Research focus aims to identify if a contribution tries to address one of the following (risk analysis, security or privacy issues, and security or privacy solutions) whilst research method describes the technique applied to achieve the goal of every work. The findings of this comparison show a need of empirical methods to support risk evaluation, secure design, clearly specify risks to user privacy and the inclusion of best practices and standards for industry developers. Later, risks are classified (in five categories i.e. software, hardware, information, communication and human-related), measured, prioritized and mapped against the four major needs identified previously. In summary, this evaluation allows to corroborate that security and privacy are issues that are not properly handled and need to be taken into consideration from the design; and that risk evaluation is a tool to identify issues that might not be considered [23]. Although this work does not propose a secure mechanism to mitigate or close breaches within software, hardware and communications, it does provide a support tool to identify possible issues that could be solved in early stages.

Enhancing software security and user privacy are the most important requirements to be addressed within SHAS. Hence, a model to support such needs from the design stage is discussed in [24]. First of all, analyzing and categorizing information is required in terms of contents, structure, but most important personal privacy; nevertheless, the context plays a particular role as it will help to identify types of data (i.e., metadata). In addition, design principles should include security requirements in terms of availability, confidentiality and integrity. In fact, these requirements need to consider hardware limitations (CPU, RAM), diversity of devices, different types of information which are part of IoT context. Finally, privacy-awareness should be in place to reduce information sensitivity (data minimization, anonymity and linkability). The aforementioned considerations would help customers to be more in control of their personal identifiable information and have safer software tools that guarantee security and privacy of information.

4 Discussion

The proposal discussed in [2] considers basic security mechanisms that might not guarantee an appropriate level of security. Besides, this approach does not focus on the communications from several devices to the gateway.

Biometric approaches might help to reduce the use of credentials for authenticating users. However, these approaches are resource demanding (RAM, CPU and energy) and apparently not all devices come with this built-in solution; it has to be built out-of-the-box and delegate functionality to other elements in order to be efficient (i.e., sent data to a server that process recognition).

Considering security as part of the design is clearly stated and supported by several works listed previously. It corroborates the fact that SHAS devices are being built to fulfill customers' needs rather than protecting their information. This requirement would have to become a standard for producing IoT devices, otherwise, all the work that is being conducted by researchers would be no more than only not realized intentions.

Building a new network layer architecture would definitely address several security issues; however, it would require a huge effort from industries and research community as they will have to focus on securing data and then start thinking on how to improve the product from the functional perspective overview.

Energy efficiency is probably a restriction and hence an advantage as it might boost the design and construction of more powerful microchips to support robust cryptography standards like AES-256, PKI among others. However, this constraint is being addressed by delegating functionality to external entities outside of the device such as cloud environments. Undoubtedly, this constraint also leads to improve and develop cryptography algorithms for IoT.

It is contradictory that industries propose solutions for securing homes while not proving security features in their products. One possible solution could be the creation of a security standard/certification that guarantee the security of a SHAS' device that contemplates at least basic mandatory security requirements.

Most of the solutions reviewed, are out-of-the-box; hence, it should be agnostic to technology to easily implement them within any SHAS environment. Some changes will have to be made for a clean integration and support.

Several works are using devices like Arduino and Raspberry to act as nodes or controllers [25]. Hence, these devices would be helpful when building Wireless Sensor Networks (WSNs) as they have the ability to connect modules and turn it into a SHAS device. A good point is that the operating system (OS) could be hardened before being deployed; nonetheless, firmware of additional controllers should also be hardened.

The use of digital certificates and nonces are initiatives that have been proposed for the Wi-Fi environment such as [26]. Such solutions are being used to connect devices and they comply with the energy-efficiency directive.

In summary, there is still a lot of work to do in regards of data privacy, authentication, software security, patch management, standards and policies. Researchers are already working on initiatives to address such issues and it is time to consolidate efforts with the industry to secure, protect and potentiate this field that is reaching millions of users. Besides, it might be time to start thinking of a near future where IPv6 would not be enough to support all the connected stuff.

5   Conclusions

There are several technological manufacturers that are betting in the SHAS and IoT market. However, there are no standards or policies in regards of properly handling security and privacy. There is still the need to come up with a common framework that adapts to all manufacturers and developers in order to grow in an organized way.

IoT is a new challenge for security; therefore, manufacturers need to improve their building processes and start considering security and privacy as part of the product rather than out-of-the-box solution or patch. Future work, will focus on finding and classifying common vulnerabilities among commercial solutions in regards of CIA triad and propose secure mechanisms to address them by focusing on user-privacy as well.

References

1. Gartner Says 4.9 Billion Connected Things Will Be in Use in 2015. https://www.gartner.com/newsroom/id/2905717. Accessed 27 Nov 2017
2. Ul Rehman, S., Manickam, S.: A study of smart home environment and its security threats. Int. J. Reliab. Qual. Saf. Eng. 23(3), 1–9 (2016)
3. Tank, B., Upadhyay, H., Patel, H.: A survey on IoT privacy issues and mitigation techniques. In: Proceedings of Second International Conference on Information and Communication Technology for Competitive Strategies, ICTCS 2016, pp. 1–4 (2016)
4. Pirbhulal, S., et al.: A novel secure IoT-based smart home automation system using a wireless sensor network. Sensors 17(1), 69 (2016)
5. Barcena, M.B., Wueest, C.: Insecurity in the Internet of Things (2015)
6. Gheorghe, A.: The Internet of Things: Risk in the Connected Home (2016)
7. Salman, T., Jain, R.: Networking protocols and standards for internet of things. In: Geng, H. (ed.) Internet of Things and Data Analytics Handbook, pp. 215–238. Wiley, Hoboken (2017)
8. Rayes, A., Salam, S.: IoT protocol stack: a layered view. In: Rayes, A., Salam, S. (eds.) Internet of Things from Hype to Reality, pp. 93–138. Springer, Cham (2017)
9. Lin, H., Bergmann, N.: IoT privacy and security challenges for smart home environments. Information 7(3), 44 (2016)
10. Reuter, T.: Security analysis of wireless communication standards for home automation, p. 92 (2013)

11. Mendes, T., Godina, R., Rodrigues, E., Matias, J., Catalão, J.: Smart home communication technologies and applications: wireless protocol assessment for home area network resources. MDPI **8**(7), 1–33 (2015)
12. Schwarz, D.: The Current State of Security in Smart Home Systems (0-14) (2016)
13. Wurm, J., Hoang, K., Arias, O., Sadeghi, A.R., Jin, Y.: Security analysis on consumer and industrial IoT devices. In: 2016 21st Asia and South Pacific Design Automation Conference, pp. 519–524 (2016)
14. S. Avizheh, T. T. Doan, X. Liu, nd R. Safavi-Naini: A secure event logging system for smart homes. In: Proceedings of 2017 Workshop on Internet Things Security and Privacy - IoTS&P 2017, pp. 37–42 (2017)
15. Iuchi, K., Matsunaga, T., Toyoda, K., Sasase, I.: Secure parent node selection scheme in route construction to exclude attacking nodes from RPL network. In: 2015 21st Asia-Pacific Conference on Communications, APCC 2015, vol. 4, no. 11, pp. 299–303(2016)
16. Hafeez, I., Ding, A.Y.: IOTURVA: Securing Device-to-Device (D2D) Communication in IoT Networks, pp. 1–6 (2017)
17. DeMarinis, N., Fonseca, R.: Toward usable network traffic policies for IoT devices in consumer networks. In: Proceedings of 2017 Workshop on Internet Things Security and Privacy - IoTS&P 2017, pp. 43–48(2017)
18. W. Jang, A. Chhabra, A. Prasad: Enabling multi-user controls in smart home devices, In: Proceedings of 2017 Workshop on Internet Things Security and Privacy - IoTS&P 2017, pp. 49–54 (2017)
19. Ishengoma, F.: Authentication system for smart homes based on ARM7TDMI-S and IRIS-fingerprint recognition technologies. CiiT Int. J. Program. Device Circ. Syst. **6**(6), 64–69 (2014)
20. Qiu, Y., Ma, M.: An authentication and key establishment scheme to enhance security for M2 M in 6LoWPANs. In: 2015 IEEE International Conference on Communication Workshop (ICCW), pp. 2671–2676(2015)
21. Hsueh, S.-C., Li, J.-T.: Secure transmission protocol for the IoT. In: Proceedings of the 3rd International Conference on Industrial and Business Engineering - ICIBE 2017, pp. 73–76 (2017)
22. Chatterjee, U., Chakraborty, R.S., Mukhopadhyay, D.: A PUF-based Secure Communication Protocol for IoT, V(212), (2015)
23. Jacobsson, A., Boldt, M., Carlsson, B.: A risk analysis of a smart home automation system. Futur. Gener. Comput. Syst. **56**, 719–733 (2016)
24. Jacobsson, A., Davidsson, P.: Towards a model of privacy and security for smart homes. In: 2015 IEEE 2nd World Forum Internet Things, pp. 727–732 (2015)
25. Chandramohan, J., Nagarajan, R., Satheeshkumar, K., Ajithkumar, N., Gopinath, P.A., Ranjithkumar, S.: Intelligent smart home automation and security system using Arduino and Wi-Fi. Int. J. Eng. Comput. Sci. **6**(3), 20694–20698 (2017)
26. Yoo, S.G., Barriga, J.J.: Privacy-aware authentication for Wi-Fi based indoor positioning systems, vol. 719 (2017)

Participatory Sensing in Sustainable Mobility: biciLAB Model

José María Díaz-Nafría[1,2] and Teresa Guarda[3,4,5](✉)

[1] BITrum-Research Group, C/San Lorenzo 2, 24007 León, Spain
jdian@unileon.es
[2] Munich University of Applied Sciences,
Dachauerstr 100a, 80636 Munich, Germany
[3] Universidad Estatal Península de Santa Elena – UPSE, La Libertad, Ecuador
tguarda@gmail.com
[4] Algoritmi Centre, Minho University, Guimarães, Portugal
[5] Universidad de las Fuerzas Armadas-ESPE, Sangolqui, Quito, Ecuador

Abstract. This project aims to develop a bicycle mobility promotion program and studies to improve urban life through intervention in urban planning and the development of intelligent territories based on bicycle lending systems with the ability to collect statistical information from mobility and other parameters of interest for sustainability and health. The aim is to achieve a long-term operation of the biciLAB under the auspices of the universities and local entities linked to its application within the scope of public policy.

Keywords: Participatory sensing · Viable System Model
Participatory sensing network

1 Introduction

Displacement is inherent in human activity; however, the way in which it takes place constitutes a cultural and technical feature that determines to a large extent the space of possibilities in which human activity unfolds. The use of pedestrian, public transport, bicycle, motor vehicle or animal traction vehicles is different in each culture, depending on multiple factors such as population density, road infrastructure, terrain, population purchasing power, customs … but also the will of the citizens and the groups of power.

Regarding bicycle transport, the Netherlands stands out with respect to the rest of the countries, being among the ones of less energy use and less environmental pollution per capita for reasons of transport [1]. Here we also observe that the current use of the bicycle is the result of multiple factors, on the one hand, the country has an urban structure and orography appropriate to this mode of transport; but historically significant trends have also been perceived since World War II. These have been conditioned by industrial interests, the oil crisis of the 1970s and most especially "the national demand in the form of huge demonstrations for the safety of cyclists", which led to the government taking the decision in the 1990s as a popular mandate, to develop a comprehensive Bicycle Master Plan. This mode of mobility (the pedestrian and cyclist) has a very relevant impact on sociability, sustainability, health and public welfare [2, 3].

© Springer International Publishing AG, part of Springer Nature 2018
Á. Rocha and T. Guarda (Eds.): MICRADS 2018, SIST 94, pp. 97–104, 2018.
https://doi.org/10.1007/978-3-319-78605-6_8

The general objective of biciLAB project is to increase the use of bicycles as a means of urban mobility and to use information obtained from bicycle mobility for the development of scientific, technical and community projects aimed at improving urban life and the development of smart territories linked to an active participation of citizens. The development of biciLAB is proposed as a pilot for its application in other cities, particularly in Castilla y León and Portugal (center-north). biciLAB project will be operationalized through the following specific objectives: (i) the development of information and communication systems for the collection of mobility information and the sustainable management of biciLAB according to the Viable System Model (VSM) [4], based on user-friendly technologies and massive open data processing (open data); (ii) the promotion of bicycle mobility within the university community and citizenship, including training activities for the development of a cycling culture and responsible mobility; (iii) the development of public bicycle in the university and in the urban environment, including public infrastructures to cover bicycle mobility; (iv) contribution to the creation of intelligent territories, development and fulfillment of Sustainable Energy and Climate Action Plans (SECAP); (v) the development of interdisciplinary scientific applications based on the biciLAB and devoted the improvement of urban life, which will also be connected to the education of inter-disciplinary capacities and community engagement; and (vi) a sustainable management of inter-institutional work based on the VSM.

The achievement of the objectives of the project and the continuity of the biciLAB are based on three fundamental pillars:

(1) The involvement of the university community, citizens and institutions to promote a change in modes of mobility in the area of intervention;
(2) The development of a system of collective bicycles with the capacity to gather information applicable to studies related to sustainability, urban planning, social welfare, etc.
(3) The development of capacities to integrate scientific and community knowledge aimed at coping with complex problems of social concern.

At the same time, as long as the number and diversity of actors and actions is high, a fundamental component for the achievement of results lies in the way of managing organizational complexity. The mode of organization that will be the basis of the whole biciLAB will be the VSM proposed by Staford Beer as a generic mode of organization [9]. In fact, Beer's model is based on the necessary and sufficient conditions that guarantee the organizational sustainability in an adaptive way with respect to the variations of the (external and internal) environment.

2   Background

2.1   Participatory Sensing

Participatory sensing is a personal centric participation technique with the inclusion of the citizens in the process, enabling the collection of environmental data with high granularity in space and time [1, 2]. Such process requires the active participation of people (citizens) to voluntarily share contextual information and sensory data, but differently to big-data approaches, to create sense of such data at the operative level of the agents that are cooperating to perform their own a work [3]. This means that the information that is used to handle the problems at a given level is absorbed by the own level in so far as the related issues are fully solved at that level, while the excess information corresponding to issues that require the attention at higher levels percolate upwards.

Participatory sensing networks (PSN) have become popular thanks to the increased use of portable devices, such as smartphones, tablets, iPads, as well as the massive adoption of social networks [3].

The central element of a participatory sensing network is the existence of the citizen capable of performing the sensing with a portable computational device. In this scenario, people participate as social sensors, voluntarily providing data on a particular aspect of a site that implicitly captures their experiences of daily living [4]. These data can be obtained with the aid of sensing devices, such as sensors embedded in smartphones, or through human sensors (subjective observations produced by users).

2.2 Viable System Model (VSM)

The Viable System Model was initially developed by Stafford Beer based on an analysis of the necessary and sufficient conditions of viability of organisms as a paradigm of viability and adaptation to their environment. The model is based on three fundamental principles [4]:

(1) The principle of recursion, according to which any Viable System (VS) is composed of VS (Fig. 1);
(2) The principle of requisite variety according to which the variety of a system must be greater than that of the problem which it affects;
(3) The principle of subsidiarity, according to which the variety is resolved at the lowest (recursive) level, so that only the residual variety percolated above (in the first instance, to the meta system or management bodies of the system; in second instance, to the upper recursive level).

The viability of each nested system means that it is able to autonomously manage the variety of its operational context (solving problems related to its own activity).

In the first approach, three basic elements can be distinguished: the set of operational units that perform the organization's primary activities, those for which the organization is constituted (Fig. 1, 'operations', S1); the meta system or meta-operational level, which is responsible for ensuring that the operating units function in an integrated and harmonic mode (M := {S2, S3, S3*, S4, S5}); and the environment constituted by all the components of the outside world that are of direct relevance to the system and in which the system is immersed.

According to the analysis of the necessary and sufficient conditions of system sustainability, it must be composed of 5 subsystems that interact with each other, represented in Fig. 1: (S1) every VS contains several primary activities. Each system 1, linked to a particular primary activity, is itself an VS according to the principle of recursion, and performs at least one of the fundamental functions of the organization. (S2) represents the information channels and functions that allow the primary activities

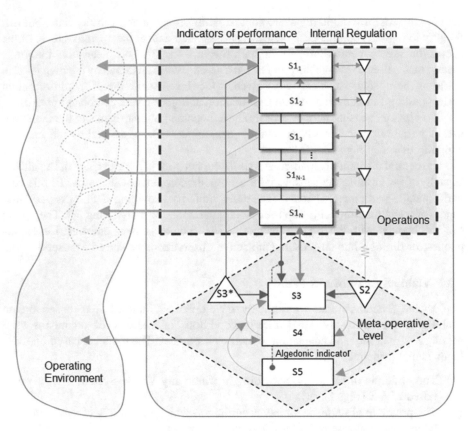

Fig. 1. Viable System Model (MSV) for an adaptive and sustainable organization. The flows of interaction with the outside are represented in green and the information flows in blue. In S2 and S3* the direction of attenuation and amplification are represented symbolically. These have a critical function when it comes to avoiding oscillatory behavior.

in S1 to communicate with one another while facilitating S3 to supervise and coordinate activities in S1. It is responsible for the programming and sharing of resources to be used by S1, conflict resolution and stability. (S3) encompasses the structures and controls arranged to establish S1 rules, resources, rights and responsibilities, guarantee internal regulation, optimize capacities and resources and synergy at the operational level. It represents the panoramic view of the processes developed in S1 while offering an interface for S4/S5. Within S3, an audit subsystem (sporadic), System 3* (S3*) can be distinguished. (S4) is composed of those responsible for looking ahead to take charge of changes in the environment and supervise how the organization has to adapt to maintain its viability in the long-term. It has to carry out, therefore, a prospective planning. (S5) is responsible for political decisions in the organization as a whole, balancing the demands of different parties and guiding the organization as a whole.

These subsystems respond to a triple role in the dynamics of operational adaptation: systems 1–3 deal with the "here and now" of the operations of the organization; system

4 deals with "there and then" as a strategic response to external, environmental and future demands; and system 5 deals with balancing the "here and now" and the "there and then" with political and axiological directives that maintain the identity of the organization as a viable entity.

According to the principle of recursion, VS is composed of VS that can be described according to an equivalent systemic description.

$$VS := \{\{S1\},\ M\ |\ S1 := SV;\ M\ S2,\ S3,\ S3^*,\ S4,\ S5\} \tag{1}$$

3   biciLAB Model

In biciLAB project, a sustainable mobility laboratory is proposed based on the capture and processing of mobility information and other parameters of interest for sustainability and health, captured by applications installed on the mobile devices of users of bicycle-sharing systems, and treated in a massive way.

As illustrated in Fig. 2, the biciLAB (located in the center) offers a basic infrastructure for the development of scientific, technological and social innovation projects.

The activity areas corresponding to the specific objectives 1 to 4 constitute an object of "continuous activity", while that corresponding to objective 5 consists of an open framework for the planning and development of interdisciplinary (ID) projects that would be chosen in annual calls, based on quality criteria and adaptation to objectives and priorities.

Thus, the component Information Systems (objective 1) is oriented to the development of the technological infrastructure for the acquisition, management and processing of information and the continuous development of applications that meet the needs of the other components aimed at: Citizen Participation in the context of Participatory sensing (objective 2); the development of Public Bike-Sharing System (objective 3); Urban Planning (objective 4) and the development of interdisciplinary scientific research linked to the objectives of biciLAB through eligible projects (objective 5).

3.1   Organization and Qualification of Knowledge Integration

The variety and diversity of actors participating in the biciLAB is represented in the figure by the number of actors involved in each component, Ni, and the number of scientific, technical or administrative disciplines involved in the respective activities, Mi. For the orchestration of this variety and multiplicity of actors and disciplines, a model of sustainable management of autonomous work is proposed (based on the Viable System Model) in which the coordination and integral planning of bike-based resources is ensured. Real-time availability of performance information of work teams distributed at various levels of organization, from the eligible projects running in a yearly basis, to the management level of biciLAB as a whole. This structure would be developed within the framework of the project aimed at constituting a pillar for the long-term operation of biciLAB under the auspices of the interested parties. In order to

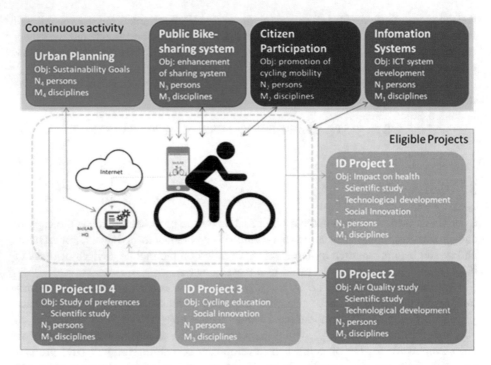

Fig. 2. Basic infrastructure for the development of scientific, technological and social innovation projects.

qualify the performance of the biciLAB in terms of its knowledge integration capacity, an innovative approach will be applied, based on the measurement of diversity and the integration of knowledge networks, contributing to the global challenge of qualifying interdisciplinarity.

From this conception, biciLAB represents altogether, on the one hand, a laboratory of social innovation oriented to propitiate a change towards a sustainable mobility from the citizen participation, on the other, an opportunity to develop the interdisciplinarity and the integration of knowledge coming from diverse scientific, technological, technical, administrative areas and from the own citizenship. This makes it an ambitious project of integral transfer of knowledge for the improvement and sustainability of urban life.

3.2   Technological Solution

The technological infrastructure is basically constituted by ad hoc mobile applications operating in users' own mobile devices and the system for acquiring, managing and processing information for its subsequent availability by citizens, public and scientific managers. Figure 3 offers some additional details about the architecture, highlighting the management of components of mobile devices (top right) that will be managed from the mobile applications of biciLAB.

On the one hand, the sensors available on any smartphone offer basic information for the collection of user mobility data. These would be stored in memory and transferred while the device has Wi-Fi access (to avoid the cost to the user caused by the transfer of data under mobile connection with her operator). In order to increase the capacity of gathering information of other types not provided by the sensors available in the user's telephone (e.g. concentration of certain gases, noise, and pressure on the pedal axis, etc.) an extension could cover the needs anticipated in specific activities. Its

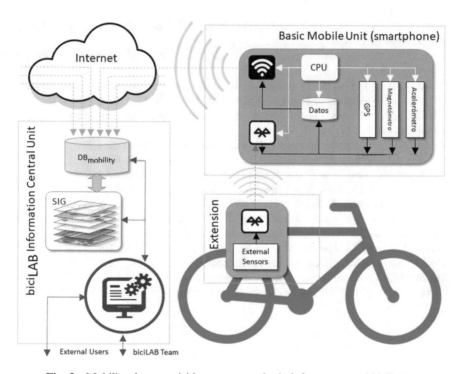

Fig. 3. Mobility data acquisition system as basic infrastructure of biciLAB.

basic structure would simply consist of a Bluetooth module attached to a sensor device. The Bluetooth port, available on the user's phone and managed by the biciLAB application, allows the transfer of information to be subsequently stored and transferred using the same protocols and data structure as for the rest of the information from the own sensors.

4 Conclusions

This document describes the biciLAB project which aims at developing a promotion program of cycling mobility and scientific-technical studies to enhance urban life based on a cycle sharing system with capacity to gather statistical information of mobility and other parameters of interest for the study of health and sustainability. The project includes the intervention in urban planning and the development of intelligent territories oriented to the achievement of Sustainable Energy and Climate Action Plans. According to current planning the biciLAB will be implemented in several metropolitan areas of Spain, Portugal and Ecuador.

References

1. MREC: En uso masivo de la bicicleta para el Buen Vivir. Embajada de Ecuador en los Países Bajos, Amsterdam (2012)
2. Litman, T.: Whose Roads? Evaluating Bicyclists' and Pedestrians' Right to Use Public Roadways (2012)
3. Molina-García, J., Castillo, I., Queralt, A., Sallis, J.: Bicycling to university: evaluation of a bicycle-sharing program in Spain. Health Promot. Intern. **30**(2), 350–358 (2013)
4. Walker, J.: The Viable Systems Model - a guide for cooperatives and federations. ICOM, CRU, CAG & Jon Walker (1998)
5. Beer, S.: Diagnosing the System for Organizations. Wiley, New York (1985)
6. Guarda, T., León, M., Augusto, M.F., Pérez, H., Chavarria, J., Orozco, W., Orozco, J.: Participative sensing in noise mapping: an environmental management system model for the province of Santa Elena. In: Rocha, Á., Guarda, T. (eds.) International Conference on Information Theoretic Security, pp. 242–251 (2018)
7. Oldenburg, J.: Griskewicz: Participatory Healthcare: A person-Centered Approach to Healthcare Transformation. CRC Press, Boca Raton (2016)
8. Burke, J., Estrin, D., Hansen, M., Parker, A., Ramanathan, N., Reddy, S., Srivastava, M.: Participatory sensing. UCLA: Center for Embedded Network Sensing (2006)
9. Giannetsos, T., Dimitriou, T., Prasad, N.: People-centric sensing in assistive healthcare: privacy challenges and directions. Secur. Commun. Network **4**(11), 1295–1307 (2011)

Territorial Intelligence in the Impulse of Economic Development Initiatives for Artisanal Fishing Cooperatives

Teresa Guarda[1,2]([✉]), José María Díaz-Nafría[3,4],
Maria Fernanda Augusto[5], and José Avelino Vitor[6]

[1] Universidad Estatal Península de Santa Elena – UPSE, La Libertad, Ecuador
tguarda@gmail.com
[2] Algoritmi Centre, Minho University, Guimarães, Portugal
[3] BITrum-Research Group, C/San Lorenzo 2, 24007 León, Spain
jdian@unileon.es
[4] Munich University of Applied Sciences,
Dachauerstr 100a, 80636 Munich, Germany
[5] Universidad de las Fuerzas Armadas-ESPE, Sangolqui, Quito, Ecuador
mfg.augusto@gmail.com
[6] Instituto Politécnico da Maia, Maia, Portugal
javemor@iiporto.com

Abstract. Sustained globalization by digital technologies has dramatically increased the capacity of the capitalist environment, putting at risk the preservation of cultural and community identities, interfering with their ability to act and adapt in a sustainable way to their environments. This paper presents a sustainable management strategy for the promotion of territorial economic development initiatives for artisanal fishing cooperatives of the province of Santa Elena, so that it can be effectively constituted as a political strategy for the country's development, seeking not only macroeconomic goals for stability and productivity excellence, but also socio-economic goals aimed at preserving natural resources, the redistribution of social wealth and the reduction of social inequalities.

Keywords: Territorial intelligence · Viable System Model · Artisanal fishing Cooperatives

1 Introduction

Intelligent territory is formed through learning environments, which function as collectors and repositories of knowledge and ideas, which are responsible for innovation and its diffusion in the conduct of the regional development process.

In the context of the solidarity economy, the construction of new labor relations through the processes of productive restructuring, highlighting the sustainability of development projects that are not always supported by the necessary mechanisms and instruments, both at the level of technologies and at the level of their articulation with macroeconomic planning and thus are verified in a set of fragmented policies.

© Springer International Publishing AG, part of Springer Nature 2018
Á. Rocha and T. Guarda (Eds.): MICRADS 2018, SIST 94, pp. 105–115, 2018.
https://doi.org/10.1007/978-3-319-78605-6_9

Contradictorily, the possibilities of solution exist through the challenges and appear at the local level, where alternative, associative and cooperative forms are built for the confrontation of the problems, starting in the initiative of the working classes. It shows the possibilities and challenges of sustainable development and solidarity economy as instruments of democratic and participatory management of workers who build networks of solidarity through new forms of organization, where social and collective interests are respected.

The aim of this project is to achieve a coordinated, decentralized, democratic, adaptive, sustainable and efficient management of cooperatives in the province of Santa Elena, integrating artisanal fishing activities and possibly other sectors of production and services, as a means of articulating solidarity activities in the province of Santa Elena. The objective includes the coordinated management of existing cooperatives as well as the cooperative formation of economic associations within the integrated sectors and the resulting grouping in a cooperative way. For the integration of the artisanal fishing activity, the creation of a Fisheries Development Unit is sought.

2   Territorial Intelligence

Territory can create an environment conducive to innovation, sustained by technology and technological innovation. Within this context, the concept of intelligent region appears as an alternative to assist the territorial organization of space, in order to generate an innovative environment capable of triggering a process of sustainable development in the precursor sites.

The discussion about technological innovation and its role in the promotion of economic development appears in the literature from the classical economists. Two types of innovation are presented: radical innovation and incremental innovation. Radical innovation is associated with the paradigm shift and brings something totally new, in incremental innovation, technique or equipment are maintained with their essential characteristics, it is an improvement of something existing. Regardless of the type of innovation, when this occurs, it brings a perspective of greater return on the volume of investments and higher rate of profit and virtuous circle of growth of the level of employment and income in the territory [1]. Support for territorial growth can only be achieved through investments in research and technical training, thus stimulating innovation and diffusion of technologies for the entire regional system.

This institutional environment, which stimulates innovation, is conceptualized as a regional innovation system, stimulating the interaction of basic research and applied research, researchers and entrepreneurs, innovation, diffusion and incorporation of new technologies, as well as increasing qualification of the workforce. In the environment, organizational and institutional structures that enable cooperation should be predominant, as well as fostering the efficient use of information flows and the means to create knowledge. It should be remembered that a regional innovation system will only be established if there is interest in organized society and local public power, since the success of this type of growth strategy depends to a large extent on the degree of interaction of local organizations and institutions.

Thus, we can deduce that the intelligent territory are regions that constitute privileged territorial contexts of interaction, learning and innovation that are configured in relational spaces between actors that intersect by cultural and economic affinities. Intelligent regions are true learning environments, whose information and knowledge are easily propagated. These regions function as collectors and repositories of knowledge and ideas, whose infrastructure and institutional environment facilitate the flow of ideas, knowledge and learning [2]. In the intelligent region, the capacity for innovation and assimilation of new techniques, technologies and knowledge is urgent. In this type of region, the institutional base must favor learning and, in turn, the accumulation of knowledge, which are the main vectors of the regional development process.

3 Relevance and Potentiality of the Santa Elena Fisheries

Among the private economic activities of the province is the fishery, with its extensive coast being one of the main centers of fishing activity in the country, along with Manabí, Guayas and Galapagos. According to archaeological evidence, fishing activity is recorded even in the early formative period (3,990–2,300 BC), which makes this activity one of the most important cultural heritage of the province and the country. But today, according to information from the Food and Agriculture Organization of the United Nations [3], the province is a privileged place for global fishing activity.

This is shown by the fact that Ecuadorian fishing and aquaculture accounts for about 14% of the country's Gross Value Added and a share of 9.6% in shrimp exports, 1.2% in the rest of the fish, 4.2% in canned fish and 0.3% fish meal. In terms of labor activity, the fisheries sector was, according to FAO estimates, occupying about 85,000 people in 2009 (of which about 6,500 were in the industrial subsector) [3]. For the Ecuadorian artisanal fishing sector, FAO highlights the favorable opportunities it offers to become an "organized, productive, highly competitive, dynamic and integrating sub-sector of social, economic and cultural development, with integrated and sustainable management of fishery resources".

In order to materialize these opportunities, the same international organization emphasizes the capacities of the cooperative way of organization integrated in Community Fisheries Centers that, in turn, sponsor Fisheries Development Units [4]. However, according to information requested from the Ministry of Agriculture, Livestock, Aquaculture and Fisheries, in relation to the cooperative articulation of the fisheries of Santa Elena, it should be noted that of the 35 fishery groups registered in the province, only 37% are cooperatives.

This outstanding fishing capacity of the region of interest; and of Ecuador as a whole, contrasts with the role of fish consumption in food sovereignty (of which the article 281 of the constitution is a specific object). Ecuadorians devote on average only 1.4% of their current expenditure to fish consumption, compared to 4.7% for meat, 4.5% for bread and cereals, and 3.4% for milk and by-products [5]. This consumption of fish is equivalent to 5.7% of the total used in food, using this source 8.1 kg per inhabitant per year (a reduction of 55% compared to the consumption estimated for 1980 [3]. In comparison, a country such as Spain (main customer of Ecuadorean fish exports), whose fishing capacity is somewhat higher than Ecuador's (around 50% higher), uses 12% of the total fish consumption in fish consumption (2.1 times more than the Ecuadorian), taking advantage of this source 43 kg per inhabitant per year (5.3 times more than in Ecuador) [3]. Since this is a double population, the balance of production and consumption makes Spain a net importer. However, without the need to reverse the positive economic balance of fishing activity in Ecuador, there is considerable scope for simultaneously favoring food sovereignty and improving health through increased consumption. On the other hand, the vertebration of artisanal fishing activity in Spain in the 20th century shows that improving the living conditions of fishers is compatible with preserving the environment for the benefit of long-term food sovereignty. The creation – before mentioned – of Community Fisheries Centers based on cooperative organization sponsored by Fisheries Development Units is a very appropriate strategy to achieve this purpose [6, 7].

However, according to the National Federation of Fishing Cooperatives of Ecuador, the artisanal fishing sector now has a regulatory framework for the defense of fishermen's rights and environmental protection, while presenting other structural problems related to: rational management of fishery resources; the implementation of sustainable development projects in the sector; the inadequacy of fishing, conservation, processing and distribution infrastructures; low level of training; and the limited applied research capacity in this subsector [8, 9]. An adequate structure of artisanal fisheries, which includes everything from the coordinated and scientific management of resources to distribution and marketing, offers a great potential for the sustainable development of the province in its social and environmental dimensions.

Despite its prominent role both in food sovereignty [10] and in employment or in the preservation of cultural identity, artisanal fishing – as reflected in the study by Benavides, García, Lindao and Carcelén (2014) – " is developed with high levels of inequality, being the most vulnerable, artisanal fisherman who lacks a boat, who is part of a crew and does not handle any other resource than his skill." [11]. The level of income significantly indicates the level of precariousness of life of artisanal fishers, which means that "in synthetic terms, 6 or 7 out of 10 fishermen live in very poor households benefiting from the Human Development Bonus" (ibid). According to these authors in their study about marketing strategies in this fishing subsector: "precariousness and inequity is explained by the way in which the activity operates. Whether or not the fisherman owns a vessel, the great beneficiary of it is the merchant who, at the beachfront level, buys fishery products, setting the price at his discretion. "These actors are closely linked with who hire the crew for the fishing task, and sometimes it is the same person adopting the figure of shipowner. In this way, "in the fishery the old figure of the" developer "is reedited, a person who by means of continuous loans commits the fisherman". The authors conclude that a serious intervention proposal in the sector is due to affect the commercialization phase, which, as it is observed, is linked to the development – credit for fishing.

4   Sustainable Management Strategy

Cooperative activity, despite its social importance and its potential for meeting the goals set by the Ecuadorian constitution, has limitations in the province of Santa Elena, as in the rest of the country, in terms of its ability to structure socio- economic scale on a scale that requires confrontation with important economic, social and environmental challenges.

In the province of Santa Elena, the artisanal fishing sector now has a normative framework for the defense of fishermen's rights and environmental protection, but at the same time presents relative problems: the rational management of fishery resources; to the implementation of sustainable development projects in the sector; insufficient fishing, conservation, processing and distribution infrastructure; low level of training; and the limited capacity for applied research [8, 9].

However, an adequate and integrated management of fishery resources could make artisanal fisheries an "organized, productive, highly competitive, dynamic and integrating subsector of social, economic and cultural development" [3].

A suitable cooperative structure in a sector of activity of socio-economic and cultural importance such as artisanal fishing could serve as an articulating axis for other socio-economic activities, in particular those that can contribute an additional value to the efficiency of the contemplated activities.

With the objective of presenting a sustainable management strategy for the promotion of territorial economic development initiatives for artisanal fishing cooperatives in the province of Santa Elena, during the year 2015 meetings were held with various fishing groups. The participation of a sufficient number of these allowed establishing the joint objective of articulating a cooperative associative structure that allows an improvement of the capacity of the associates based on a model of sustainable management, without impairing the autonomy of the operative groupings in their context of performance.

4.1   Santa Elena Fisheries Organizations

According to the information provided by the Ministry of Agriculture, Livestock, Aquaculture and Fisheries, the number of fishing organizations in the intervention area amounts to 35 (of which 20 are classified as associations, 2 as pre-associations and 13 as cooperatives), so this subgroup accounts for 25% of the total groupings.

It should be noted that among the participating (non-cooperative) associations of fishermen the cooperative way of organizing, according to their representatives' statement, had not been chosen for reasons of administrative complexity and for lacking a clear idea of their organizational peculiarities, obligations and benefits. However, in all cases the interest of developing a cooperative group constitution was manifested. The geographical distribution of these fishing groups, which, as can be appreciated, offers a pillar of provincial articulation around which eventually more clusters could be integrated (see Fig. 1).

Fig. 1. Distribution of fishermen's groups. Associations and cooperatives.

The fact that these are artisanal fisheries groups allows to focus the operational management in this type of activity whose socio-economic and cultural impact presents special advantages with respect to other fisheries subsectors with regard to food sovereignty, employment, cultural identity, socio-community structure and sustainable management of resources. The benefit of the integration of research, education and coordinated and sustainable management of resources – according to the analysis of the artisanal fishing organizations themselves and according to the studies and international recommendations mentioned above – strengthens the objective of creating a Development Unit Fisheries with a cooperative base and links to academic institutions [6–8]. On the other hand, this goal is aligned with the Millennium Development Goals, as global fisheries are considered to be a highly effective instrument for their achievement [7].

4.2   Conformation of the Fisheries Development Unit

For the articulation of the other components required for the Fisheries Development Unit (FDU), it is considered that the distribution of components considered for this project correspond to the components required for its long-term sustainability, i.e.: (1) socioeconomic observatory to component 100); (2) cooperative integration (dedicated to the extension of the cooperative grouping or the intensification of its level of integration, akin to component 200); (3) development of information systems for the structuring of cooperative activities (related to component 300); (4) Training and research (related to component 400); (6) Adaptive management of FDU (related to component 500). To these components it would be necessary to add, at the operational level, the cooperative activity itself (or integrated sectorial activity). Beyond the horizon of the project, this unit – temporarily operative – could come together with the second component in that this integration activity is something that must be addressed – according to the principles of cooperativism – from the grouping of cooperatives. Likewise, assuming that the grouping transcended the sectorial level, the Fisheries Development Unit would be transformed into a Cooperative Development Unit dedicated to the integral management of capacities and resources with the relevant academic units.

The technical feasibility of the project rests on the use of the Viable System Model (VSM) [12] applied to the coordinated organization of the objective sectorial activity (articulated by cooperatives and local and provincial cooperative groups) of the appropriate information tools that meet the sustainability conditions of the VSM. The same organizational model will be used for the formation of the FDU and for the management of the project itself, so that the structures created for it will be a substrate for the sustainability of the long-term project results.

VSM offers the basic organizational scheme to which all the organizational structures foreseen in the project will be accommodated: management of the activities foreseen in the project; re-organization of fishing cooperatives to improve their operational effectiveness while preserving compliance with their own cooperative principles; the sectorial or inter-sectorial grouping of cooperatives; and the fishery development unit.

Figure 2 illustrates the composition of the FDU in the different levels of organization considering a median size of cooperatives. For larger cooperatives, an additional level of organization would have to be added, so that in total the planned structure includes four levels of organization, which according to the initial forecast would allow efficient and democratic articulation of up to 100 000 people. If this organizational complexity is compared with that of the Mongragón Group, it is observed that they are corresponding and that the second articulates at present, in a very efficient way, the activity of 74 335 people distributed in 261 groups [13].

However, taking into account that the upper level in our case corresponds to the UDP, the equivalent in the case MCC would be that of the cooperative movement of Euskadi, and therefore would be 5 organizational levels. On the other hand in the cooperative integration of the MCC are represented a multiplicity of sectors of activity (financial, productive, distribution, services, knowledge). The MCC's organizational capacity would therefore be of a higher order and could correspond to the case of adding a level for the integrated management of several sectors.

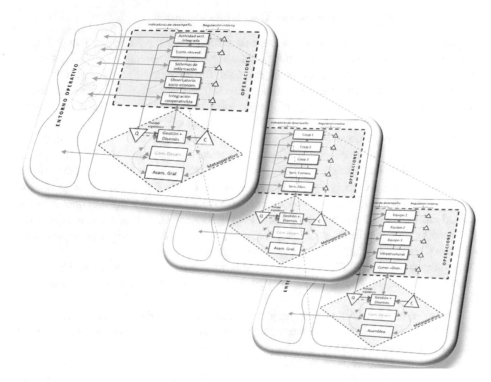

Fig. 2. Nesting of organizational structures from FDU to grassroots cooperatives.

Artisanal fishing sector is weakly structured and has deficient infrastructures. According to statistics provided by FAO [3], based in turn on information from the Secretariat for Fisheries Resources (SFR) 2009, artisanal fishing activity in the old province of Guayaquil – which included the present provincial territory of Santa Elena – was distributed in 41 coves where 17 819 fishermen worked, representing around 30% of the national total.

In the Santa Elena region there are 41 coves, 17816 fishermen and a total of 4070 boats. The composition of the artisanal fishing fleet of this region its composed by fiberglass vessels (32.6%) and wooden boats (27%) stand out [3].

In Santa Rosa and Anconcito are located the main fishing ports of the province, Santa Rosa being the main provincial port in volume of catches of artisanal fishing. The two ports are within the scope of intervention of the government project that intends to build – among other basic infrastructures – 27 docks along the entire Ecuadorian coast until the end of 2017 for the modernization and strengthening of this activity.

The new Anconcito port infrastructures undoubtedly represent a major substrate for the improvement of the subsector in the province, which, however, operational level should be complemented by extractive, productive and distributive infrastructures as well as organizational structures.

However, although both ports concentrate the largest artisanal fishing activity in the province, the rest of the 31 coves distributed along the entire coast of Santa Elena have a significant volume of fishing and occupation, infrastructure and vertebration is even more remarkable [11]. Organizationally, the artisanal fishing activity of the province, as indicated above, is supported by some 7 475 people that are articulated in 35 groups, of which 20 are associations, 2 pre-associations and 13 cooperatives.

As indicated above, the fishing sector, and in particular the artisanal sector, is not very structured, but it has entities that group – generally at national level – the sector with different purposes. In general, it can be said that for any sub-sector, there are no intermediate instances (local and regional) that allow the decentralized management of issues of common interest, while there is a lack of representative national, regional or local bodies that allow comprehensive management of common problems. In the case of fishing cooperatives, this problem is visualized by the existence of a single national representation body, the National Federation of Fishing Cooperatives of Ecuador. For national associations and fishing companies, the National Chamber of Fisheries, the National Chamber of Aquaculture, the Association of Shipowners of Fishing Boats, the Tuna Association of Ecuador and the Association of Exporters of White Fishing.

It is interesting to carry out integration in a sector from which significant wishes and benefits can be obtained in the area described. The fishing activity undoubtedly represents a very appropriate sector for the proposed objectives. On the one hand, there are enough cooperatives and groups of fishermen to tackle the proposed objective, but it is also an economic sector of the highest importance in the area of intervention that is affected by important problems of inequality.

The World Conference on Small Scale Fisheries, "Ensuring Small-scale Fisheries: Responsible Fisheries and Social Development," convened by the Food and Agriculture Organization of the United Nations (FAO) in Bangkok, noted the need to achieve "A comprehensive and coordinated strategy to ensure and expand the capacity and freedom enjoyed by fishing communities, including civil and political freedom to participate meaningfully in the processes that shape their lives." In this sense, a general recognition of the mode of cooperative organization in its capacity was pronounced: (1) Increase the stability and responsiveness of fishing communities; (2) To increase the capacity of fishermen to negotiate prices with intermediaries, to help stabilize markets, to improve post-harvest practices and facilities, and to sustain marketing logistics and market information; (3) Encourage higher levels of commercial competition by establishing auctioning systems, contributing to market information and, where appropriate, procuring supplies or investing in common structures such as refrigeration plants and fish processing facilities; (4) Save-by increasing bargaining power-in wholesale purchases of rigs, engines, equipment and fuel, and gain more political influence and bargaining power with the government and (5) Also provide microcredit schemes for fishermen, which would reduce their dependence on intermediaries and give them more flexibility in the choice of buyers, which would favor small-scale fish buyers and sellers.

Taking into account the low penetration of social insurance, it is expected that the population over 65 (5.6%) will present significant cases of vulnerability. In this sense, the cooperative mode of organization is in itself an adequate measure to deal with these situations. As an example, the cooperative experience of Mondragón can be considered, which is deeply illustrative in this sense: in its origins the population lacked basic social protection which was solved by the creation of the Lagun Aro Social Provision

Service that until the day of today it constitutes one of the pillars of cohesion of the group Mondragón and in fact between the services of this nature is leader in satisfaction of its clients [13].

5 Conclusions

The aim of this project is to achieve a coordinated, decentralized, democratic, adaptive, sustainable and efficient management of cooperatives in the province of Santa Elena, integrating artisanal fishing activities and possibly other sectors of production and services, as a means of articulating solidarity activities in the province of Santa Elena; in order to a sustainable and local development to be effectively constituted as a political strategy for the country's development, seeking not only macroeconomic goals for stability and productivity excellence, socio-economic goals should be set for the preservation of natural resources, the redistribution of social wealth and the reduction of social inequalities. As well as transforming solidarity economy into public policy, revolutionizing not only local structures, the emergence of a network that shelters all productive chains, from its origin and the chain of all phases, taking advantage of existing experiences.

References

1. Oliveira, G.: O desenvolvimento das regiões: uma iniciação às estratégias de desenvolvimento regional e urbano. Protexto, Curitiba (2008)
2. Palacios, T.B., Galván, R.S.: La estrategia basada en el conocimiento en el ámbito territorial Revisión teórica. Pensamiento & Gestión 25, 58–77 (2008)
3. FAO: Report of the Global Conference on Small-Scale Fisheries Securing Sustainable Small-Scale. Fisheries: Bringing Together Responsible Fisheries and Social Development. FAO., Roma (2009)
4. Ben-Yami, M., Anderson, A.M.: Centros comunitarios de pesca: pautas para su fundación y operación (1987)
5. INEC: Proyecciones a nivel de provincias por grupos de edad. 2010–2050, Quito (2013)
6. Wilson, D., Nielsen, R.J., Degnbol, P.: The fisheries co-management experience: accomplishments, challenges and prospects, vol. 26. Springer, Netherlands (2003)
7. Pomeroy, R., Andrew, N.: Small-scale fisheries management: frameworks and approaches for the developing world. Cabi, London (2011)
8. FENACOPEC: Situación actual y perspectivas de los derechos de acceso de la pesca artesanal en Ecuador (2009)
9. Darricau, D., Marugán, L.: Proyecto para el desarrollo de las comunidades pesqueras de la República del Ecuador., Guayaquil (2012)
10. AN: Ley orgánica del régimen de la soberanía alimentaria. Registro Oficial Suplemento 583 de 05-May-2009 (2009)

11. Benavides, A., Espinoza, L.G., Lindao, C.A., Carcelén, F.: El sector pesquero de Santa Elena: análisis de las estrategias de comercialización. Revista Ciencias Pedagógicas e Innovación **2**(2), 79–86 (2014)
12. Walker, J.: The Viable Systems Model – a guide for cooperatives and federations. ICOM, CRU, CAG & Jon Walker (1998)
13. Arregui, N., Antonio, P.: La Experiencia Cooperativa de Mondragón: estudio de su viabilidad organizacional en el contexto de Euskadi. Revista de economía pública, social y cooperativa 54 (2006)

Assistive Devices and Wearable Technology

Use of Drones for Surveillance
and Reconnaissance of Military Areas

Carlos Paucar[1], Lilia Morales[1], Katherine Pinto[1], Marcos Sánchez[1],
Rosalba Rodríguez[1(✉)], Marisol Gutierrez[1], and Luis Palacios[2]

[1] Universidad de las Fuerzas Armadas-ESPE, Sangolqui, Quito, Ecuador
{cmpaucar2,lmmoralesl,kjpinto,mgsanchez7,mrodriguez5,
megutierrez2}@espe.edu.ec
[2] Fuerza Aérea Ecuatoriana, Quito, Ecuador
lpalacios@fae.mil.ec

Abstract. This research is intended to contribute to the design of control algorithms for static cameras and drones. These were modelled on the quadcopter Parrot Bebop 2 by using the communication software Robot Operating System ROS to provide data and recognize different drone stages (landed, on flight, and its maneuvers: yaw, throttle, roll, pitch). The controller was developed after conducting several tests regarding the ideal distance for effective drone operations which include image processing for target detection or tracking in real time. This study also analyzes the benefits on the implementation of this technology in the Ecuadorian Armed Forces for surveillance and reconnaissance operations. Based on the results, it can be concluded that the use of drones in aspects of national security would have a positive impact because it allows to reduce costs as well as to optimize human resources in military operations.

Keywords: Drones · Images · Detection · Surveillance · Reconnaissance

1 Introduction

The globalization and the technological advance in different academic areas has not been the exception in the aeronautical field. Unmanned Aerial Vehicles (UAVs) are a clear example of continuous development, they are more commonly known as drones, "aircrafts able to fly without having a pilot on board, consisting on an aerial platform handled from the ground with the capacity to obtain and transmit information immediately through the use of sensors" [1]. These devices can be reused and are capable to perform a controlled flight hold by a combustion engine. Moreover, "UAVs are considered as tactical, autonomous and are connected to a ground control station (GCS); in addition, drones are able to perform surveillance and reconnaissance tasks as well as sending live information" [2].

Nowadays, these technological devices are used for commercial, military, and personal purposes. "At the moment, the most frequent uses are: aerial photography, cinematography, monitoring, surveillance, inspection of facilities, search and rescue missions, emergency management, and land mapping" [3]. It has been observed an increased number of drone users in fields such as geology, wealth management,

© Springer International Publishing AG, part of Springer Nature 2018
Á. Rocha and T. Guarda (Eds.): MICRADS 2018, SIST 94, pp. 119–132, 2018.
https://doi.org/10.1007/978-3-319-78605-6_10

hydrology and security and border control. Likewise, there are different types of UAVs, classified as: fixed wing & rotary wing drones. "Fixed wing drones are mostly used by militaries for intelligence, reconnaissance and attack missions due to their superiority in terms of autonomy, scope, service ceiling, speed and loading capacity" [4].

As stated in the previous paragraph, drones are also used during military operations; Armed Forces of various countries such as the United States have included these technological devices for several years; for example, in 1983, UVAs were employed by the American militaries in Lebanon. After the terrorist attack occurred on September 11th of 2000, the United States increased the production of drones as well as their use in the Pakistani confrontation. "The US Government led by President Barack Obama systematically intensified drone attacks in Pakistan due to their effectiveness in Afghanistan. It can be concluded that the main American war strategy in Pakistan was based on the use of unmanned aerial vehicles" [5]. UAVs were employed in numerous operations and "the idea of using drones MQ1-1 and MQ-9 in surprised attacks against military leaders without having great losses was generalized" [6].

Moreover, the use of these technological aircrafts, have started to take place in South America, especially in Colombia, which is a country that has been battling for more than 50 years an internal conflict. In order to combat this situation, most public institutions have incorporated UAVs for reconnaissance, surveillance, photography and communication missions; this decision have brought favorable results against armed illegal groups [7].

Apart from that, Ecuador has also entered in the use of this technology, particularly the Armed Forces, which have conducted several researches in this field. An example, presented by the Ecuadorian Air Force, is the design and implementation of a pre-flight automated system for the "UVA-1 prototype Phoenix which is able to perform operations like landing, taking off, and autopilot. It can also transmit real time data through the use of a video camera with an electro-optical system; this feature contributes to the mission of surveillance and reconnaissance which is a responsibility of the nation's Defense area [8]. The Geographic Military Institute, which belongs to the Ecuadorian Army, is in charge of the nation's mapping service, and acquired a drone to "conduct research studies related to the geography and mapping of a specific zone in the Antarctica called Pedro Vicente Maldonado Station..." [9].

Based on what has been previously explained, it can be deducted that the use of drones in surveillance and reconnaissance missions is feasible, however, this technology is not applied with the frequency it should be by the Armed Forces. Some Ecuadorian universities have also centered new academic programs in this field by offering degrees in various areas associated to technology. Students who get their majors in these careers possess the necessary knowledge to design a software capable of controlling UAVs. The fact of having the opportunity to apply drones for military operations as well as the contribution of professionals in technological fields, are the main reasons of this research. It is aimed at the designing of control algorithms for static cameras and drones, which were modelled in Parrot Bebop 2 quadcopter, for efficient real-time people and objects detection.

This article has been structured in four sections: the first part is the introduction; the second, describes the process of capturing images; the third, presents the analysis and procedures for tracking implementation, additionally, it is briefly examined the applicability of drones for military operations. Finally, the research conclusions are presented in the last part of this work.

2   The Use of Computer Vision for Image Processing

The methodology applied for modelling the algorithms, used for the communication system as well as for the detection and tracking of the drone Parrot Bebop 2, is the Pre-Experimental design based on the design "One group Pre-test, Post-test". Several tests were done to determine the distance needed for proper robot operation. Visual Servoing (VS), also known as Vision-Based Robot Control, is the technique used during this research to monitor movements and actions of the drone [10]. A camera is employed to provide data related to the location of an object in relation to the collected image in order to control the robot's movements [11].

Through image-based visual servoing (IBVS) technique, the error is estimated among current and desired characteristics in the image plan; characteristics might be the coordinates of visual characteristics, lines, or regional moments [11].

For a precise recognition of the target that will be tracked, through a quadcopter type Dron Bebop 2, the following stages are considered: image acquisition, image preprocessing, noise reduction, segmentation, characteristics extraction, and recognition & interpretation.

The first step in the process of image capturing, is the implementation of (ROS) Robot Operating System which captures the image and saves it as a message format sensor_msgs/image. The use of the CvBridge library allows to change the format and convert it from ROS to OpenCv (cv:: Mat) [12]. The process is illustrated in Fig. 1.

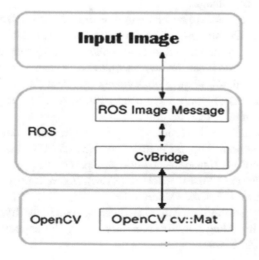

Fig. 1. Conversion process from ROS image to OpenCv image.

Once converted to OpenCv, this file allows the use of pre-trained classifiers and other resources such as specialized libraries in computer vision, the algorithm Histogram of Oriented gradients (HOG) among the most well-known [13]. It is based on the orientation of gradients within localized portions of an image. This image is divided into several cells which contain addresses of the gradient histogram or edge orientation pixel cells, as illustrated in Fig. 2.

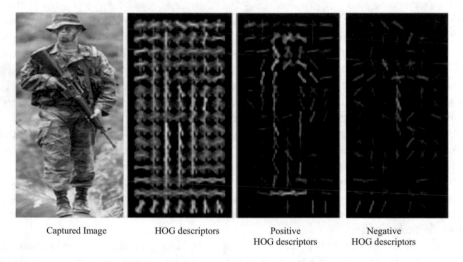

| Captured Image | HOG descriptors | Positive HOG descriptors | Negative HOG descriptors |

Fig. 2. Captured image and HOG feature descriptor.

HOG descriptors measure the intensity changes based on the borders of an image which are computed by calculating image gradients that capture contour and silhouette information of grayscale images.

It is convenient to standardize histogram values, to minimize the difference on image captions with the purpose of achieving similar gradient magnitudes in both images. Standardization is performed in zones called blocks (group cells). These blocks of standard descriptors are what the authors called HOG descriptors.

As a result of standardization, global values (gradients) of histograms are equal, therefore, the differences of values have been reduced to a final representation among similar images, obtaining a final descriptor [14].

Finally, standardized gradients on each portion of the image are used as input data for a classification system, Support Vector Machine (SVM), these sets of organized vectors in a n-dimensional space will build a separation hyperplane on that space, which is known as Support Vector Machine (SVM). Figure 3 shows the whole process from image capture to vector data storage.

It is considered that the best data classification tool is the hyperplane that maximizes the distance regarding the closest points. Being the support vectors the points which are close to the edge.

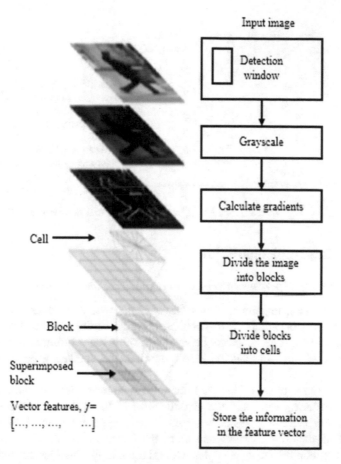

Fig. 3. Process of extraction, gradients calculation and characteristics storage.

The use of support vector machines, for detection, enables the separation of positive and negative class, this means that the data that corresponds to the *person* class are positive samples and the *non-person* class represents negative samples, as it is illustrated in Fig. 4.

The concept of optimal separation is where the fundamental characteristic of SVM is located; this type of algorithm allows the hyperplane to have the maximum distance (edge) with the points closer to them. In this way, the vector points are separated when they are labeled in a category on each side of the hyperplane.

In order to implement these categories, previous training of the machine is needed by providing it examples of persons or *positives* as well as examples of negatives or *non-persons*. Once the examples have been given during training, the algorithm of SVM classification elaborates a M-dimensional curve that divides both groups, obtaining the kernel (function that allows to convert a nonlinear classification problem in the original dimensional space to a simple linear classification problem in a greater dimensional space) of the machine.

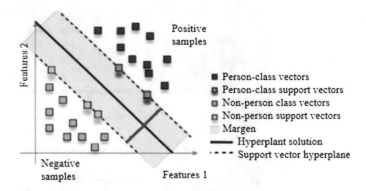

Fig. 4. Classification of samples person and non-person.

3 Tracking Implementation

A. Monitoring of persons or objects in a specific external environment.

For the tracking implementation ROS is employed, it includes packages for the development of applications: its main characteristic is coding reuse and development of robot applications. The package bebop_autonomy has communication controllers among the ROS platform and the Parrot Bebop2, the updated version is called parrot_arsdk [15].

The first step is to set up the package and the libraries, it is essential to create the *workspace* (working area) and the classification of the different packages. Once the package bebop_autonomy has been set up, it will be permitted the access to communicate and get data from various sensors of the Parrot Bebop 2, this includes sending commands to control the quadcopter. Complementarily, the files manifest.xml and CMakeList.txt must be set up for the compilation that is done by the CMake.

The next step is the *launch* files setting, which are needed to execute effectively the nodes that have been created. In this research two launch files have been executed: (1) the first executes the *bebop_driver node*, opening the image flow from the camera of the quadcopter Parrot Bebop 2; and (2) the second the execution of the node for visualizing the user interface.

Then, the ROS system is executed to implement internal communication among the nodes through subscriptions and publications of topics that are in charge of transmitting and receiving data; in this case there must be communication between the nodes *bebop_driver* and *run_bebop*.

The subscriptions are used to receive information that is sent from the node: the video flow obtained from the camera of the quadcopter by using the topic called/ *bebop/image_raw* along with the position and the speed obtained from the topic/ *bebop/odom*. To these topics and some others is added the node *run_bebop* which receives all the necessary data required to be published in the user interface.

There are also the publishing nodes, these are in charge of sending data from the node *run_bebop* to the node *bebop_driver* which in this case will be commands for the quadcopter: for taking off is the topic/*bebop/takeoff* and for landing the topic/*bebop/land.*

The following step is the execution of the launch files through the terminal for the nodes activation and the communication between the quadcopter and the computer. Table 1 shows the control that can be sent to the quadcopter by using the keyboard of the manual control.

Table 1. Orders of the keyboard to control the debop 2.

Key	Command	Description
T	Take off	Send the order of taking off to the quadcopter
L	Landed	Send the order of landing to the quadcopter
R	Reset	Resets the parameters of the Bebop 2
F	Flattrim	Calibrates the camera in frontal position, in relation to the orientation of the Bebop 2
P	Permit tracking	Allows to start the manual mode
H	Home	Sends the order of going home

Considering that the system that flight tracking is in real time, images processing must be fast; taking this into consideration, it is used the resolution of 640×280 pixels to process data. Consequently, the control orders will be executed adequately at the moment of sending them to the drone for tracking, without delay as it may happen when using a higher resolution.

In Fig. 5, it is illustrated the visual field that will cover the frontal camera of the drone. Within this field, reference axis of the drone are established: roll (y axis) and throttle (z axis). Based on this, the body describes a continuous path that will occupy a position (y, z), being an irregular path due to the movement of the target as well as the drone when trying to track.

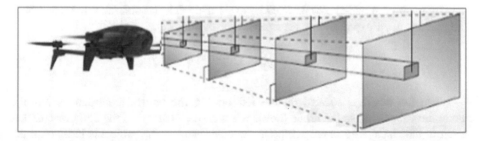

Fig. 5. Communication between nodes, subscription and publication of topics.

The most relevant aspect is the determination of the *position error*. During the tracking, the main objective is to locate and keep the body or object in the center of the received image, for that reason a new point is established in the center of the received image, this is (320; 240) which is illustrated in Fig. 6.

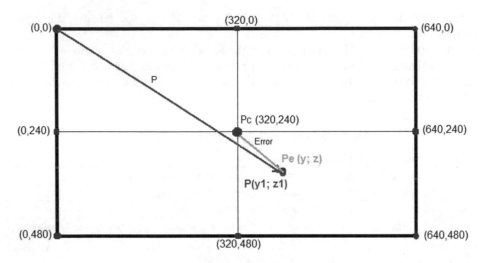

Fig. 6. Center point and position error

The following equation is used to determine the position error of the target in regard to the center of the camera.

$$Pe(y,z) = P(y1,z1) - Pc\,(320,240) \tag{1}$$

Where, Pe = Calculation error point; P = Subject center point; PC = Image center point.

The position error calculated may show values within a range of ±320 pixels for the y axis (roll) and as for the z axis (throttle) it will be of ±240 pixels; as a result, a new point of origin in the center of the image will be created automatically.

Next, the *conversion of the position error at speed* takes place, and in order to fulfill this task, flight commands are published through topics (themes), in this case, it is used *cmd_vel*. The conversion of values related to position errors to speed values is done by scaling, in which 320 px is equal to −1 and −320 px and takes the value of 1. The mathematical expression is represented on the Eq. (2).

$$Vy = -\frac{Epy}{320} \tag{2}$$

When the target is located on the left border, the control command sends the maximum value of 1, so the drone moves left until the position of the body reaches the center of the visual field or set point. On the other hand, if the subject is located on the right border (320), it sends the maximum value of −1.

As for the pitch axis (x axis, depth), is a different kind of analysis, which is done by the compilation of the areas obtained from the target when it is positioned within the visual field at certain distance. Once the distance of the target is verified and safe for tracking, it is selected and the area is saved. The area is highlighted in the shape of a rectangle pointing the target, obtaining the maximum value of the area when the target is closer to the quadcopter, whereas, when the target moves off, the size of the area will decrease until reaching the maximum and minimum values of depth position.

To determine the maximum and minimum values of depth position, the reference area or set point can be established by using the Eq. (3).

$$Arefx = \frac{A_{max} - A_{min}}{2} + A_{min} \tag{3}$$

Where Arefx: reference area for the distance of the target; A_{max} = maximum detection area 60000px^2; A_{min} = maximum detection area 10000px^2.

When the target is close to the frontal camera and gets the maximum value of the area, the command value will be sent as −1, so the drone moves backwards. When the target moves off from the drone and gets the minimum value of the area, the command value will be sent as 1, so the drone moves forward directly to its target.

The values that make the drone moves forward or backwards can be obtained by using the following Eq. (4).

$$Vx = \left[\left(1 - \frac{Area}{Arefx} \right) * K \right] \tag{4}$$

K is a compensation constant which is obtained by tracking tests, it is applied in order to make the drone receive the adequate commands for throttle.

Once the previous activities have been achieved, the modelling of the quadcopter *Bebop 2* must be completed to establish the location of the quadcopter at any point in the space; its position must be established regarding a fixed reference point and its orientation about an inertial frame [16]. When the quadcopter moves on any of its axis, the turning speed of its rotors changes applying more or less force on these to reach the desired movement in throttle, yaw, pitch and roll.

To know the orientation of the quadcopter, the Euler angles must be obtained regarding C, being C the frame that indicates the rotation of the quadcopter on the tridimensional space [16]. Where Φ is the roll angle (rotation regarding Xb); θ is the pitch angle (rotation regarding Yb); Ψ is the yaw angle (rotation regarding Zb) [17]. See Fig. 7.

To get the C complete rotational matrix regarding I, the product of the three rotational matrices is calculated [18], getting as a result the rotation matrix of the C reference frame concerning the fixed reference frame I.

$$R_I^C = \begin{bmatrix} \cos\Psi\cos\theta & \cos\Psi \sin\theta\sin\varnothing - \sin\Psi\cos\varnothing & \cos\Psi\sin\theta\cos\varnothing + \sin\Psi\sin\varnothing \\ \sin\Psi\cos\theta & \sin\Psi\sin\theta\sin\varnothing - \cos\Psi\cos\varnothing & \sin\Psi\sin\theta\cos\varnothing + \cos\Psi\sin\varnothing \\ -\sin\theta & \cos\theta\sin\varnothing & \cos\theta\cos\varnothing \end{bmatrix} \tag{5}$$

The mathematical modelling of a quadcopter expressed in the Eq. (5) shows that it can be decomposed into each of its axis to carry out the control and analysis respectively; in order to achieve this modeling has to be based on its state variables.

This system is similar to a variable-mass system without damping, which can be controlled by the input, having as a result the transfer function [19]. See Table 2.

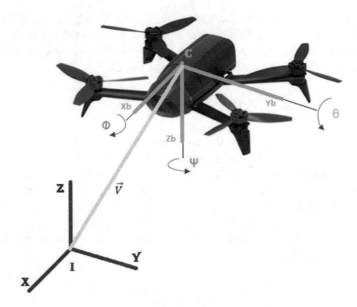

Fig. 7. Reference frames for modelling

Table 2. Keyboard commands to control debop 2

Axis	Transfer function
Throttle	$G(s) = \frac{1}{m.s^2}$
Pitch	$G(s) = \frac{l1}{I_x.s^2}$
Roll	$G(s) = \frac{l2}{I_y.s^2}$
Yaw	$G(s) = \frac{1}{I_z.s^2}$

Next, it is needed to calculate the quadcopter inertial matrix, which is done considering that it is symmetrical [20] according to Eq. (6).

$$I = \begin{bmatrix} I_{xx} & 0 & 0 \\ 0 & I_{yy} & 0 \\ 0 & 0 & I_{zz} \end{bmatrix} \tag{6}$$

The inertial calculation for each axis is the following:
X-axis

$$I_{x1} = I_{x3} = \tfrac{1}{12} m_m \left(l_y^2 + l_z^2 \right) \qquad I_{x2} = I_{x4} = \tfrac{1}{12} m_m \left(l_y^2 + l_z^2 \right) + m_m d_{cg}^2 \tag{6.1}$$
$$I_{xx} = 2I_{x1} + 2I_{x2}$$

Y-axis

$$I_{y2} = I_{y4} = \tfrac{1}{12} m_m \left(I_x^2 + I_z^2 \right) \qquad\qquad I_{y1} = I_{y3} = \tfrac{1}{12} m_m \left(I_x^2 + I_z^2 \right) + m_m d_{cg}^2 \qquad (6.2)$$
$$I_{yy} = 2I_{y1} + 2I_{y2}$$

Z-axis

$$I_{z1} = I_{z2} = I_{z3} = I_{z4} = \tfrac{1}{12} m_m (I_x^2 + I_y^2) + m_m d_{cg}^2$$
$$I_{zz} = 4I_{z1} \qquad\qquad (6.3)$$

On Table 3 are presented the necessary inertial values to calculate the gain of the Proportional-integral-derivative (PID), when replaced in transfer functions of each axis from Table 2.

Table 3. Values to calculate inertia on different axis.

Variables	Values
Drone mass	0.536 kg
Motor mass (m_m)	0.04275 kg
Rotor length on X (I_x)	0.023 m
Rotor length on Y (I_y)	0.023 m
Rotor length on Z (I_z)	0.0113 m
Distance from the motor to the quadcopter's gravity center (d_{cg})	0.143 m

Once the quadcopter Bebop 2 modeling has been completed, and for having a more suitable result on flight, is recommendable to control each axis separately, in the case of roll and pitch the mathematical expression is the following.

Pitch Axis

$$G(s) = \frac{l1}{I_x . s^2} = \frac{0.114}{0.00175 . s^2} \qquad fx(s) = \frac{0.3s^2 + 0.603s + 0.006}{s} \qquad (7.1)$$

Roll Axis

$$G(s) = \frac{l2}{I_y . s^2} = \frac{0.088}{0.00175 . s^2} \qquad fy(s) = \frac{0.12s^2 + 0.1206s + 0.0024}{s} \qquad (7.2)$$

The easiest way to generate the controller is through a transfer function and then apply the Laplace transform [10]. Getting to the Eq. (8).

$$G(s) = K_p + \frac{K_i}{s} + K_d s = \frac{K_d s^2 + K_p s + K_i}{s} \qquad (8)$$

It is important to remember that the maximum values accepted by the quadcopter and that are managed by the bebop driver, fluctuate between −1 and 1, therefore, PID control cannot send higher values than the established.

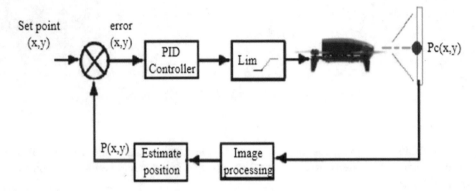

Fig. 8. PID Function Blocks of control diagram.

On Fig. 8, it is illustrated the quadcopter's control blocks diagram; the feedback is done with the obtained image processing since this one brings a point with coordinates and an area that are mandatory for the roll and throttle movement of the drone.

B. Use of drones for reconnaisance and surveillance of Military areas.

The important modernization process that is taking place among humanity, particularly in the Ecuadorian Army, has caused that its members show interest for technological innovation, advances that have forced the modification of the characteristics and management of military operations.

The use of drones to improve public safety is having more relevance, especially in border's control, illegal businesses, facilities surveillance and control of marine spaces.

The Armed Forces possess facilities in all the Ecuadorian territory and most of them are large areas that must be watched by military personnel, for this reason, several guard shifts must be conformed to satisfy this necessity. For example, the Jaramijo Naval Base is a warehouse where weapons and ammunitions of the Armed Forces are stored. The logistic support is ruled by manuals that specifies: the entry, exit, transport, storage, distribution, safety rules as well as removal processes in case of expired ammunitions to prevent accidents. There are also rules for the transportation of this material, these stipulates the vehicle characteristics for each kind of weapon and procedures for safety. There are precise policies for the storage and protection of weapons and ammunitions; in addition, these manuals described how war material must be stowed, labeled by color codes depending on the type of material, and its level of danger. [21].

The advantages of applying this study for surveillance missions of The Jaramijo Naval Base, which has about 11 thousand hectares, are the following:

- Planning Rounds in specific timetables.
- Ease access to dangerous or inaccessible places for human beings.
- Cover wide extensions of territory in a short period of time to achieve better search results.
- Risk reduction for human beings avoiding exposition to tasks considered as dangerous.

- Precision in detection, classification and tracking of people or objects in a limited environment in real time.
- Reduction of time in decision making processes.
- Considerable cost reduction on implementation of UAVs by using open software such as: Linux, ROS, OpenCv.

Optimal computer vision system for light and movement changes which enables the target tracking in low light environment.

4 Conclusions

Image processing for location and tracking of a person or object needs a great amount of technological resources, as much as memory and speed processing. The use of Robot Operating System (ROS) and the Parrot Bebop 2 Quadcopter on this research has represented considerable benefits; the most important, was the access to flight data which made easier the development of applications, and by having its own Wi-Fi signal, the programs were executed directly from the computer by sending commands and receiving different flight states.

When transforming the video image to OpenCv format, some characteristics were lost because of the way these were stored by ROS; due to the camera's resolution of the quadcopter Parrot Bebop 2, this loss was relative, therefore, it did not modify image processing and final results.

The use of HOG algorithm provided solid characteristics to support different light changes in non-controlled environment, and through the use of SVM classifier, to categorize person from non-person, was possible to detect targets effectively when it was applied in the control algorithms of the static camera and drone.

Once detected the person or object, commands in charge of flight control are sent, these are: taking off and landing as well as throttle and pitch movements. It is necessary to point that different tests (not included in this article) have been successfully applied. This fact guarantees the appropriate functioning of the mathematical model selected for this research.

The implementation of this technology would allow more control and coverage of the Armed Forces strategic assets. It would also optimize the resources and minimize the risk in Internal Defense operations including the monitoring of the state dependencies like the Presidential Palace or the National Assembly offices. The Air Force as responsible of the Ecuadorian airspace could implement this device to combat illegal acts on border controls such as: drug-trafficking, arms and explosives trafficking; illegal mining; clandestine drug laboratories.

References

1. Arias, G., Cristina, D., Sánchez, P., Fernando, D.: Diseño e implementación de un sistema automatizado de prevuelo para el prototipo UAV1 Fénix de la Fuerza Aérea Ecuatoriana, » Latacunga/ESPE/2014, Latacunga (2014)

2. Ministerio de Defensa Nacional: Dirección de Comunicación Social. Boletin de prensa 48, Quito (2012)
3. Addati, G., Pérez, G.: Introducción a los UAV's Drones o VANTs de uso civil. Seire Documentos de Trabajo, Universidad de CEMA, No. 551, Buenos Aires (2014)
4. Golmayo, S.: Cálculo y selección de sistema de propulsión para mini UAV de apoyo a pequeñas unidades de Infantería de Marina, Centro Universitario de la Defensa en la Escuela Naval Militar, España (2015)
5. Correa, N.: Estrategia de intervención militar de estados unidos en medio oriente: dinámicas y consecuencias del uso de drones en Yemen y Pakistán entre 2009 a 2013, Universidad Colegio Mayor Nuestra Señora del Rosario, Bogotá (2015)
6. Williams, B.: The CIA's cover Predator drone war in Pakistan 2004–2010: the history of an Assassination compaign, Studies in Conflict & Terrorism, Miami (2010)
7. Brito, M.: Los drones, un nuevo socio en el espacio aérco colombiano, Universidad Militar Nueva Granada, Bogotá (2014)
8. Arias, G., Cristina, D., Sánchez, P., Fernando, D.: Diseño e implementación de un sistema automatizado de prevuelo para el prototipo UAV1 Fénix de la Fuerza Aérea Ecuatoriana, Latacunga - ESPE, Latacunga (2014)
9. Morán, C.: El drone del Instituto Geográfico Militar, uso específico y su operación en áreas estratégicas requeridas por diferentes organizaciones, Universidad Politécnica Salesiana Sede Quito, Quito (2016)
10. Jácome, C., Teresa, M., Andrade, M., Javier, D.: Control de un cuadricoptero usando ROS, Quito: Escuela Pólitecnica Nacional (2014)
11. Hutchinson, S., Hanger, G., Corke, P.: A tutorial on visual servo control. IEEE Trans. Robot. Autom. **12**(5), 651–671 (1996)
12. ROS.org: cv_bridgetutorialesUsingCvBridgeToConvertBetweenROSImagesAndOpenCV Images, 05 2015. [En línea]. http://wiki.ros.org/cv_bridge/Tutorials/UsingCvBridgeTo ConvertBetweenROSImagesAndOpenCVImages. (Último acceso: 10 2016)
13. Betancourt, G.: Las Máquinas de Soporte Vectorial (SVMs), Pereira – Colombia: Universidad Tecnológica de Pereira – Scientia et Technica Año XI (2005)
14. Valveny, E.: Lecture 29 – L4.4. HOG – Cálculo del descriptor, (2015). [En línea]. https:// es.coursera.org/learn/deteccion-objetos/lecture/uAEKB/l4-4-hog-calculo-del-descriptor. (Último acceso: 20 03 2017)
15. Mani, M.: AutonomyLab, Simon Fraser University (2015). [En línea]. http://bebop-autonomy.readthedocs.io/en/latest/. (Último acceso: 11 2016)
16. Barrientos, A., Peñin, L., Balaguer, C., Aracil, R.: Fundamentos de robótica. Mcgraw-Hill/ Interamericana de España, Madrid (2007)
17. Reinoso, M.: Diseño de un sistema de control por regimen deslizante para el seguimiento de trayectoria lineal de un quadrator. Universidad Politecnica Salesiana, Cuenca (2014)
18. Barrientos, A., Peñin, L., Balaguer, C., Aracil, R.: Fundamentos de robótica, Mcgraw-Hill/Interamericana de España, Madrid: S.A. (2007)
19. Bristeau, P., Callou, F., Vissiere, D., Petit, N.: The navigation and control technology inside the AR. Drone micro UAV, Milano: IFAC (2011)
20. Brito, J.: Quadrator Prototype, Lisboa- Portugal: Instituto Superior Tecnico Portugal (2009)
21. Sanchez, E.: Gestión Logística en el Ambito Militar. [En línea]. http://businessmanagement. globered.com/categoria.asp?idcat=34. (Último acceso: 15 03 2011)

Engineering Analysis and Signal Processing

The Electroencephalogram as a Biomarker Based on Signal Processing Using Nonlinear Techniques to Detect Dementia

Luis A. Guerra[1(✉)], Laura C. Lanzarini[2], and Luis E. Sánchez[3]

[1] Universidad de las Fuerzas Armadas-ESPE, Sangolqui, Ecuador
laguerra@espe.edu.ec
[2] Universidad Nacional de la Plata, La Plata, Argentina
lidi@lidi.info.unlp.edu.ar
[3] Universidad de la Mancha, La Mancha, Spain
luisenrique@sanchezcrespo.org

Abstract. Dementia being a syndrome caused by a brain disease of a chronic or progressive nature, in which the irreversible loss of intellectual abilities, learning, expressions arises; including memory, thinking, orientation, understanding and adequate communication, of organizing daily life and of leading a family, work and autonomous social life; leads to a state of total dependence; therefore, its early detection and classification is of vital importance in order to serve as clinical support for physicians in the personalization of treatment programs. The use of the electroencephalogram as a tool for obtaining information on the detection of changes in brain activities. This article reviews the types of cognitive spectrum dementia, biomarkers for the detection of dementia, analysis of mental states based on electromagnetic oscillations, signal processing given by the electroencephalogram, review of processing techniques, results obtained where it is proposed the mathematical model about neural networks, discussion and finally the conclusions.

Keywords: Biomarker · Dementia · Electroencephalogram · Signal processing
Neuronal network

1 Introduction

The understanding of how the brain works with the appearance of symptoms of syndromes of progressive pathological deterioration, physiological aging and brain diseases will allow the development of new therapeutic and rehabilitative approaches; therefore, it is necessary to focus the research in this field in order to contribute in the identification of quantitative and specific biomaterials [1], which will allow understanding the

This work was supported by the Universidad de las Fuerzas Armadas, Sangolquí – Ecuador. Luis A. Guerra work at the University of the Fuerzas Armadas at the campus in Latacunga City. Laura C. Lanzarini work at the National University of the Plata – Argentina and Luis E. Sanchez work at the University of the Mancha - Spain.

© Springer International Publishing AG, part of Springer Nature 2018
Á. Rocha and T. Guarda (Eds.): MICRADS 2018, SIST 94, pp. 135–150, 2018.
https://doi.org/10.1007/978-3-319-78605-6_11

development of cognitive processes, and therefore know the link between structural, functional changes and brain dysfunction [2]. Dementia is a degenerative disease of the central nervous system, which can be described clinically as a syndrome of progressive pathological deterioration that causes a decrease in the cognitive domain centered on attention, memory, executive function, visual-spatial ability and language [3].

Currently, the interest of the research with the support of the electroencephalogram has allowed the detection of cortical anomalies associated with cognitive deterioration and dementia [6, 7]. An electroencephalogram marker provides signals to be processed and analyzed by nonlinear techniques [3] such as support vector machines and neural network.

1.1   Types of Dementia and Cognitive Spectrum

Dementia occurs when the brain has been affected by a specific disease or condition that causes cognitive impairment [8]. According to its cause, there are different types of dementia; Alzheimer's disease, Lewy body, fronto temporal dementia, Parkinson's disease [4, 9], vascular dementia [10, 11]. In Fig. 1, the advance of cognitive deterioration is illustrated until reaching the spectrum of dementia, which can be seen as a sequence in the cognitive domain that starts from mild cognitive impairment and ends with severe dementia, and the period in which the brain is at risk of reaching cognitive impairment not dementia [3].

In reference to Fig. 1, clinically, mild cognitive impairment is the transition stage between early normal cognition and late severe dementia and is considered

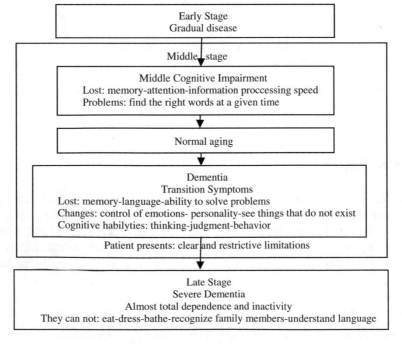

Fig. 1.   Spectrum of dementia

heterogeneous because some patients with mild cognitive impairment develop dementia, although others remain as deteriorating patients cognitive mild for many years [12].

1.2 Biomarkers for the Detection of Dementia

Biomarker is an objective measure, which is related to molecules, biological fluids, anatomical or physiological variables that are concentrated in the brain, which help diagnose and evaluate the progression of the disease, as well as, the response to therapies. They analyze the pathogenesis of dementia, predict or evaluate the risk of disease by providing a clinical diagnosis [13]. For the early detection of dementia, biomarkers include studies divided into four categories: biochemistry [4, 15–17], genetics [4, 5, 9, 18], neuroimaging [19], and neurophysiology [4, 5, 9, 14].

1.3 Analysis of Mental States in Function of Electromagnetic Oscillations

Alpha oscillations appear in healthy adults while they are awake, relaxed, and with their eyes closed. It fluctuates in a frequency range from 8 Hz to 13 Hz with a voltage range of 30 µV to 50 µV. It decreases when the eyes are opened, by the hearing of unknown sounds, anxiety or mental concentration [22, 23]. The alpha rhythm is composed of sub spectral units [24, 25]. It is defined in the occipital1 region, occipital2 [26] and on the frontal cortex.

The frequency of the beta waves oscillates between 13 Hz and 30 Hz, with a range of voltage of 5 µV to 30 µV. They appear by excitation of the central nervous system, increasing when the patient puts attention and control, is associated with thought and active attention, in the solution of concrete problems, reaches frequencies close to 50 Hz; they replace the alpha wave during cognitive decline [22]. They are delimited in the parietal7, parietal8 and frontal7, frontal8 [26] regions.

The frequency of the theta waves oscillates between 4 Hz to 7 Hz. They predominate during sleep, emotional stress, creative inspiration states in children and adults, deep meditation. They are recorded in temporal region7, temporal8 and parietal7 region, pariental8 with an amplitude of 5 µV to 20 µV [22]. In adults, two types of theta wave are presented according to their activity; the first is associated with decreased alertness, drowsiness, deterioration and dementia; the second is linked to activities of mental effort, attention and stimulation [26].

The frequency range of the gamma wave varies from 30 Hz to 100 Hz [22]. It is recorded in the somatosensory cortex, reflects the mechanism of consciousness, states of short-term memory to recognize objects, sounds, tactile sensation, particularly when it is related to the theta wave [26].

The frequency of the delta waves oscillates between 0.5 Hz to 4 Hz, its amplitude goes from 20 µV to 200 µV. It is generated during deep sleep, waking state and with serious diseases of the brain, never reach zero because that would mean brain death. This wave tends to confuse signals from the artifacts caused by the muscles of the neck and jaw, which is because the muscles are close to the surface of the skin, while the signal that is of interest originates from the inside of the brain. Before entering the dream state the brain waves pass successively from beta to alpha, theta and finally delta [22].

2   Materials and Method

Figure 2 shows the steps of the processing of the signals given by the electroencephalogram; the acquisition of biological signals, reduction of artifacts, extraction of characteristics, classification and presentation of the signal.

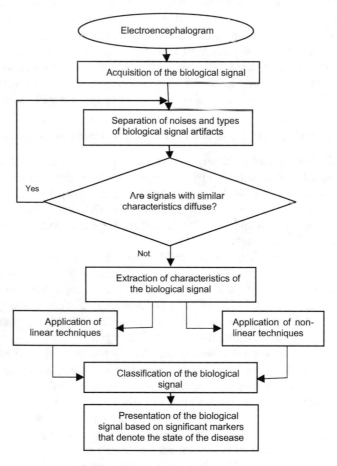

Fig. 2. Stages of signal processing

In Fig. 2, the electroencephalogram needs to record successive stages of biological signal processing to extract significant markers from patients with dementia, so that these markers reflect the pathological changes in the brain.

The stage of signal acquisition reflects the electrical activity of neurons in the brain [29]. The referencial assembly of the electroencephalogram is done based on the international 10/20 system [30–35].

The artifact reduction stage verifies that the recorded signal is affected by the noise factors. The artifacts are superimposed on the frequencies of the electroencephalogram signals. These artifacts are divided into physiological: muscle activity, pulse and ocular flicker [37–39] and non-physiological artifacts: interference noise from the electric line, sweat [38, 40]: background noise. Processing techniques allow us to overcome this problem and extract relevant information from the recorded signal. We disseminate the methods used for the elimination of artifacts: analysis of independent components [41, 42], wavelet transform [43–45], combined technique of component analysis with the wavelet transform [46, 47].

The stage of extraction and selection of the characteristics separates the useful information by means of linear and non-linear techniques. Linear techniques based on coherence [52] and spectral calculations [36, 48–51] facilitate finding the anomalies provided by the electroencephalogram [28]. Dynamic non-linear techniques analyze dynamic information [53–56]. The non-linear methods used are: the correlation dimension and the Lyapunov exponents [57, 58], to quantify the number of independent variables [59]; the fractal dimension, in terms of the waveform used to measure the structure of the signal [20, 60]; Lempel-Ziv-Welch [61], metric to evaluate the complexity of the signal and its recurrence rate; Entropy methods [20, 21, 56, 62], analyze the ability of the system to create information.

The techniques for classifying dementia are scenarios that predict the qualitative properties of the mental state. In Fig. 3, vectors of similar characteristics extracted from the previous stage are classified into three categories: cognitive impairment, not dementia, mild cognitive impairment, and dementia. Vectors with similar characteristics are analyzed before being applied to the classifier to avoid overloading the classifier and reducing computational time, increasing the accuracy of the classification. These vectors are processed using dimensionality reduction techniques based on the analysis methods: principal and independent components [63–65].

In Fig. 3, the efficiency of the classification is related to the extracted characteristics, the classifiers and the reduction of dimensionality. Classifiers: linear discriminant analysis and support vector machines are efficient methods for classifying brain disorders, such as dementia and epilepsy [63, 64]. The linear discriminant [66] creates a new variable that combines the original predictors by finding a hyperplane that separates the data points representing different classes.

On the other hand, support vector machines, integrate feature vectors with many components [3], is a problem of convex optimization. They are based on an algorithm that establishes a hyperplane that optimally separates the points of one class from the other that have been previously projected in a space of superior dimensionality [67], has the ability to build the model with a subset of training data. The support vector machines maximize the marginal discriminant, formulating a solution by Lagrange methods, the output of its classifier has the following expression.

$$y = \text{sgn}(\sum_{i=1}^{N} \alpha_i y_i k(x_i, x_j) + b) \tag{1}$$

Where, it y represents the exit; sgn is the signum function; Σ is the summation; N constitutes a set of training patterns; i is the input unit; (x_i, y_i) are training samples, with input vectors x_i and classes $y_i = [-1, 1]; \alpha_i \geq 0$, are Lagrange multipliers; b is the bias;

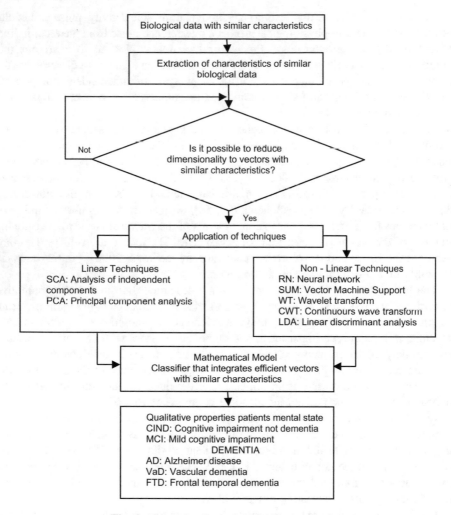

Fig. 3. Scenario about classification processing

$k(x_i, x_j)$ is the Kernel function. The two classes are mapped with kernel methods in a new space of greater dimension characteristics through non-linear measurements [67].

$$k(x_i, x_j) = \exp\left(\frac{-\|x_i - x_j\|^2}{2\sigma^2}\right) \tag{2}$$

Where, $k(x_i, x_j)$ is the function of the Gaussian kernel; exp is the exponential function; $\|x_i - x_j\|^2$ is the Euclidean distance squared; σ is the kernel extension parameter, which expresses a measure of similarity between vectors.

The technique of support vector machines is based on the principle of structural risk minimization, where the mathematical kernel classification function tends to minimize the error by separating the data, minimizing the error in the classification, and maximizing the margin of separation. This technique is used for the classification of signals, due to its high precision which makes it insensitive to over training and dimensionality [64, 68].

The neural network technique is used to: develop non-linear classification boundaries, information processing, resolution of classification problems, modeling, association, mapping, interpretation of spectra, calibration and pattern recognition. In Fig. 4, the artificial neural network is presented.

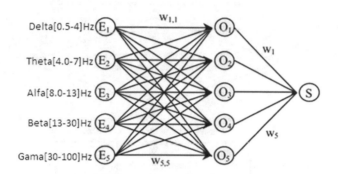

Fig. 4. Neural network of multilayer type

Figure 4 is based on multilayer models: the one is an input unit (E1, E2, E3, E4, E5) and no mathematical operations are performed on this unit, it distributes the information to the first hidden layer, the next one is the hidden layer (O1, O2, O3, O4, O5) and the last one is the output layer (S). The mathematical technique of multilayer modeling of the neural network is explained in the results section obtained.

To evaluate the capacity of neural networks, it is permissible to use architectures corresponding to a multilayer model with a single hidden layer. First, some different transfer functions should be used in the hidden layer node, two sigmoidal functions, a logarithmoid function, and the linear function. Secondly, the use of different numbers of neurons (from one to three) should be analyzed.

Figure 5 shows the Architecture of the technique of a Neural Network, the performance of the network is dependent, among other variables, on the choice of the processing elements (neurons), the architecture and the learning algorithm.

In relation to the architecture of Fig. 5, the objective of the neural network technique is to provide a tool that can be used to select the optimal design, which is obtained by optimizing the neural network itself through the implementation of the algorithm backpropagation based on available training data.

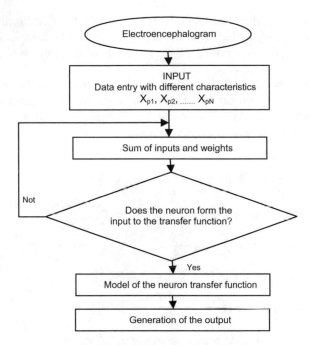

Fig. 5. Architecture of the neural network technique

3   Results

As work in the future, for the electroencephalogram project based on the processing of biological signals, we will focus the experimental phase on the development of the backpropagation type neural network model, the data of the frequencies of the biological signals alpha, beta, theta, delta, the range provided by the electroencephalogram will act as training inputs for the neural network.

The mathematical technique of multilayer modeling of the neural network of Fig. 5, depends on the values considered as input patterns provided by the electroencephalogram, which enter each of the neurons that make up the input unit, the unit that distributes the information to the next layer called hidden and the last one is the output layer. Result of the mathematical modeling study, we distinguish the following steps that demonstrate the validity of the training algorithm: (1) initialization of the weights of the network with small random values, (2) reading of an input pattern x_p : $(x_{p1}, x_{p2}, \ldots, x_{pN})$ and specification of the desired output that it must generate the network $d : (d_1, d_2, \ldots, d_M)$ associated with said input, (3) calculation of the current output of the network for the input presented; to do so, the inputs to the network are presented and the output of each layer is calculated until reaching the output layer, this is the output of the network; the substeps to be followed are described; the hidden net entries from the input units are calculated for a hidden neuron O_j, with the following formula.

$$net_{pj}^h = \sum_{i=0}^{N} w_{ji}^h x_{pi} + \theta_j^h \tag{3}$$

Where, *net* is the function that is transformed to obtain the output signal of the i-th node by means of the sigmoid transfer function; h are the magnitudes of the hidden layer; is the p-th training vector; it is the j-th occult neuron; it is the $\sum$ summation; it N constitutes a set of training patterns being the number of processing units of the hidden layer, it is the i input unit, it is the w_{ji} synaptic weight of the connection between E_i and O_j; is the x_{pi} input pattern of the connection between E_i and the p-th training vector and is the θ_j minimum threshold that the neuron must reach for its activation, acting as an input unit. Based on these inputs, the outputs y of the hidden neurons are calculated using the sigmoid activation function f to minimize the error:

$$y_{pj}^h = f_j^h (net_{pj}^h) \tag{4}$$

Where, y is the transfer function of the output units of the neuron; p is the p-th training vector; it j is the j-th occult neuron; they h are the magnitudes of the hidden layer; f_j^h sigmoid activation function; *net* is the function that is transformed to obtain the output signal of the i-th node by means of the sigmoid transfer function.

Analyzing is that the activation function integrates:

$$f(net_{jk}) = \frac{1}{1 + e^{-net_{jk}}} \tag{5}$$

Equation (5), is the sigmoid function, calculates that the output neurons are binary; it is net_{jk} the neuron, where it is j the j-th occult neuron; it is k the neuron of the output layer; e mathematical constant of irrational numbers equal to 2,71828.

To obtain the results of each neuron in the output layer, the same is done:

$$net_{pk}^o = \sum_{j=1}^{L} w_{kj}^o y_{pj} + \theta_k^o \tag{6}$$

Where, the function of transformation net_{pk}^o from the p-th training vector and the output of the neuron of the hidden layer, is calculated by adding the multiplication between the weight between the neuron of the output layer k and the j-th is neuron is hidden w_{kj}^o by the transfer function which integrates the output units y_{pj} of the neuron; L is a set of training patterns being the number of processing units of the hidden layer, j is the j-th occult neuron; plus the minimum threshold θ_k^o that the neuron of the hidden layer must reach for its activation.

Based on these inputs, the outputs of the hidden neurons y are calculated using the sigmoid activation function to minimize the error f:

$$y_{pk}^o = f_k^o(net_{pk}^o) \tag{7}$$

In which, the transformation function y_{pk}^o from the p-th training vector and the output of the neuron from the hidden layer, is calculated by multiplying the sigmoidal activation function f_k^o; is net_{pk}^o the function that is transformed to obtain the output signal from the p-th training vector and the output of the neuron from the hidden layer.

Once all the neurons have an activation value for a given input pattern, (4) the algorithm continues calculating the error for each neuron, except those of the input layer. For the neuron k in the output layer, if the answer is $(y_1, y_2, \ldots, y_M)$, the error δ is expressed as:

$$\delta_{pk}^o = (d_{pk} - y_{pk})f_k^{o'}(net_{pk}^o) \tag{8}$$

Where, it δ_{pk}^o calculates the error in the output layer; $(d_{pk} - y_{pk})$ represents the linear output; the f_{pk}^o function is derivable; it's net_{pk}^o the neuron in the output layer.

For the neuron k in the output layer with respect to the sigmoid function in particular:

$$\delta_{pk}^o = (d_{pk} - y_{pk})y_{pk}(1 - y_{pk}) \tag{9}$$

Where, is $y_{pk}(1 - y_{pk})$ the result of the derivative of the function $f_k^{o'}$.

If the neuron j is not an output, then the partial derivative of the error can not be calculated directly. Therefore, the result is obtained from known values and others that can be evaluated. The resulting formula is:

$$\delta_{pj}^h = x_{pi}(1 - x_{pi}) \sum_k \delta_{pk}^o w_{kj}^o \tag{10}$$

Where, the calculation of the error between the p-th training vector and the j-th occult neuron δ_{pj}^h is obtained by multiplying $x_{pi}(1 - x_{pi})$ which is the derivative of the function $f_k^{o'}$ by the sum of the multiplication between the error of the output layer δ_{pk}^o by the weight comprised between the neuron k of the output layer and the j-th hidden neuron w_{kj}^o.

Also, the error in the hidden layers depends on all the error terms of the output layer; so it is determined that the error must be propagated from the output layer to the input layer, which is possible using the backpropagation algorithm.

The error in a hidden neuron is proportional to the sum of the known errors that occur in the neurons connected to its output, each multiplied by the weight of the connection. The internal thresholds for the neurons adapt in a similar way, considering that they are connected with weights of auxiliary inputs with constant value.

To update the weights, (5) a recursive algorithm is used, starting with the output neurons and working backward until the input layer is reached. These weights are adjusted in the neurons of the output layer with:

$$w_{kj}^o(t+1) = w_{kj}^o(t) + \Delta w_{kj}^o(t+1) \tag{11}$$

Where, is w_{kj}^o the weight of the neuron from the output layer k to the j-th neuron of the hidden layer; it is $(t+1)$ the entry from the last layer; represents (t) the current entry; is Δw_{kj}^o the variation of the weight of the output layer k towards the j-th hidden neuron, which is calculated as follows:

$$\Delta w_{kj}^o(t+1) = \alpha \delta_{pk}^o y_{pj} \tag{12}$$

Where, the variation of the weight of the output layer Δw_{kj}^o is calculated by multiplying the learning rate α by the sigmoid function δ_{pk}^o by the transfer function y_{pj} which integrates the output units of the neuron.

The weights of the neurons of the hidden layer are calculated with:

$$w_{ji}^h(t+1) = w_{ji}^h(t) + \Delta w_{ji}^h(t+1) \tag{13}$$

Where, is w_{ji}^h the weight of the neuron of the j-th neuron of the hidden layer towards the input unit (i); is $(t+1)$ the entry from the j-th neuron of the hidden layer; represents (t) the current entry; is Δw_{ji}^h the variation of the weight of the neuron of the j-th neuron of the hidden layer towards the input unit (i), which is calculated as follows:

$$\Delta w_{ji}^h(t+1) = \alpha \delta_{pj}^h x_{pi} \tag{14}$$

Where, the variation of the weight of the output layer Δw_{ji}^h is calculated by multiplying the learning rate α by the sigmoid function δ_{pj}^h by the input unit x_{pi}.

In both cases, to accelerate the learning process, a learning rate α equal to 1 is included, and to correct the error address, the following term is used for the case of an output neuron:

$$\gamma(w_{kj}(t) - w_{kj}(t-1)) \tag{15}$$

And to correct the address of the error of a hidden layer, with a rate γ less than 1, the following term is used:

$$\gamma(w_{ji}(t) - w_{ji}(t-1)) \tag{16}$$

This process (6) is repeated until the error term E_p is acceptably small for each of the learning patterns:

$$E_p = \frac{1}{2} \sum_{k=1}^{M} \delta_{pk}^2 \tag{17}$$

Where, it M constitutes a set of patterns that accelerates the learning process; is p the p-th training vector; it k is the neuron of the output layer.

4 Discussion

From the analysis of the biological signals emitted by the electroencephalogram with respect to the mental states, it is concluded that until adulthood, the frequencies of the delta and theta waves decrease, while those of the alpha and beta waves increase linearly [27].

We have investigated functional connectivity with electroencephalogram and brain networks in patients with neurological diseases in general, arriving to determine that certain patients show different characteristic patterns of functional connectivity and alterations of the network; more specifically, in the future we will focus on the application of the algorithm based on the massive parallelization of neural networks to model and process the data obtained with the electroencephalogram, in such a way that the extracted information contained in the signal can be compared to classify the signal between two or more classes, which allows to determine patterns that facilitate the identification of signs of Alzheimer's disease called dementia.

From results obtained, what can be evidenced is that the computational learning method of vector support machines allows finding a hyperplane that separates multi-dimensional information perfectly into two classes. However, because the data is not generally linearly separable, this learning method introduces the notion of kernel-induced characteristic space, which transforms the information into a higher dimensional space where the data is separable. The key to vector support machines is that this higher dimensional space does not need to be treated directly, so the resulting error must be minimized.

On the other hand, the development of neural network models of backpropagation type, is based on a structure of neurons joined by nodes that transmit information to other neurons, which give a result by means of mathematical functions. The neural network learns from existing information through some training, a process by which its weights are adjusted, in order to provide an approximate result close to the desired one. In addition, it is noted that the backpropagation algorithm uses the descent method by the gradient and performs an adjustment of the weights starting with the output layer, according to the error committed, and proceeds by propagating the error to the previous layers, from back to forward, until you reach the layer of the input units, denoting your ability to organize the knowledge of the hidden layer so that any correspondence between the input unit and the output layer can be achieved.

5 Conclusions

To understand or design a learning process in any of the cases, method of vector support machines or neuron network, it is necessary: (1) a learning paradigm supported by the information provided by the electroencephalogram, (2) learning rules that govern the process of weight modification, (3) a learning algorithm.

The vector support machines are constructed based on a convex function, so that a global optimum is obtained and allows the construction of its dual formulation. Another great advantage is the ability to model non-linear phenomena by using a transformation of the space of origin to a larger one. On the other hand, it has limitations, when working with numerical data it is necessary to transform the nominal attributes to a numerical format. There is no kernel function that is better than all. The use of different kernel functions can determine different solutions, so it is necessary to be determined to solve each particular problem.

Result of the analysis of the readings on neural networks as a work in the future, the backpropagation algorithm will be applied, with a layer of five input units, a hidden layer with five neurons and an output layer with a neuron. For the tasks in the hidden and output layers, the classifier of the neural network will be integrated by bipolar and unipolar sigmoid functions, respectively, as a decision function; on the other hand, we will normalize the weights and the entries. We will determine the most effective set as well as the optimal length of the vector for the high precision classification. By minimizing the error, we optimize the number of neurons in the hidden layer to five. The weights and the slope of the sigmoid function will be trained. Appropriate times and an appropriate learning rate will be considered, in order to reach an optimal classification accuracy, that are within the ranges of the frequencies of the biological signals alpha, beta, theta, delta, gamma.

References

1. Griffa, A.: Structural Connectomics in Brain Diseases. Neuroimage. **80**, 515–526 (2013)
2. Sporns, O., Tononi, G., Kötter, R.: The human connectome: a structural description of the human brain. PLoS Comput. Biol. **1**(4), e42 (2005)
3. Al-Qazzaz, N.K.: Role of EEG as biomarker in the early detection and classification of dementia. Sci. World J. **2014**, 16 (2014)
4. Cedazo-Minguez, A., Winblad, B.: Biomarkers for Alzheimer's disease and other forms of dementia: clinical needs, limitations and future aspects. Exp. Gerontol. **45**(1), 5–14 (2010)
5. Hampel, H.: Biomarkers for Alzheimer's Disease: academic, industry and regulatory perspectives. Nat. Rev. Drug Discov. **9**(7), 560–574 (2010)
6. Vialatte, F.B.: Improving the specificity of EEG for diagnosing Alzheimer's Disease. Int. J. Alzheimer's Dis. **2011**, 7 (2011)
7. Hampel, H.: Perspective on future role of biological markers in clinical therapy trials of Alzheimer's disease: a long-range point of view beyond 2020. Biochem. Pharmacol. **88**(4), 426–449 (2014)
8. Borson, S.: Improving dementia care: the role of screening and detection of cognitive impairment. Alzheimer's Dement. **9**(2), 151–159 (2013)

9. DeKosky, S.T., Marek, K.: Looking backward to move forward: early detection of neurodegenerative disorders. Science **302**(5646), 830–834 (2003)
10. Román, G.C.: Vascular dementia may be the most common form of dementia in the elderly. J. Neurol. Sci. **203**, 7–10 (2002)
11. Thal, D.R., Grinberg, L.T., Attems, J.: Vascular dementia: different forms of vessel disorders contribute to the development of dementia in the elderly brain. Exp. Gerontol. **47**(11), 816–824 (2012)
12. Petersen, R.C.: Mild cognitive impairment as a diagnostic entity. J. Intern. Med. **256**(3), 183–194 (2004)
13. Dorval, V., Nelson, P.T., Hébert, S.S.: Circulating MicroRNAs in Alzheimer's Disease: The Search for Novel Biomarkers. Frontiers in Molecular Neuroscience **6**, 24 (2013)
14. Poil, S.S.: Integrative EEG biomarkers predict progression to Alzheimer's disease at the MCI stage. Front. Aging Neurosci. **5**, 58 (2013)
15. Mattsson, N.: CSF biomarkers and incipient Alzheimer disease in patients with mild cognitive impairment. JAMA **302**(4), 385–393 (2009)
16. Paraskevas, G.: CSF biomarker profile and diagnostic value in vascular dementia. Eur. J. Neurol. **16**(2), 205–211 (2009)
17. Frankfort, S.V.: Amyloid beta protein and tau in cerebrospinal fluid and plasma as biomarkers for dementia: a review of recent literature. Curr. Clin. Pharmacol. **3**(2), 123–131 (2008)
18. Folin, M.: Apolipoprotein E as vascular risk factor in neurodegenerative dementia. Int. J. Mol. Med. **14**, 609–614 (2004)
19. Schneider, A.L., Jordan, K.G.: Regional attenuation without delta (RAWOD): a disqtinctive EEG pattern that can aid in the diagnosis and management of severe acute ischemic stroke. Am. J. Electroneurodiagn. Technol. **45**(2), 102–117 (2005)
20. Henderson, G.: Development and assessment of methods for detecting dementia using the human electroencephalogram. IEEE Trans. Biomed. Eng. **53**(8), 1557–1568 (2006)
21. Zhao, P., Ifeachor, E.: EEG assessment of Alzheimers diseases using universal compression algorithm. In: Proceedings of the 3rd International Conference on Computational Intelligence in Medicine and Healthcare (CIMED2007), Plymouth, UK, 25 July (2007)
22. Ochoa, J.B.: EEG signal classification for brain computer interface applications. Ec. Polytech. Federale de Lausanne **7**, 1–72 (2002)
23. Guérit, J.: EEG and evoked potentials in the intensive care unit. Neurophysiol. Clin. Clin. Neurophysiol. **29**(4), 301–317 (1999)
24. Moretti, D.: Quantitative EEG markers in mild cognitive impairment: degenerative versus vascular brain impairment. Int. J. Alzheimer's Dis. **2012**, 12 (2012)
25. Moretti, D.: Vascular damage and EEG markers in subjects with mild cognitive impairment. Clin. Neurophysiol. **118**(8), 1866–1876 (2007)
26. Pizzagalli, D.A.: Electroencephalography and high-density electrophysiological source localization. In: Handbook of Psychophysiology, vol. 3, pp. 56–84 (2007)
27. John, E.: Developmental equations for the electroencephalogram. Science **210**(4475), 1255–1258 (1980)
28. Jeong, J.: EEG dynamics in patients with Alzheimer's disease. Clin. Neurophysiol. **115**(7), 1490–1505 (2004)
29. Taywade, S., Raut, R.: A review: EEG signal analysis with different methodologies. In: Proceedings of the National Conference on Innovative Paradigms in Engineering and Technology (NCIPET 2012) (2014)
30. Husain, A., Tatum, W., Kaplan, P.: Handbook of EEG Interpretation. Demos Medical, New York (2008)

31. Punapung, A., Tretriluxana, S., Chitsakul, K.: A design of configurable ECG recorder module. In: Biomedical Engineering International Conference (BMEiCON) IEEE (2012)
32. Klem, G.H.: The Ten-Twenty Electrode System of the International Federation
33. Anderson, C.W., Sijercic, Z.: Classification of EEG signals from four subjects during five mental tasks. In: Solving Engineering Problems with Neural Networks: Proceedings of the Conference on Engineering Applications in Neural Networks (EANN 1996), Turkey (1996)
34. Müller, T.: Selecting relevant electrode positions for classification tasks based on the electro-encephalogram. Med. Biol. Eng. Compu. **38**(1), 62–67 (2000)
35. Sanei, S., Chambers, J.A.: EEG Signal Processing. Wiley, Chichester (2013)
36. Moretti, D.V.: Individual analysis of EEG frequency and band power in mild Alzheimer's disease. Clin. Neurophysiol. **115**(2), 299–308 (2004)
37. Jung, T.P.: Removal of eye activity artifacts from visual event-related potentials in normal and clinical subjects. Clin. Neurophysiol. **111**(10), 1745–1758 (2000)
38. Núñez, I.M.B.: EEG Artifact Detection (2011)
39. Guerrero-Mosquera, C., Trigueros, A.M., Navia-Vazquez, A.: EEG Signal Processing for Epilepsy, in Epilepsy-Histological, Electroencephalographic and Psychological Aspects, InTech (2012)
40. Molina, G.N.G.: Direct brain-computer communication through scalp recorded EEG signals. École Polytechnique Fedérale de Lausanne (2004)
41. Naït-Ali, A.: Advanced Biosignal Processing. Springer Science & Business Media, Berlin (2009)
42. McKeown, M.: A new method for detecting state changes in the EEG: exploratory application to sleep data. J. Sleep Res. **7**(S1), 48–56 (1998)
43. Zikov, T.: A wavelet based denoising technique for ocular artifact correction of the electroencephalogram. In: Proceedings of the Second Joint Engineering in Medicine and Biology, 24th Annual Conference and the Annual Fall Meeting of the Biomedical Engineering Society EMBS/BMES Conference. IEEE (2002)
44. Krishnaveni, V.: Removal of ocular artifacts from EEG using adaptive thresholding of wavelet coefficients. J. Neural Eng. **3**(4), 338 (2006)
45. Jahankhani, P., Kodogiannis, V., Revett, K.: EEG signal classification using wavelet feature extraction and neural networks. In: IEEE John Vincent Atanasoff 2006 International Symposium on Modern Computing, JVA 2006. IEEE (2006)
46. Akhtar, M.T., James, C.J.: Focal artifact removal from ongoing eeg–a hybrid approach based on spatially-constrained ICA and wavelet denoising. In: Annual International Conference of the IEEE Engineering in Medicine and Biology Society, EMBC 2009. IEEE (2009)
47. Inuso, G.: Wavelet-ICA methodology for efficient artifact removal from electroencephalographic recordings. In: International Joint Conference on Neural Networks, IJCNN 2007. IEEE (2007)
48. Jelles, B.: Global dynamical analysis of the EEG in Alzheimer's disease: frequency-specific changes of functional interactions. Clin. Neurophysiol. **119**(4), 837–841 (2008)
49. Escudero, J.: Blind source separation to enhance spectral and non-linear features of magnetoencephalogram recordings: application to Alzheimer's disease. Med. Eng. Phys. **31**(7), 872–879 (2009)
50. Hornero, R.: Spectral and nonlinear analyses of MEG background activity in patients with Alzheimer's disease. IEEE Trans. Biomed. Eng. **55**(6), 1658–1665 (2008)
51. Markand, O.N.: Organic brain syndromes and dementias. Curr. Pract. Clin. Electroencephalogr. **3**, 378–404 (1990)
52. Dauwels, J., Vialatte, F., Cichocki, A.: Diagnosis of Alzheimer's disease from EEG signals: where are we standing? Curr. Alzheimer Res. **7**(6), 487–505 (2010)

53. Jeong, J.: Nonlinear dynamics of EEG in Alzheimer's disease. Drug Dev. Res. **56**(2), 57–66 (2002)
54. Subha, D.P.: EEG signal analysis: a survey. J. Med. Syst. **34**(2), 195–212 (2010)
55. Abásolo, D.: Analysis of EEG background activity in Alzheimer's disease patients with lempel-ziv complexity and central tendency measure. Med. Eng. Phys. **28**(4), 315–322 (2006)
56. Escudero, J.: Analysis of electroencephalograms in Alzheimer's disease patients with multiscale entropy. Physiol. Meas. **27**(11), 1091 (2006)
57. Grassberger, P., Procaccia, I.: Measuring the strangeness of strange attractors. Phys. D **9**(1–2), 189–208 (1983)
58. Wolf, A.: Determining lyapunov exponents from a time series. Phys. D **16**(3), 285–317 (1985)
59. Hamadicharef, B.: Performance evaluation and fusion of methods for early detection of Alzheimer disease. In: International Conference on BioMedical Engineering and Informatics, BMEI 2008. IEEE (2008)
60. Henderson, G.T.: Early Detection of Dementia Using The Human Electroencephalogram (2004)
61. Ferenets, R.: Comparison of entropy and complexity measures for the assessment of depth of sedation. IEEE Trans. Biomed. Eng. **53**(6), 1067–1077 (2006)
62. Costa, M., Goldberger, A.L., Peng, C.-K.: Multiscale entropy analysis of biological signals. Phys. Rev. E **71**(2), 021906 (2005)
63. Subasi, A., Gursoy, M.I.: EEG signal classification using PCA, ICA, LDA and support vector machines. Expert Syst. Appl. **37**(12), 8659–8666 (2010)
64. KavitaMahajan, M., Rajput, M.S.M.: A comparative study of ANN and SVM for EEG classification. Int. J. Eng. Res. Technol. IJERT **1**, 1–6 (2012)
65. Vialatte, F.: Blind source separation and sparse bump modelling of time frequency representation of eeg signals: new tools for early detection of Alzheimer's disease. In: IEEE Workshop on Machine Learning for Signal Processing. IEEE (2005)
66. Besserve, M.: Classification methods for ongoing EEG and MEG signals. Biol. Res. **40**(4), 415–437 (2007)
67. Garrett, D.: Comparison of linear, nonlinear, and feature selection methods for EEG signal classification. IEEE Trans. Neural Syst. Rehabil. Eng. **11**(2), 141–144 (2003)
68. Lehmann, C.: Application and comparison of classification algorithms for recognition of Alzheimer's disease in electrical brain activity (EEG). J. Neurosci. Methods **161**(2), 342–350 (2007)

Data Link System Flight Tests for Unmanned Aerial Vehicles

Anibal Jara-Olmedo[1,2], Wilson Medina-Pazmiño[1],
Eddie E. Galarza[1,2(✉)], Franklin M. Silva[1,2], Eddie D. Galarza[1,2],
and Cesar A. Naranjo[2]

[1] Centro de Investigación y Desarrollo FAE (CIDFAE), Ambato, Ecuador
egalarzaz@gmail.com
[2] Universidad de las Fuerzas Armadas - ESPE, Sangolquí, Ecuador

Abstract. This document establishes the considered procedures for UAV - F communication system's flight tests, considering the compliance with the requirements of international standards showing the results of applying those procedures. In the paper there is a description of the aircraft's communications system as the starting point of analysis. The tests were done considering each of the airworthiness requirements of the communications system to establish the way for data gathering. It is also described the related procedures and test parameters which allow to develop a new system as described in the general methodology. The outcomes of this work demonstrated the system's ability to extend the communication to the compliance of the USAR of the STANAG4701 standard of NATO, in order to establish procedures for validation and certification of an "in development" aircraft and to build communication over obstacles.

Keywords: Unmanned Aerial Vehicle · Unmanned aerial systems
Ground Control Station · Ground Data Terminal · Air Data Terminal
Flight test

1 Introduction

P Unmanned aircrafts (UAV), commonly known as drones, are a technology that develops quickly and worldwide as established by the International Civil Aviation Organization (ICAO) [1]. ICAO states that the operation of UAVs must be as safe as manned aircrafts, more so when some test could represent dangers for people or goods on ground or air, similar to the operation of manned aircrafts. It is necessary to establish parameters for flight tests that consider the particularities of these aircrafts, however, it should be considered that the requirements, procedures and regulations for unmanned aircraft which are in development process, as established by Dalamagkidis et al. [2].

The development of drones should be framed as ingenious aerial devices, hence, the regulation should consider the aspects related with production and certification of UAVs oriented to their integration in the controlled airspace. The European Union members countries have already made considerable progress in the regulations for this kind of industry, thus, the UK policy for lightweight UAV systems, presented by

© Springer International Publishing AG, part of Springer Nature 2018
Á. Rocha and T. Guarda (Eds.): MICRADS 2018, SIST 94, pp. 151–161, 2018.
https://doi.org/10.1007/978-3-319-78605-6_12

Haddon and Whittaker [3], establishes the need for a certification of the parameters in design, production, maintenance and operation processes.

Austin in his book, "Unmanned aircraft systems: UAVS design, development and deployment", identifies that an unmanned aircraft has complex systems such as propulsion [4], navigation, instrumentation, communication among others. It is necessary to understand the unmanned aircraft as part of an Unmanned Aerial System (UAS), which integrates two large components that are the Unmanned Aerial Vehicle (UAV) and the Ground Control Station (GCS). The communication between the UAV and the GCS basically consists of an "up link" for the commands transmission and the operator aircraft control and a "down link" that returns the aircraft's conditions as well as images of the electro-optical sensor or the useful load. In this situation, as analyzed by Stansbury et al. [5], a large part of the work in autonomy of the Unmanned Aircraft (UAV) depends critically on communication links and data transmission, unlike traditional aviation where there is a crew in the aircraft's cabin which controls the flight. The communications system becomes critical for the aircraft's operation as well, especially when the aircraft must be monitored and controlled remotely.

The importance of establishing a methodology to test the UAV communications system within the flight tests is very important. Hirling and Holzapfel [6] highlight the applicability of military airworthiness standards for UAVs that require a certificate from the aeronautical authority. Therefore, based on the airworthiness requirements of a military regulation established by NATO (North Atlantic Treaty Organization), the methodology for testing the communication systems applied in the experimental UAV-F complies with the required airworthiness parameters.

As stablished by Jain in his work [7] "Wireless Datalink for Unmanned Aircraft Systems: Requirements, Challenges and Design Ideas", the two key challenges for datalinks are the long distances and the high-speed of the aircraft. Covering these distances and aircraft speeds affects the efficiency of the datalinks.

Numerous researches have been performed in UAV data link communication, in [8] Haw et al., presents a flight demonstrations of cooperative control for UAV teams, Haala [9] describes the performance test on UAV-Based Photogrammetric Data Collection. Edrich [10] analyzes the aperture radar based tests on a 35 GHz Frequency Modulated Continuous Wave (FMCW) sensor concept and online raw data transmission. Pinkney et al. [11] make the data link tests for the communication relay of Unmanned Aerial Vehicle (UAV) and Cetin and Zagli [12] analyzes the continuous airborne communication relay approach using UAVs.

The purpose of this work is to assist people who works in UAV communication to understand the structure of the datalink system and the procedures to tests the operation of the datalink communication system. In this paper there is a description of the test requirements of the aircraft data link system, the flight tests, data link tests results, and conclusions that we get from the work.

2   Test Requirement of the Aircraft Data Link System

Within the development of unmanned aircrafts in the CIDFAE (Ecuadorian Air Force Research and Development Center), the prototype called UAV-F has been built, for which the compliance with airworthiness requirements based on the Standardization Agreement of the Treaty Organization of the North Atlantic STANAG 4671 [13] has been set as an starting reference. This document contains a set of technical airworthiness requirements aimed primarily in the certification of military UAVs, with a maximum takeoff weight of 20.000 kg, which are intended to operate regularly in non-segregated airspace. As established by ICAO in Annex 8 [14, 15], Airworthiness is defined as the measure of the aircraft's suitability for a safe flight. The prototype UAV-F has been established as a test platform for the communications system. The design characteristics of this aircraft are detailed in Table 1. Annex 8 establishes, within the flight tests, the basic applicable parameters to the UAV's communication system, considering the technical and operational aspects that allow verification of compliance of the airworthiness requirements, considering the particularities of unmanned aircraft [16], as proposed by Pontzer et al.

Table 1. UAV characteristics.

Characteristic	Description
Type	Fixed wing
Wingspan	5,5 m
Dimensions	4.15 m × 1.35 m (long, height)
Maximum gross weight	>150 kg
Cruiser Speed	135 km/h

The flight test planning procedures begin with the elaboration of a Test Base Plan for the UAV that is based on the design characteristics of the aircraft, as presented by Johnson et al. [17] and Williams and Harris [18]. Considering the previous flight tests experience prototypes for UAV-F, the following analysis phases have been established: aerodynamic behavior; automatic flight; autonomy, scope and limitations and application. Within the plan, the correct operation of the communications system will be examined, through which the aircraft is monitored and controlled from ground. The STANAG 4671 for airworthiness requirements, establishes that UAS systems are related to each phase of normal flight tests. According to the agreement, the UAV Communications Systems are related with the aircraft in general, the limitations in operation and information, and the data link for command and control. This document presents the analysis focused to the data link for command and control.

The data link to be tested includes several configurations depending on the test phase. Commercial off-the-shelf (COTS) communication equipment is integrated with hardware and software developed for the UAV. The line-of-sight (LOS) digital data link provides range to the horizon and integrates compressed video, metadata and serial port data from different UAV systems. Initial data link configurations start with industrial, scientific and medical (ISM) radio bands with redundant command and control (C2) backup. After initial test phases the data link changes frequencies to C band and provides GMSK modulation and modem bit rate up to 11 Mbps considering the applications and requirements of the UAV, especially the transmission of NTSC/PAL video from the electro-optical system (Fig. 4).

The data link system for command and control is one of the main differences of an UAV regarding a manned aircraft, considering the particularities of this system as established by Stansbury et al. [5]. The architecture of the system for the UAV-F is presented in Fig. 1, it considers as main components: the Ground Control Station (GCS), the Ground Data Terminal (GDT), the Air Data Terminal (ADT) and the Antenna System. For the UAS system of the UAV-F aircraft, the GCS is constituted by a mobile, and a self-sustaining and rugged station that houses the system where the aircraft is controlled and monitored.

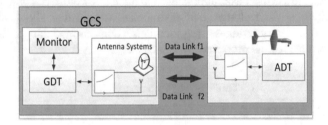

Fig. 1. UAS system architecture.

The Ground Data Terminal (GDT) contains data link receivers and transmitters, signal conditioners, and an alternate monitoring station. The GDT has been adapted to a mobile station that allows the location of the antennas in best positioning for testing. The Air Data Terminal (ADT) is an aircraft on board equipment that completes the communications system. The antenna system includes on-board antennas that consider the limitations of weight, energy consumption and space conditions of the aircraft. They are related with the omnidirectional and the directional antennas that are located in Earth. These subcomponents allow a link that includes the control data, the monitoring data, as well as the video down-link of an electro-optic sensor, as shown in Fig. 2.

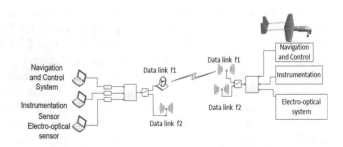

Fig. 2. Communications systems components.

The structure of the Basic Test Plan covers all test sessions, with which a matrix is formed with the different systems and parameters to be tested. The general objectives of each phase, specifically oriented to the data link system, are detailed in Table 2.

Table 2. Basic tests planning.

Phase	General objective	Tests location
1. Aerodynamic behavior	Architecture tests and back up	CIDFAE Airfield
2. Automatic Flight	Automatic guiding monitoring and control	CIDFAE Airfield
3. Autonomy, span and limitations	Specific link parameters determination	CIDFAE Airfield and external airfield tests
4. Application	Data reception and useful load control	CIDFAE Airfield and external airfield tests

Table 3 establishes a summary of the Unmanned System Airworthiness Requirements (USAR), which will be considered within the Basic Test Plan of the UAV. The procedures discussion in flight tests is established below.

Table 3. Unmanned System Airworthiness Requirements (USAR).

USAR classification	Description of the airworthiness requirement
USAR.U1601	General
USAR.U1603	Command and Control Data Link Architecture
USAR.U1605	Electromagnetic compatibility and interference
USAR.U1607	Monitoring and performance
USAR.U1611	Latency
USAR.U1613	Data Link Loss strategies
USAR:U1615	Data Link antenna masking

3   Flight Tests Discussion

For flight tests, the airworthiness requirements established in STANAG 4671 are related to the data link command and control with the specific phases and tests established in the Test Plan of the UAV-F. For each test, the procedures, configurations and tools to be used are established as well as the recommendations obtained from the tests.

Regarding the general requirement of the USAR.U1601, the communication system, as established in the architecture described above, includes the uplink for transmitting the commands from ground to the UAV, as for the downlink, the data transmission of the UAV status. The USAR.U1603 considers that avoiding a single failure can lead to a dangerous event, the architecture of the UAV communication system is scaled from a basic configuration to the final configuration, as the milestones of the basic plan are fulfilled.

Figure 3 shows the UAV-F designed for testing purposes at the CIDFAE. The basic architecture complies with Phase 1 requirements that are visual monitoring of the aircraft and a RPA control type (the aircraft is controlled by an operator from ground).

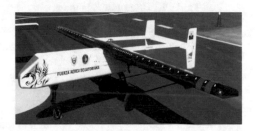

Fig. 3. UAV-F

The evaluation of the aerodynamic behavior includes the basic maneuvers such as straight and leveled flight, controlled turns, ascents and staggered descents as established by Cotting et al. [19]. This evaluation, done by the UAV operator, is supported by the telemetry data obtained from the interfaces implemented in the Command and Control Station. Data gathered from telemetry is analyzed after the flight has been completed for a more detailed qualities evaluation of the aerodynamic response of the aircraft. The data link to be tested includes several configurations depending on the test phase. Commercial off-the-shelf (COTS) communication equipment is integrated with hardware and software developed for the UAV. The line-of-sight (LOS) digital data link provides range to the horizon and integrates compressed video, metadata and serial port data from different UAV systems. Initial data link configurations start with industrial, scientific and medical (ISM) radio bands with redundant command and control (C2) backup. After initial test phases the data link changes frequencies to C band and provides GMSK modulation and modem bit rate up to 11 Mbps considering the applications and requirements of the UAV, especially the transmission of NTSC/PAL video from the electro-optical system.

Once the aerodynamic flight conditions of the aircraft have been determined in Phase 1, the staggered tests proposed for Phases 2, 3 and 4 are continued. System redundancy is necessary to allow the UAV aircraft to maintain the flight conditions or to be recovered even in the event of the failure of the main system. The system redundancy must be maintain in air and in ground segments. During Phases 1 and 2, the main communications system is supported by an alternate equipment, this includes controllers, operators and frequencies. However, for Phases 3 and 4, which establish the limitations and applications of the aircraft, due to weight and space conditions, a complete back-up system of the main link can't be implemented. Therefore, the redundancy of the automatic navigation system is established as a risk management measure.

The compatibility and electromagnetic interference, referred in USAR.U1605, is considered in the pre-flight procedures, where the UAV communications system is tested, in parallel with the ATC (Air Traffic Control) communication links, as well as the possible combinations of the ascending and descending links. For UAV

Fig. 4. UAV telemetry interface

communication systems, verification of electromagnetic compatibility and interference on the aircraft and in the test area is considered as essential, which is established in the NATO Electromagnetic Compatibility Procedures detailed in the RTO-AG-300-V27.

Considering that the UAV aircraft has equipment of smaller dimensions than a manned aircraft, the pre-flight tests review that situation in detail, as analyzed by Deng and Yuan [21]. A sweep of signals in the station area is also necessary, which verifies the absence of unwanted signals, such as those established by Faughnan et al. [22]. For this, the NI PXI-8108 unit is used, which analyzes the ranges of frequencies in which the tests are carried out. The data link system of the UAV-F could counteract a dedicated interference system based on its transmission power from ground, the transmission in C-band frequencies and the directionality of its main link due to it is difficult to insert the power of the signal interferer, as argued by Austin [4]. However, the UAV communication system is additionally tested with local signals, without any interference to the system.

The flight tests, as stablished by USAR.U1607, it tests the effective maximum range of each command and control data link in Phase 3: autonomy, scope and limitations. The maximum data link range depends on the data link frequency, the bit error rate for the digital link, the relevant parameters of the climate zone, the UAV altitude, among others. For this reason, to establish an acceptable levels of signals, the monitoring interfaces of the communications system consider the real-time visualization of the BPS transfer rate, RSSI signal level, synchronization parameters as the bit, frequency and mux, BER error rate, signal carrier band, geo-referenced location of the antenna on ground, geo-referenced location of on-board antennas, distance between antennas, addressing the antenna on ground and the temperature of the on board and on land equipment (Fig. 5).

The relationship with the UAV conditions are established using the control interface of the automatic navigation system, which presents information about the height of the aircraft, geo-referenced location and the attitude of the aircraft (inclination in bank and warp).

The masking of the antennas for different attitudes and UAV relative orientations with respect to the signal source are considered in the USAR.U1615. The conditions of masking are simulated during the aircraft's design process and the outcome of this

Fig. 5. Transfer rate analysis interface

condition regarding the links are tested in Phase 1, where the flight tests are performed at lower heights and distances. The structure of the aircraft produces the masking of on board antennas. The technical data and the position data of the flight tests aircraft are obtained from the previously indicated same interfaces.

The data link signal levels are fundamental, considering that based on its levels, the system's alert signal and the procedures of the activation should be established in case of a link loss. The behavior simulation of the link is done during the flight tests and the parameters for the alerts of the interfaces are obtained. In Phase 3, where the specific parameters of the link are determined, the minimum signal levels established as triggers of the signal alarm are verified. Table 4 establishes the parameters considered for the first tests of phase 3.

Table 4. Minimum parameters levels to keep the link between GCS and UAV

Parameter	Minimum value
RSSI (Received Signal Strength level)	−85 dB
BLER	Marginal
VER	Marginal
Temperature	<60°C

4 Data Link Tests Results

With the signal parameters, the data link ranges for the different power and height levels are verified. The establishment of the range curves are based on the results of the different tests. Table 5 establishes the results of the first scope test.

The loss of the data link between the UAV and the ground control station (GCS), must be considered for flight tests, and therefore a strategy must be established before the loss of command and control data, as required by USAR.U1613. Although the GDT, by design, once the link is lost, establishes an automatic search of the UAV, taking as reference the last location of the received signal. It is necessary to establish the response of the automatic navigation system, including the emergency form to finish the flight (Flight Termination), aspects that was included in the Stansbury et al. work [23].

The flight finishing routines are established considering the tests scenarios, the coverages and areas of operation, as well as the prediction of sensitivities, which are

Table 5. Data link between GCS and UAV.

Operating height [m]	Sensed minimum level [dBm]	Distance [km]	Comments
2733	−60	0	Optimal link
3103	−80	20	Slight loss of signal
3139	−86	23	Partial loss of signal
3218	−97	41	Total loss of signal

Table 6. Results for keeping the range curves

Condition	Procedure	Observation
Main link loss	Automatic activation of the backup link	Link activated at station
Link loss at times higher than t1 s	Navigation in "holding state" until link recovery	UAV last position stay showing at navigation map
Link loss at times higher than t2 s	Navigation return to the last contact point	Returning follows the last navigation points trying not to keep a straight line to the link loss point
Link loss at times higher than t3 s	Navigation in "holding state" over the point	External searching of the UAV must be activated

simulated by the Systems Tool Kit STK, in order to establish the safest conditions in the flight test, as already stated by Charlesworth [24]. The t1, t2 and t3 times established for the activation routines are determined for each flight test, considering the risk conditions of the work area (Lum et al. [25]). Therefore, the emergency recovery capacity of the UAV is implemented through commands programmed into the autonomous navigation system, in order to mitigate the effects of the critical failures of the data link. The strategies for link loss in flight tests are detailed in Table 6, where t1 < t2 < t3.

5 Conclusions

This document establishes the flight tests for the data link system of the UAV-1 that allows to determine the compliance of the USAR of the STANAG4701 standard of NATO, in order to establish procedures for the validation and certification of an aircraft that is in development. However, regulations changes according to countries and continents, the methodology used in this work compliance with requirements that are considered mandatory in any regulation

The carried out tests analyze the automatic navigation, link redundancy and configuration of the interfaces that were developed for the UAV navigation in order to comply with the flight tests in adequate safety conditions. The analysis establishes the procedures, how to obtain data, and additional considerations for testing the data link system, as reference of a suggested methodology for other similar tests.

References

1. Organización de Aviación Civil Internacional: Unmanned Aircraft Systems (UAS). https://www.icao.int/Meetings/UAS/Documents/Circular%20328_en.pdf. Accessed 20 Oct 2017
2. Dalamagkidis, K., Valavanis, K., Piegl, L.: On Integrating Unmanned Aircraft Systems into the National Airspace System: Issues, Challenges, Operational Restrictions, Certification, and Recommendations, 2nd edn, pp. 503–519. Springer, Munchen (2010)
3. Haddon, D., Whittaker, C.: UK-CAA policy for light UAV systems. UK Civil Aviation Authority, United Kingdom, pp. 79–86, May 2004
4. Austin, R.: Unmanned Aircraft Systems: UAVS Design, Development and Deployment, 1st edn, pp. 10–11. Wiley, Chichester (2010)
5. Stansbury, R., Vyas, M., Wilson, T.: A survey of UAS technologies for command, control, and communication (C3). In: Unmanned Aircraft Systems, Netherlands, pp. 71–78, June 2008
6. Hirling, O., Holzapfel, F.: Applicability of Military UAS Airworthiness Regulations to Civil Fixed Wing Light UAS in Germany, p. 25. Bonn, Deutsche Gesellschaft für Luft-und Raumfahrt-Lilienthal-Oberth eV (2012)
7. Jain, R., Templin, F.: Requirements, challenges and analysis of alternatives for wireless datalinks for unmanned aircraft systems. IEEE J. Sel. Areas Commun. **30**(5), 852–860 (2012)
8. How, J., Kuwata, Y., King, E.: Flight demonstrations of cooperative control for UAV teams, - AIAA 3rd, "Unmanned Unlimited", Technical Conference, Workshop and Exhibit, Infotech@Aerospace Conferences Technical, pp. 505–513, September 2004
9. Haala, N., Cramer, M., Weimer, F., Trittler, M.: Performance Test on UAV-Based Photogrammetric Data Collection in International Archives of the Photogrammetry, Remote Sensing and Spatial Information Sciences, vol. XXXVIII-1/C22, Switzerland, pp. 1–6, September 2011
10. Edrich, M.: Ultra-lightweight synthetic aperture radar based on a 35 GHz FMCW sensor concept and online raw data transmission. IEE Proc. Radar Sonar Navig. **153**(2), 129–134 (2006). Germany
11. Pinkney, F., Hampel, D., DiPierro, S.: Unmanned aerial vehicle (UAV) communications relay. In: Proceedings of the Milcom 96, McLean, VA, USA, pp. 47–51, October 1996
12. Cetin, S., Zagli, I.: Continuous airborne communication relay approach using unmanned aerial vehicles. J. Intell. Rob. Syst. **65**(1–4), 549–562 (2012)
13. STANAG 4671: Unmanned Aerial Vehicles Systems Airworthiness Requirements (USAR). NSA/0976, NATO Naval Armamental Group (2009)
14. NATO: RTO-AG-300-V27: Annex A-Typical Tactical Class EMC soft procedure. Annex A – Typical Tactical Class EMC Soft Procedure, NATO, April 2010
15. ICAO 2010: Annex 8 – Airworthiness of Aircraft. http://www.icao.int/secretariat/postalhistory/annex_8_airworthiness_of_aircraft.htm. Accessed 12 Nov 2017
16. Pontzer, A., Lower, M., Miller, J.: Unique aspects of flight testing unmanned aircraft systems (No. RTO-AG-300 AC/323 (SCI-105) TP/299), NATO Research and Technology Organization Neuilly-Sur-Seine (France), pp. 15–16, April 2010
17. Johnson E., Schrage D., Prasad, J., Vachtsevanos, G.: UAV flight test programs at Georgia Tech. In: Proceedings of the AIAA Unmanned Unlimited Technical Conference, Workshop, and Exhibit, pp. 1–13, September 2004
18. Williams, W., Harris, M.: The challenges of flight-testing unmanned air vehicles (Doctoral dissertation, Systems Engineering Society of Australia), Sidney Australia, October 2002

19. Cotting, M., Wolek, A., Murtha, J., Woolsey, C.: Developmental flight testing of the SPAARO UAV. In: 48th AIAA Aerospace Sciences M & E, Orlando, FL, pp. 103–118, October 2010
20. Medina-Pazmiño, W., Jara-Olmedo, A., Tasiguano-Pozo, C., Lavín, J.M.: Analysis and implementation of ETL system for unmanned aerial vehicles (UAV). In: International Conference on Information Theoretic Security, pp. 653–662. Springer, Cham, January 2018
21. Deng, D., Yuan, H.: UAV flight safety ground test and evaluation. In: IEEE AUTOTESTCON, New Harbord, USA, pp. 422–427, November 2015
22. Faughnan, M., Hourican, B., MacDonald, G., Srivastava, M., Wright, J., Haimes, Y., White, J.: Risk analysis of unmanned aerial vehicle hijacking and methods of its detection. In: Systems and Information Engineering Design Symposium (SIEDS), Charlottesville, USA, pp. 145–150, April 2013
23. Stansbury, R., Wilson, T., Tanis, E.: A technology survey of emergency recovery and flight termination systems for UAS. In: Proceedings of AIAA Infotech@ Aerospace Conference and AIAA Unmanned, Unlimited Conference, Seattle, OR, pp. 2375–2382, April 2009
24. Charlesworth, P.: Simulating missions of a UAV with a communications payload. In: IEEE UKSim 15th International Conference on Computer Modelling and Simulation (UKSim), pp. 650–655, April 2013
25. Lum, C., Gauksheim, K., Kosel, T., McGeer, T.: Assessing and estimating risk of operating unmanned aerial systems in populated areas. In: 11th AIAA Aviation Technology, Integration, and Operations (ATIO) Conference, Virginia Beach, USA, pp. 1–13, September 2011

Statistical Study to Determine the Sample Size to Define a Propagation Model Adjusted to the Equatorial Jungle Environment: A Proposal to Optimize Telecommunications Resource I

V. Stephany Cevallos, C. Manolo Paredes[✉], B. Federico Rodas,
and Rolando Reyes

Universidad de las Fuerzas Armadas, ESPE, Sangolquí, Ecuador
{spcevallos3, dmparedes, rpreyes1}@espe.edu.ec

Abstract. This work shows a methodology to determine the sample size required to define a propagation model adjusted to a specific and uncharacterized jungle region, based on the Longley Rice (L-R) propagation model. The analysis is performed in the HF frequency band and using high performance military tactical communication equipment. Initially, a simulation of the telecommunications system to be implemented were developed using the Radio Mobile platform. Subsequently, through the field measurements done insitu, we estimated the interval of confidence to get an accurate model and the data size to get such interval. The research includes a set of meteorological measurements to validate and characterize the environment under which the optimization is going to be applied. The results obtained yields a set of protocols for determining propagation models to be used on voice, video and data transmission in hostile and uncharacterized areas.

1 Introduction

The increasing demand for robust, low cost, scalable, and highly flexible telecommunications systems required a constant research aimed at seeking efficiency in the use of available means to provide voice, video and data services.

Currently, the vertiginous evolution of telecommunications networks has reached a fourth generation 4.5G. However, it is expected that the world would be using the fifth-generation cellular phone technology-5G by the year 2020. Therefore, it is essential to optimize software and hardware resources in such a way that existing technology provides a better service to their users.

In the area of data access and transport networks, systems are deployed based on a previous study of efficiency, coverage and scope, based fundamentally on simulating the performance of the system using generic propagation models and computational tools. As modernity needs accurate systems, those models and software have to properly converge regarding geography, orography and environmental conditions under which the system would be implemented.

© Springer International Publishing AG, part of Springer Nature 2018
Á. Rocha and T. Guarda (Eds.): MICRADS 2018, SIST 94, pp. 162–170, 2018.
https://doi.org/10.1007/978-3-319-78605-6_13

Considering that the propagation model of radio signals used worldwide is generic for urban and rural regions, it is necessary to establish a model that is more in line with the terrain where the telecommunications system will be deployed; however, this is not a simple job, since it requires an appropriate methodology to define the maximum likelihood model, it is necessary to use inferential statistics to get the best fit result.

In our proposal, the experimental protocol to follow during the data gathering must be set. Subsequently, the reliable data set should be discriminate from the whole sample. Finally, based on the previous stages and according to inferential statistical analysis, the experimental process is validated or not, so the data is considered as inputs for further treatment to determine the propagation model more suited to implement in the field.

The contributions generated by this research are structured from an introduction exposed in Sect. 1. The Sect. 2 briefs the existing propagation models. The Sect. 3 provides an approximation to the inferential statistics used. Next, Sect. 4 shows an analysis and discussion of results. Finally, in Sect. 5, we expose the conclusions and future work.

2 Model of Propagation of RF Signals

The propagation model is responsible for predicting the attenuation of a signal. The present research considers the L-R propagation model, also called ITM (Irregular Terrain Model) [1], which reaches a distance of 1 to 2000 [Km] and it is applied to several communication systems. that one of them is point-to-point communication. You can determine the specific parameters of the trajectory in the frequency range from HF to EHF with values from 20[MHz] to 20[GHz] [2].

2.1 Radio Frequency (RF) Propagation in High Frequency (HF) Band

Hertzian signals in this band are used for long-range communications, since the ionosphere functions as a reflective layer, and the refractive index depends on the weather conditions, variances in temperature and humidity benefits or harms the propagation of the signal, so the acceleration and reflection achieved by the wave in the ionosphere allows the wave to return to the terrestrial surface [3].

This band presents some difficulties for electromagnetic propagation. For instance, it requires a high level of transmission power. Likewise, due to the characteristic of the wavelength, the size of the antenna is also affected, requiring antenna dipoles with larger dimensions, in contrary the transmission frequency decreases. Besides, ionospheric propagation allows that part of the atmosphere where there are free electrons and ions in large quantities can affect radio propagation. All frequencies below 30 [MHz] are generally used by amateur radio and mostly by military forces.

3 Inferential Statistics to Determine the Sample Size

Inferential statistics allows an analysis of data easing decision making and estimations [4]. Through the estimation by interval, the limit of the sample tends to sample distribution of probabilities. The interval calculated is called the "Interval of confidence" and it represents the set of values that is formed from a sample of data, thus there is the possibility that the population parameter occurs with a specific probability named as confidence level [5].

3.1 No Probabilistic Sample

The non-probabilistic sampling establishes the choice of the intentional elements, that it does not depend on the probability but on the conditions that allows sampling, they do not ensure informal mechanisms and total representation of the population [6].

Generally, this type of studies exempts sampling procedures, being the researcher and his knowledge a fundamental actor, since he is the one who selects the sample, the units of analysis and the conditions of the population, therefore, the importance of all these variables depend of the researcher's criterion. This particular type of sampling is known as quota sampling.

The sample allows the estimation of the population parameters to be acceptable. To make this decision, factors such as: error range, confidence level and population dispersion are analyzed. For the present analysis, the variables and thresholds depend on the generic conditions of the transmission and reception equipment used, which are represented in the Table 1.

Table 1. Initial conditions for sampling

Name	Type	PowerTx	Threshold Rx	Gain	A. Height
Militar	Dipole	High = 20 W	−107 dBm	6 dBi	5 m
Militar	Dipole	Medium = 5 W	−107 dBm	3 dBi	5 m
Militar	Dipole	Low = 1 W	−107 dBm	1 dBi	5 m

Error margin. - corresponds to the maximum admissible error that is tolerated when estimating a population parameter. It is denoted by the letter E. If you want to obtain a small margin of error, the sample size must be large, however, it represents a longer time and money to get it [7].

Confidence level. - depends on the value of z corresponding to the level of confidence you want to obtain from 0 to 100%. If large samples are used, higher confidence level is required.

For the calculation of the sample size, qualitative variables are used in proportion estimation, that it is the percentage of measures providing acceptable or unacceptable values on the sample made [8]. The value that is calculated in the sample will be the most likely from our universe and as we move away from this value, it will be less likely. For the estimation of a proportion the following equation is used:

$$n = \frac{z^2 p(p-1)}{x^2} \tag{1}$$

Regarding the data size, an extended formula oriented to global data is used to calculate a chosen level of confidence neglecting the total size of the population.

$$n = \frac{z^2 pq}{x^2}$$
(2)

where:

- n: number of samples obtained.
- z: confidence level.
- p: probability of failure.
- q: probability of success.
- x: error marge.

4 Reception Power Measurement

One of the fundamental variables that must be consider when calculating the performance of a communications link is the weather. For the present research, a Wheatear Hawk-Pro equipment was used to collect variables that temperature, humidity, wind, solar radiation. The data obtained by the weather station is shown in the Fig. 1. As it shows, the image confirms the extreme conditions of humidity, solar radiation and temperature that influenced the coverage of the deployed telecommunications systems.

The weather conditions are typical in this Ecuadorian zone during September, with humidity of 98.9%, solar radiation of 19[watts/m^2] and barometric pressure of 28.9 [inHg]. It is important to cite that density of vegetation was upper to 5[m]. This is a factor varying the performance of the link.

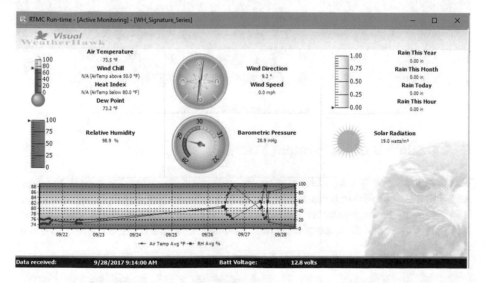

Fig. 1. Meteorological conditions observed during field testing

To determine the performance of a wireless access communication system, the physical field must be analyzed to establish a theoretical estimate of the link calculation. Next the band of frequency, climatic characteristics of the zone and technical specifications of the equipment of transmission and reception, besides the gain of the antennas, sensitivity of the receiver, error rate and power of transmission and reception.

In this testing, 18 units were located in the jungle area of Ecuador. A military radio equipment was used with the following characteristics: multiband, portable HF-SSB / VHF-FM, with frequency range up 60[MHz], high precision of frequency stability. This equipment is considered as a high-performance radio.

The verification of the radio link parameters as well as the calculation made are established by means of paths, allowing the analysis of the basic losses in the radio communication systems. Using Radio Mobile software, you can compare the reception power in each implemented link. According to the level of coverage, the graphical interface assigns a green color to the graphic when there is communication between the transmitter and the receiver. Otherwise, the link is red as shows in the Fig. 2.

Fig. 2. Radio link network

5   Calculation of the Sample Size

Through the technique of discretionary sampling or sampling by quotas the study population will be divided into different subgroups, the choice of subjects in each quota is made based on the researcher's criteria [8]. The characteristic to define each quota is the signal power and distance of each link, establishing a margin of error of 1 to 6% for each quota.

The level of confidence was defined by estimating the probability of success or failure of the measured values at each reception point. It will be considered a probability of success if the reception power is considered as acceptable (based on other experiences where it was found that a voice communications is received with a minimum power of 108[dBm], it means that taking into account the color level showed by Radio Mobile software is based on the Longley Rice propagation model where the power is acceptable if the range is −60[dBm] to −105[dBm], and it will be considered a probability of failure if the power is in the range of −114[dBm] to −150[dBm] as explained in the Fig. 3.

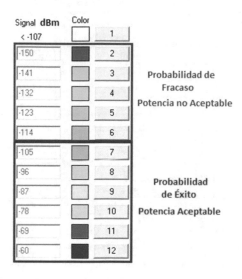

Fig. 3. Levels of probability of success and failure

The level of confidence will be selected according to the probability of failure, if this is low, that is in the range of 0 to 25%, it means that the probability of success or percentage of acceptable power will be 75%. Therefore, the level of confidence that will be taken should be required can be low, since there are no major complications for the link to be unsuccessful. The Table 2 indicates the assumptions percentage of confidence that will be taken according to the probability of failure.

Table 2. Percentage of power not acceptable according to the level of confidence

P—Failure probability for not acceptable power	25%	50%	75%	100%
Confidence level	68%	80%	95%	99%
K	1	1,28	1,96	2,58

Note that for a probability of failure of 25%, the level of confidence has been considered at 68%, given that favorable conditions is observed, consequently lower demands on the reliability are imposed. Whereas the probability of error is 99%, the totally unfavorable conditions for the link is evident, in contrary, a confidence level of

99% is imposed, if this confidence level is not reached, the data is considered valid. This is summarized in the fact that the higher the probability of error, the higher levels of confidence are imposed. All in all, the sample size must be increased.

By means of Eq. (2), the calculation of the sample size for each link was determined. Establishing the confidence levels for each probability of failure, the final results are detailed in Table 3.

Table 3. Final result of sample size

Frequency [MHz]	Distance [m]	TX Power [Watts]	Simulated power [dBm]	Average	Sample size	Name
20	1810	5	−68,2	−85,821	13	M1
20	1810	1	−77,2			M1
20	2060	20	−74,1	−77,21	3	M2
20	2060	5	−83,1			M2
20	2060	1	−92,1			M2
25	1810	20	−62,1	−110,5	142	M1D
25	1810	5	−71,2			M1D
25	1810	1	−80,2			M1D
25	2610	5	−78	−127,353	1305	J8
25	2860	5	−74,2	−110,63	202	J7
25	11650	5	−122,3	−128,1	1305	V5
25	11650	1	−131,3			V5
30	2400	20	−120	−121,95		V2

6   Result Analysis

According to the results obtained it can be verified that, the higher the level of reception power, the lower the number of samples that are required to be taken in the field. Regarding our experiment, we could say that the smallest number of samples required corresponds to point M2 with a total of 3 samples, whose acceptance percentage is 100%. It means that all reception power values were admissible, therefore, an adjustment to the propagation model is relatively simple under these conditions.

On the other hand, the points with the highest number of samples required are the point J8, V5 and V2 with a quantity of 1305 samples in which it can be verified that its percentage of inadmissible powers receptions are of 99%. This indicates that it did not exist at least one correctly accepted power, therefore, its number represents a very difficult value to experiment in the field, and finally this conditions not allow the transmission of any kind of information.

At the point M1D located at a distance of 1810[m], the number of samples calculated is 142; while at point M1, located at the same distance, but with a different orientation angle, the number of samples is much smaller considering the 95% of reliability as the percentage of samples acceptable achieved is 13; this happens, because the density of vegetation in the two points is different.

It can be verified that while increasing the transmission frequency, the reception power decreases, this is a phenomenon that normally occurs, given that the higher the frequency of transmission, the link requires direct sight access (not guaranteed in this testing) between the transmitter and the receiver, otherwise, the attenuation of the signal is very significant, being able to be total. According to the measurements obtained in the field, it was verified that at 30[MHz] frequency and 20[Watts] transmission power, the reception power is at level of −120 dBm, while at 20[Watts] and 20 [MHz] the power of reception is −74,1[dBm] regardless of the distance.

The Longley Rice propagation model has been defined as one of the most common for bond yield calculation, which is why it has been considered in this research as a reference model. Therefore, the conditions required by the Longley Rice model and the configurations under which the field measurements were made, are shown in Table 4, as a reference that the test conforms to what is established in the scientific literature.

Table 4. Performance of the Longley Rice propagation model

Parameters	Research	L-R Parameters
Frequency [MHz]	20, 25, 30, 40	20-20000
Distance [Km]	1-11,7	1-2000
A. Height [m]	1,5-5	0,5-3000
Polarization	Vertical	Vertical-Horizontal
Refractivity	250	250-400
Weather	Equatorial	Equatorial Continental [B]

The results detailed here gives rise to several interpretations, initially claiming that the propagation model used is adjusted to the defined frequencies, despite the weather conditions, it is also observed that the central limit theorem is fulfilled, which affirms that for a population whose distribution of data is asymmetric and not normal, an approximate size of 50 samples is required.

However, the results highlight another fundamental aspect, note that the difference between the simulated data and the data measured in the field is significant, in especial to links upper to 20[MHz], that is, the simulation alone does not guarantee that the levels of propagation and reception are similar in the countryside. This is just one of the hypotheses that can be shown in our research, that there is a need to adjust a generic propagation model to a physical, geographical and climatological reality. Adjustment that is important to optimize the resources used in telecommunications.

7   Conclusions

The present investigation showed that there is a reliable methodology to determine the sample size required to adjust a propagation model under special weather and terrain conditions, to verify the effectiveness of the results thrown by a propagation model in uncharacterized zones, such as jungle with dense vegetation and climatic conditions different from those commonly found. For the present study, it is observed that a reliable sample size for frequencies below 20[MHz], is 13 samples. While it is emphasized that for frequencies above 20[MHz], it is fundamental to perform a deeper analysis, in order to analytically define the causes of signal fading, a possible cause could be the absorption by vegetation, given the high density of trees in the test areas. In addition to other causes such as a possible unfavorable ionization to transmit on these frequencies.

Likewise, the differences found between the simulated results and the measurements obtained in the field, highlight the importance of the adjustment to the propagation model. Giving rise to a great deal of future work, as well as awakening other hypotheses, such as the following: it will be necessary to invest large amounts of money in expensive telecommunication equipment, or it will be necessary to investigate adequately what are the conditions that radiant systems must have radio equipment, to be efficient in different geographical regions.

Finally, it is important to emphasize that having an adequate telecommunications system is fundamental to maintain an efficient command and control of the actions to be taken in any event, activity or disaster, since this translates into saved lives or optimized resources. In conclusion, it is pertinent to characterize the behavior of radio systems in isolated populations or uncharacterized zones, in order to optimize the resources used in telecommunications and allow the population and the authorities to know what the performance conditions of their telecommunications equipment will be. (radio, television, cell phone) in these areas.

References

1. Salous, S.: Radiopropagation Measurement and Channel Modelling. Durham University, United king, WILEY (2013)
2. Ying, X., Cheng, X.: Propagation Channel, Characterization Parameter Estimation and Modelling Wireless Communication. Wiley, Singapore (2016)
3. Rico, M., Contreras, J., Diana, C., Urrutia, S.J., Enrique, A.: Interferencias en sistemas de radiodifusión en zonas de frontera colombiana. Ingenium Revista de la facultad de ingeniería 15(30), 50 (2014)
4. Pérez, N., Herrera, J., Uzcátegui, R., Peña, J.: Modelo de Propagación en las ciudades de Mérida (Venezuela) y Cúcuta (Colombia) Para Redes WLAN, Operando En 2.4 GHz, En Ambientes Exteriores. Inicio 16(62), 54–64 (2012)
5. Ballesteros, L., Peraza, D., Vaca, H.: Model study of spread in the environment of the University Distrital Francisco Jose de Caldas Luis. Rev. Vis. Electr. 4(2), 77–87 (2010)
6. Phillips, C., Sicker, D., Grunwald, D.: The Stability of The Longley- Rice Irregular Terrain Model for Typical Problems (2011)
7. Rappapport, T.: Wireless Communications: Principles and Practice, pp. 529–551. Prentice Hall, Upper Saddle River (1996)
8. Buettrich, S.: Unidad 06: Cálculo de Radioenlace, pp. 1–22 (2007)
9. Blake, R.: Basic Electronic Comunication, 1st edn. West Group (1993)
10. Lind, D., Marchal, W., Wathen, S.: Statistical Techniques in Business and Economics. McGraw-Hill, New York (2012)
11. Pérez, C.: Informe sobre técnicas de muestreo estadístico. AlfaOMega, México DF (2000)
12. Webster, A.: Estadistica aplicada a los negocios, pp. 5–12 (2000)
13. Martínez-Salgado, C.: El muestreo en investigación cualitativa: prin- cipios básicos y algunas controversias. Ciencia e Saúde Coletiva 17(3), 613–619 (2012)

Interference of Biological Noise in Sonar Detection

Teresa Guarda[1,2,3](✉), José Avelino Vitor[4], Óscar Barrionuevo[5],
Johnny Chavarria[2], Maria Fernanda Augusto[2], José Garcés[5],
and Luis Morales[5]

[1] Universidad de las Fuerzas Armadas-ESPE, Sangolqui, Quito, Ecuador
tguarda@gmail.com
[2] Universidad Estatal Península de Santa Elena – UPSE, La Libertad, Ecuador
johnny_chavarria@consultant.com,
mfg.augusto@gmail.com
[3] Algoritmi Centre, Minho University, Guimarães, Portugal
[4] Instituto Politécnico da Maia, Maia, Portugal
lemm47@gmail.com
[5] Armada del Ecuador, Guayquil, Ecuador
oscarbarrionuevovaca@gmail.com,
jgarces@armada.mil.ec, javemor@iiporto.com

Abstract. There are a wide range of ambient noise sources within the underwater acoustics. Acoustic noise can vary over time in intensity and spectral composition, depending on environmental conditions. In passive sonar detection the target of interest is surrounded by several noises, environmental and other radiated noise from other sources. The acoustic detection systems in underwater warfare are important for ships and submarines that have acoustic sensors to determine the presence of any contact. The lack of information on the noise that disrupts the acoustic detection sensors and interferes with the recognition, possibly mask the contacts endangering the surface and submarine operations. In order to solve this problem, it is fundamental to develop an information system model that allow the identification of the signals of interest and classify the targets.

Keywords: Sonar · Noise · Noise mapping · Information system
Data Mining

1 Introduction

Sound Navigation and Ranging (Sonar), being an element of great importance can be affected by the environmental noise which is understood by the biological, seismic, hydrodynamic and maritime traffic. Biological noise is produced by marine life, which can reduce considerably the passive an active sonar detection range at sea, and as a result the response time to an unforeseen event is considerably diminished, hampering the correct time of response in underwater warfare [1].

In the marine environment the main sources of biological noise are whales, dolphins, fish and other marine species, which increase the ambient noise level and this in

© Springer International Publishing AG, part of Springer Nature 2018
Á. Rocha and T. Guarda (Eds.): MICRADS 2018, SIST 94, pp. 171–179, 2018.
https://doi.org/10.1007/978-3-319-78605-6_14

itself affects the sonar detection range of contacts of interest and masking these contacts, worsening by the lack of information on the location, type, and frequency of sound of said species, compromising the success of the operations. Environment noise can be harmful to the sonar system, impairing the detection of the signals of interest [2].

The non-existence of information on studies, analysis, classification by types and noise areas originated by marine species of jurisdictional aquatic spaces; could be preventing the correct application of sonar functions generating a masking of contacts, making acoustic detection difficult and the development of underwater warfare proceedings.

In submarine warfare, passive acoustics detection is very important for warfare and safety procedures. In this sense, it is fundamental to create an information system that processes the frequencies of marine animals, allowing the correct application of the sonar filters, without masking the contacts. It is fundamental to create an information system that processes the frequencies of marine animals, fishes and other biological sources of noise in order to allow the sonar operators to apply the right sonar filters and use the most adequate function in the equipment, without masking the contacts. And in other hand, to stablish the most silence service mode of the engineering submarine systems to avoid self-interferences that reduce the sonar performance in biological saturated noise areas.

In the underwater warfare, acoustics is very important for warfare and safety procedures, then, the sonar performance and sonar detection range have a relevant importance, to extend the detection range and reduce the counter detection range in the operation area, optimize the sonar detection range trough adequate procedures will be decisive for officers and commander to make right decisions. The focus of this research will be on the biological noise that is produced by marine life specifically in cetaceans and fishes that affect the sonar performance and mask contacts at sea. The order Cetacea id divided in two suborders: Odontoceti y Mysticeti (Table 1) [3].

Table 1. Cetacea Suborder.

Suborder	Common name
Odontoceti	Common dolphin
	Porpoise
	Sperm whale
	Killer whale
Mysticeti	Blue whale
	Fin whale
	Bowhead whale
	Humpback whale

This underwater bioacoustics research work analyzes the biological noise and its effect on the sonar ranges, with the purpose of developing an information system model that allows assists real-time decision-making selection process of the best sound signs identifying and classifying the targets. This process will be leveraged by Data Mining (DM) process and Machine Learning (ML) algorithms to automate species identification.

2 Background

2.1 Sound Navigation and Ranging

Sonar is an instrument that nowadays has a lot of use in navigation, in fishing, in the study and research of the oceans, and in atmospheric studies; although in its origins it was only used to locate in submarines during the Second World War [2].

Sonar encompasses different devices, with different objectives and modes of operation. In general, Sonar can be considered as a system or device that uses sound to explore and/or obtain information from an underwater environment [3].

Sonar is currently used for the location and qualification of fish shoals, marine substrate mapping, study of marine sediment composition, the location of submerged objects and various other fields of application.

Comparing with civilian sonars, military sonars (MS) operated at higher power levels, being used for target detection, location and classification. Low frequency MS are used for surveillance, gathering information in large areas.

The acoustic frequencies used in Sonar systems vary from the infra-sonic (very low) to the ultra-sonic (extremely high). There are two types of Sonar technologies: Passive Sonar and Active Sonar. Passive Sonar picks up sounds from sources; and Active Sonar, emits pulses and picks up echoes [3].

Passive sonars exploit these irradiated signals, allowing the detection and location of the target of interest; and also applying spectral analysis techniques, determine the number of blades in the propeller of the ship, and the characteristics of machinery, to classify the sources of sound and ships [3]. Passive sonar has a wide variety of techniques for identifying the source of a sound that has been detected. Intermittent sound sources can also be detected by passive sonar. Passive sonar has the characteristic of not emitting any type of signal. To detect and locate targets of interest, passive sonar is based on the detection of acoustic noise emitted by the object of interest itself [2, 3]. Being the passive sonar for military applications usually used in submarines, because the fact of not emitting a signal, ends up making difficult its detection by others.

Passive sonar systems may have a large sound database, but the sonar operator is usually the one who does the final classification.

The active sonar uses a sound transmitter and a receiver. When the two are in the same place it is called a monostatic operation. When transmitter and receiver are separated, the operation is bistatic. When more than one transmitter or receiver is used, spatially separated, the operation is called multistatic. Most Sonar operates multistage, with the same array being often used for transmission and reception [2].

The active sonar creates a sound pulse (ping) and then hears the echoes of the ping. This sound pulse is usually created electronically through a sonar projector consisting of a signal generator, power amplifier and electro-acoustic transducer. A beam former is used to focus acoustic power on a beam, which can sweep the required research angles. The sonar pulse is emitted when encountering an obstacle, and returns to the emitter; being half the time in which the "ping" took to go and return, being possible to calculate the distance of the object echoed with relative precision. The sound pulse can be a constant frequency or a chirp of frequency change. Simple sonars generally use constant frequency with a filter large enough to cover possible Doppler changes due to target motion, while more complex sonars generally use the last technique [4].

Active sonar is most commonly used in civil applications for tasks such as: autonomous navigation, submarine communication and seabed mapping.

2.2   Acoustic Underwater Noise

Noise in the marine environment is important because it masks the information contained in underwater acoustic signals. The whole signal process that the sonar must perform is to extract the information from the signal-to-noise combination. The parameter of interest here will be the signal-noise ratio [6].

In the underwater acoustic environment, noise sources are classified into two categories: radiated noise and ambient noise [2]. The radiated noise comes from artificial sources (oil extraction platforms, ships, submarines, among others). In a ship or submarine, the noise comes from its equipment contained in it; which start vibrating at certain frequencies, and the vibrations propagate through the structure of the ship or submarine and from there to the sea [3].

Ambient noise comes from natural sources presented in the ocean. Among the main sources of acoustic ambient noise, we can highlight: state of the sea; marine fauna and rain [4]. Ocean ambient noise sources and characteristics can be divided in three frequency bands: low; medium and high. And each band has a different set of noise sources, and different forms of noise propagation from the source [8]. Low-frequency sources (10 to 500 Hz), have significant potential for long-range propagation because they have little attenuation. On the contrary the medium frequencies (500 Hz to 25 kHz) have a limited potential of propagation, due to their greater attenuation and, therefore, only local or regional sources contribute to the field of ambient noise. In the case of high frequencies (>25 kHz), acoustic attenuation becomes extreme so that all sources of noise are confined to an area a few kilometers from the receiver [8, 11].

Biological noise is generated from natural sources; marine animals, including fish, invertebrates and mammals that use sound, with specific frequency, to communicate [4]. The frequency spectrum of sound produced by marine animals ranges from 10 Hz to 200 kHz [4] (Table 2).

The biological activity affects the underwater acoustics detection ranges as part of ambient noise level, decreasing the value of the Figure of Merit (FOM) of a sonar system and masking some important tonal signals from the contacts: warships, maritime traffic, trawlers and another kind of boats, that are used by the sonar technician to classify and get important data for tracking, especially when passive tracking is used. All this information configures the tactical panorama that is used for making decision process of duty officers and commanders.

Table 2. Cetacea characteristic sounds frequency range.

Suborder	Common name	Sound type	Frequency range
Odontoceti	Common dolphin	Whistle	0.2 – 150
		Click	0.2 – 151
	Porpoise	Pulse	100–169
	Sperm whale	Coda	16 – 30
	Killer whale	Scream	0.25 – 35
Mysticeti	Blue wahle	Moan	0.2 – 0.02
	Fin whale	Call	0.16 – 0.75
	Bowhead whale	Call	0.1 – 0.58
	Humpback whale	Song	0.05 – 10

Adapted from [4]

3   Biologic Noise Mapping

Currently new technologies have allowed the construction of more silent ships and submarines, that is including merchant ships, the warships to avoid detection, reducing the source level and increasing detection ranges; the merchant or commercial ships to reduce the cost of fuel using propulsion systems that have a low source levels using anti-cavitating propellers making it difficult to detect and classify.

In a subaquatic complex environment with an important ambient noise level that affect the Figure of Merit (FOM) and the detection sonar range, it will be useful to have a decision support tool like a biological noise mapping. At this point, integrating sonar and non-sonar data from multiple sources across multiple platforms is shown as an option to establish a more accurate underwater scenario, improving the process of target detection and classification and, consequently, support tactical decision-making.

The proposed information system model, the aCousticR it's organized in four combined processes: (1) Data acquisition; (2) Knowledge Discovery; (3) Data analysis and (4) Distribution (see Fig. 1).

aCousticR information system has two functional modes: online and offline. In the online mode the system is synchronized in real time with the cloud server; in turn the offline system remains functional, and can be synchronized whenever there is Internet access (for example, in the ports).

Despite the use of Data Mining (DM) techniques, OLAP tools and Machine Learning (ML) algorithm's that help in the automatic classification of the signs of interest; the records can be analyzed and if necessary changed by the users, in order to guarantee the quality and accuracy of the information.

The data acquisition (1) process will be made by warships and submarines that will use the sonar systems to log the reference position, time, date, estimated deep, frequency, amplitude of the signal, including sound record.

Before the data captured in the external sources are loaded into aCousticR information system MySQL databases, these data are preprocessed by DM models for Knowledge Discovery (2) that will extract, transform and load a set of heterogeneous data, which are consolidated and cleaned. DM refers to non-trivial extraction of the

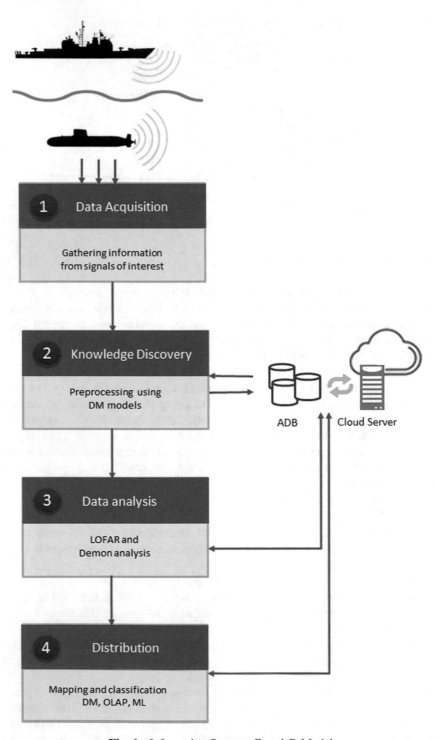

Fig. 1. Information System aCousticR Model

identification of valid, new, potentially useful and understandable data patterns from database data [9]. The DM aims to construct data models, with reference to the following models: predictive models; descriptive; dependencies and deviations [10]. The selection of DM activities is directly related to the objectives of the collection of classified biological noises (Table 3).

Table 3. DM Models.

Model	Type	Description
Predictive	Are constructed from the input data set (independent variables) for the output values (dependent variables)	
	Classification	Learning function that allows to associate to each data object one of the classes of finite set of predefined classes.
	Regression	Learning function that maps each object to a continuous value
Descriptive	Discover groups or data categories of objects that share similarities and help in describing datasets	
Dependency	Describe the dependencies or associations between certain objects	
Deviation	Describe the dependencies or associations between certain models that try to detect the most significant deviations, considered as a reference the past behavior	

The data obtained in this step will be load in an Acoustic Database (ADB). All the information loaded in the aCousticR databases must be analyzed (3) on the time, amplitude and frequency domains to be validated like useful. This step considers make spectrum analysis of sounds and noises recorded by the acoustics devices, using Spectrum Lab, a useful open source software. All the additional information gets in this step and useful to mapping the biological noise at sea will be uploaded in the system. The spectral analysis requires sonar technician prepared in Low Frequency Analysis (LOFAR) and Demodulation analysis (Demon analysis) of the sound that will be qualified by the Acoustical Analysis Center.

Finally in distribution process (4), the data will be mapped, geo-reference and share with the users using ArcMap. ArcMap is a geospatial processing program used primarily to view, edit, create, and analyze geospatial data; allowing user to explore data within a data set and create maps [8]. Using ArcMap, all the biological noise radiated and important data about the frequencies, noise level, type of fishes, will be georeferenced; it will permit to have graphical information. Users can explore and analyze large volumes of aCousticR data using OLAP tools, DM techniques and ML algorithms. OLAP tools are a combination of analytical processing procedures and graphical user interface, providing fast and flexible access to data and information and a multidimensional view of data. The DM aims to build data models. There are several algorithms available, each with specific characteristics. ML is part of an emerging Artificial Intelligence (AI) technology that over the last has been used by a growing number of disciplines to automate complex decision making. ML is a set of methods that allow machines to acquire knowledge to solve problems based on the history of past cases. At this level all activities must be developed to analyze data, search for patterns, and load information organized and encoded, and then the results will be distributed to end users for further feedback if necessary.

To share the information, users in the submarines or ships will have access to the system through intranet or Internet, to update all the shared information about biological noise and to the digital format of an atlas of biological noise in the ocean in jurisdictional waters.

Users had available an interface trough a cloud server or in standalone version, to the visualization and edition of data. In the case of ships that do not have intranet or Internet service at sea they will use a standalone version, and all the data will be uploading at the cloud server at port.

4   Conclusions

Biological noise information makes possible to make decision to configure sonar filters and functions, to optimize the submarine or ship machinery to decrease own noise level to increase the sonar detection range in complex underwater environmental. To optimize the sonar performance will make possible to classify contacts in an environmental with interferences of mammals, fishes and another kind of biological life that mask some tonal in the ships acoustic signature.

The biological noise mapping will allow to have information about areas where some species of fishes, mammals and crustaceans saturate the acoustic spectrum in the different periods of the year, to take forecasts useful for the underwater warfare and safety for ships and submarines. The use of the aCousticR information system allows the identification, classification and mapping of the signals of interest; in this particular case the cetaceans. The system databases are hosted in a cloud server, allowing data synchronization with all systems which integrate the aCousticR extranet (submarines, ships, investigation centers, and others).

aCousticR systems in underwater warfare it's extremely important for ships and submarines that have acoustic sensors to determine the presence of any contact. The lack of information on the noise that disrupts the acoustic detection sensors and interferes with the recognition, possibly mask the contacts endangering the surface and submarine operations.

With the use of mapping tool it will be possible to get the databases in order to define trawlers operation areas, to avoid these areas by the submarines or take actions to avoid collisions or problems with the trawlers nets.

References

1. Hodges, R.: Underwater Acoustics Analysis, Design and Performance of SONAR. John Wiley and Sons Ltd, UK (2010)
2. Board, O., Council, N.: Ocean Noise and Marine Mammals. National Academies Press, Washington, D.C. (2013)
3. Urick, R.: Principles of Underwater Sound, 2nd edn. (1975)

4. Wang, L., Wang, Q.: The Influence of Marine Biological Noise on Sonar Detection. Ocean Acoustics (COA), pp. 1–4 (2016)
5. Stergiopoulos, S.: Advanced Signal Processing Handbook: Theory and Implementation for Radar, Sonar, and Medical Imaging Real Time Systems. CRC Press, Boca Raton (2000)
6. Richards, C.: Sistemas electrónicos de datos: aspectos prácticos. Sistemas electrónicos de datos: aspectos prácticos, Reverté (1980)
7. Bradley, D., Stern, R.: Underwater Sound and the Marine Mammal Acoustic Environment: A Guide to Fundamental Principles, Marine Mammal Commission (2008)
8. Etter, P.: Underwater Acoustic Modeling and Simulation. CRC Press, Boca Raton (2013)
9. Hildebrand, J.: Anthropogenic and natural sources of ambient noise in the ocean. Mar. Ecol. Prog. Ser. **395**, 5–20 (2009)
10. Board, O., Council, N.: Ocean Noise and Marine Mammals. National Academies Press, Washington, D.C. (2003)
11. Ketten, D.: The cetacean ear: form, frequency, and evolution. In: Marine mammal sensory systems, pp. 53–75 (1992)
12. Fayyad, U., Smyth, P., Uthurusamy, R.: Advances in Knowledge Discovery & Data Mining. The AAAI Press/The MIT Press, Cambridge (1996)
13. Povel, O., Giraud-Carrier, C.: Characterizing data mining software. Intell. Data Anal. **5**, 1–12 (2001)
14. Minami, M.: Using ArcMap. In: ESRI (2002)
15. Pierce, A.: Acoustics: An Introduction to its Physical Principles and Applications, p. 678. McGraw-Hil, New York (1981)
16. Jefferson, T., Webber, M., Pitman, R.: Marine Mammals of the World: A Comprehensive Guide to their Identification. Academic Press, San Diego (2011)

Leadership (e-Leadership)

Mindfulness in the Dutch Military Train Your Brain

Tom Bijlsma[1(✉)], Susanne Muis[2], and Anouk van Tilborg[2]

[1] Faculty Military Sciences, Netherlands Defence Academy,
90.002, 4800 PA Breda, Netherlands
t.bijlsma.01@mindef.nl
[2] Maastricht University, Maastricht, Netherlands

Abstract. Mindfulness training (MT) programs are, apart from curative MT programs (e.g., PTSD treatment), not yet widely offered in the military. However, military (wo)men, who are often exposed to extremely stressful situations, might benefit from the "preventive" effects of MT (e.g., stress-reduction, enhanced wellbeing, increased military resilience and situational awareness). In order to meet busy military operational schedules, the current research investigates the potential effects of an economic, low dose, self-training mindfulness intervention (i.e., 10-day Mindfitness training) in a Dutch military sample (N = 173) that was subdivided into an intervention- and a waitlist-control group. By using a pre-/post-test design, the effects of our MT on mindfulness, stress, wellbeing, working memory capacity, and situational awareness were explored. Concluding from a multivariate analysis of covariance, the intervention had a negative effect on stress, and a positive effect on mindfulness, wellbeing and (self-rated) situational awareness. These results indicate the need to further explore the potential benefits of implementing (both extensive and low-dose) MT programs in other domains.

1 Introduction

Military are expected to be fit, both physically and psychologically. This means that they should not only be able to run for a few minutes without being out of breath and to carry heavy armory. What is also expected from military, is for them to go to war without losing their minds and to be able to make the right decisions in highly stressful situations. How can this be ensured? How can it be trained? A military's training schedule is filled with exercises to ensure their physical is in superb condition and they master their skills and drills. However, a military's psychological does not receive the same attention.

There is a growing interest in practicing mindfulness in order to train the mind. This study will look at the usefulness of mindfulness practices in the daily life of Dutch military. Instead of a time-intensive and costly plenary training, it will do so by using a low-threshold intervention. It will explore the effect of a short, ten-day mindfulness self-training intervention on perceived stress, well-being, working memory capacity, and situational awareness on Dutch military.

© Springer International Publishing AG, part of Springer Nature 2018
Á. Rocha and T. Guarda (Eds.): MICRADS 2018, SIST 94, pp. 183–192, 2018.
https://doi.org/10.1007/978-3-319-78605-6_15

2 Mindfulness

Mindfulness is a topic gaining more and more popularity in both the general population as the academic research field. It is a process of controlling one's attention and being aware of the moment (Kabat-Zinn 1990). It is also described as *"the awareness that arises through intentionally attending in an open, caring, and nonjudgmental way"* (Shapiro and Carlson 2009, p. 4). This capacity is developed by a form of mental training and using mediation techniques (Bishop et al. 2004). Bishop et al. (2004) approached mindfulness as involving two different components. One component is to be able to self-regulate attention in order to maintain increased focus on the present moment. The other component is characterized by an open, curious, and accepting attitude toward the present moment experiences. Mindfulness often starts as simple as being aware of one's own breathing (Shapiro and Carlson 2009). Taking the breathing as the starting point, the attention expands to being aware of the thoughts and feelings a person has right at that very moment, without evaluating or judging them. Chambers et al. (2009) stated that mindfulness exercises can be understood as attention or concentration exercises. As it can be frustrating to notice that the mind is wandering, an open and accepting attitude is necessary to prevent the practice from becoming a source of aversion (Chiesa 2013).

Mindfulness practices can focus on serving multiple purposes. For one it can focus on the development of concentration, through directing the attention on a single object such as an image or sensation, while blocking possible distractions. Another style of mindfulness practices can be to focus on developing an open monitoring of all sensory and cognitive information (Chiesa 2013). These are not opposed processes however, but rather different directions or ways of developing one's attention or concentration (Lutz et al. 2008; Chambers et al. 2009). For people new at the practice, it is often easier to concentrate on a single object. Once the practices advance, a transition into open monitoring can be made.

As the basis of mindfulness rests in Buddhist meditation techniques, mindfulness is best known as a form of therapy which reduces stress (Shapiro and Carlson 2009). Well-known examples are mindfulness-based stress reduction (MBSR; Kabat-Zinn 1990), and mindfulness-based cognitive therapy (MBCT; Segal et al. 2002). MBSR is one of, if not the, first mindfulness-based intervention, expressed in accessible language, for people who do not have any experience with the Buddhist style of mediation (Kabat-Zinn 2011). It inspired the development of a broad range of mindfulness interventions, among them MBCT, a mindfulness based intervention developed to help reduce depressions (Segal et al. 2012). There has been a growing interest in other clinical effects of mindfulness as well. For example, mindfulness has been used in the treatment for eating disorders (Kristeller et al. 2006), relationship therapy (Carson et al. 2006), and relapse prevention after addictions (Marlatt and Witkiewitz 2005).

Besides a clinical effect of mindfulness, the practice has also been researched in the healthy population, and with positive effects, regarding for example decreased levels of distress and increased levels of positive mood (Jain et al. 2007; Carson et al. 2006), a better quality of sleep (Hülsheger et al. 2015), and a reduction of chronic pain among hospital patients (Shapiro and Carlson 2009, Chapter 6).

3   Mindfulness in the Military

Mindfulness has already been studied in highly demanding jobs such as firefighters, marines, or the police. Johnson et al. (2014) showed that U.S. Marines after following an eight-week during mindfulness training had greater heart reactivity and enhanced recovery, among other bodily mechanisms associated with stress recovery. Stanley et al. (2011) found that the self-reported stress among U.S. Marines who engaged in mindfulness training prior to deployment, decreased. Not only do these jobs expose people to more stress than usual, they also require more alertness. A wandering mind is not necessarily a problem, but in highly demanding jobs in which it is important to stay alert and react quickly, it could become problematic. Jha et al. (2015) looked at the effects of mindfulness on the attention of people in the military. They even made a distinction between a training-focused mindfulness intervention and a didactic-focused mindfulness intervention. The training-focused version prioritized instruction on mindfulness exercises, while the didactic-focused version prioritized instruction on basic principles of stress and resilience, among other things. Both these interventions had a positive effect on attentional performance compared to a control group. However, the people who received the training-focused version of the intervention performed better than those who received the didactic-focused version of the intervention. This suggests that knowledge about the processes behind mindfulness can already help in improving one's attention span, but actually practicing mindfulness through exercises will increase this effect.

Another study looked at the working memory capacity of U.S. Marines, another important variable for a military (Jha et al. 2010). They found that those marines who seriously practiced over the 8 weeks of MT during their busy predeployment period improved their working memory capacity scores. The authors found a positive correlation between mindfulness practice-time and working memory capacity ($r = .37$).

A study among soldiers of the German Armed Forces suggested that a mindfulness intervention with plenary mediation training and daily homework led to increases in well-being and sleep quality, and to reduction of stress (Zimmerman 2015). Trousselard et al. (2012) found that military's mindfulness levels in two units of the French military were positively correlated with self-reported well-being and negatively with stress. A study in a Norwegian F-16 combat aircraft squadron found a reduction of anxiety, and an improvement of attention regulation and arousal regulation (Meland et al. 2015a). Another study in a Norwegian military helicopter unit revealed that the MT participants had a more positive bodily reaction to stress (related to the amount of cortisol) (Meland et al. 2015b).

However, regardless of the evidence on the benefits of MT, not much is done when it comes to researching and/or implementing MT in the Dutch military. A pilot study of the author in 2016 on the effects of a MT with half the crew of a Dutch submarine was unsuccessful and this was mainly due to practical reasons. A 2-week MT, which required participants to engage in four plenary sessions is absolutely unpractical for an operational military unit. Yet, research in a civilian population indicates that even brief, self-training in mindfulness is effective (Hülsheger et al. 2013; Hülsheger et al. 2015).

4 Mindfitness in the Dutch Military

4.1 'Train Your Brain'

Halfway 2017 the research 'Mindfitness in the Military; Train your Brain' was conducted to explore the possibility of using a short, low barrier mindfulness training. The above mentioned studies in the military's abroad have used intensive, plenary trainings of multiple weeks. As MT we use a 10-day, economic, mindfulness-based self-training program, hugely based on validated studies (Hülsheger et al. 2013, Hülsheger et al. 2015). It was called 'Mindfitness' because the military might be too masculine and task-focused to accept mindfulness, and many military might think it is too "woolly". The variables to be measured are military operational important: trait mindfulness, wellbeing, working memory capacity, situational awareness, and stress. Participants were administered to either the MT group or the waitlist-control group and completed a pre- and post-test survey. Only the MT group engaged in the training in-between both tests. The waitlist-control group was provided with the training after completing the post-test questionnaire. The questionnaires were anonymously assessed through the Qualtrics program.

The training is a short twice a day training, approximately 15 min on mindfulness exercises a day, that participants could easily do from their own home, on a mission, or anywhere else, as there were no planetary sessions and the material was distributed digitally. Participants received a digital booklet with general information about mindfulness and instructions on when and how to do the mindfulness exercises. The following six exercises were explained: the Three-Minute Breathing Space, the Body Scan, the Mindful Routine Activity exercise, the Mindful listening exercise, the Walking Meditation exercise, and the object-focus exercise.

A Dutch version of de Mindful Attention Awareness Scale (MAAS; Brown and Ryan 2003) was administered to measure mindfulness. Well-being was measured through the Personal Well-Being Index for adults (PWI-A; IWbG 2006). A Dutch version of the Perceived Stress Scale (PSS; Cohen et al. 1983) was used to measure stress among the participants (Albers 2011). As well as in Jha et al. (2010), working memory capacity was measured by means of an OSPAN. Situational awareness was measured by means of a subjective, self-report measure (the Situation Awareness Rating Technique (SART; Taylor 1990) and a more objective measure in the form of a self-made situational judgment test (SJT, only at t1). The data were analyzed with IBM SPSS Statistics version 21. ANOVA's were conducted to check for differences between complete and incomplete responses, and between the control and experimental condition. A power analysis calculated an estimated total of 96 participants as minimum necessary sample size.

4.2 Results

The final 173 participants consisted of 30 females and 143 males. The waitlist-control group consisted of 110 participants (91 males, 19 females). The experimental group consisted of 63 participants. The mean age of the participants was 38.9 ($SD = 10.6$), the average tenure was 18.1 years ($SD = 10.9$). Participants came from all four branches of the Dutch military: army ($n = 71$), air force ($n = 24$), navy ($n = 61$), and military police ($n = 17$). The ranks of the participants were categorized into personnel (private/corporal, $n = 28$), non-commissioned officers ($n = 57$), and army officers ($n = 88$). The participants in the experimental group reported to have spent an average of 6.9 days (ranging from 3 to 12, $SD = 2.4$) and an average of 103.57 min in total (ranging from 8 to 361, $SD = 67.3$) on the mindfulness exercises.

After establishing the raw scores for all outcome variables (see Table 1), and Pearson Correlation scores (see Appendix 1) were calculated, we started with a manipulation check and found a statistically significant difference in scores on the MAAS as a function of Time and Condition (F1, 170 = 7.702, p < .006; Wilk's Lambda = 0.957). Therefore, the 10-day MT intervention is effective in increasing self-rated mindfulness (see Fig. 1).

Table 1. Means and standard deviations of the main outcome variables for both groups at t0 and t1.

Condition	Scale	t0, Pre-test		t1, Post-test	
		M	SD	M	SD
Waitlist-control Group (n = 110)	MAAS	3.81	.68	3.81	.72
	PSS	2.21	.66	2.16	.67
	PWIA	7.25	1.23	7.26	1.26
	OSPAN	14.23	6.93	16.35	6.74
	SART	19.25	6.88	19.88	6.16
	SJT	–	–	8.33	1.75
	Days between t0 and t1			11.81	1.41
Experimental Group (n = 63)	MAAS	3.66	.77	3.84	.69
	PSS	2.21	.63	2.03	.56
	PWIA	7.44	.75	7.66	.91
	OSPAN	14.54	6.51	15.90	6.40
	SART	18.11	6.18	21.16	5.38
	SJT	–	–	8.16	1.41
	Days between t0 and t1			11.83	1.36
	Training time in days			6.87	2.42
	Training time in minutes			103.57	67.26

A multivariate analysis of covariance (MANCOVA; N = 171) was conducted, using pre-test measures as covariates. Concluding from a MANOVA, we did not need to include other covariates (e.g., tenure, rank, age) in the MANCOVA, since the MT intervention group and the waitlist-control group did not significantly differ on the demographic variables that were established at t0. Results of the MANCOVA were (marginally) significant (Wilk's Lambda = .931, = F6, 159 = 1.970, p = .073). With pre-test differences controlled, univariate tests support a group effect for the effect of the 10-day Mindfitness training on mindfulness (F1, 0.945 = .945, p = .022), wellbeing (F1, 1.951 = 1.951, p = .057), and stress (F1, 0.059 = .059, p = .084).

188   T. Bijlsma et al.

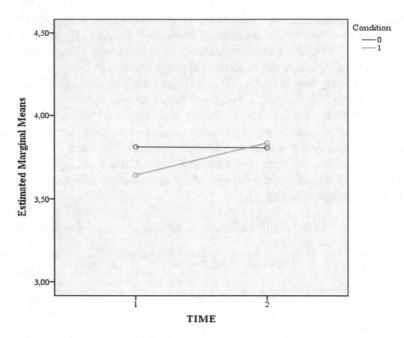

Fig. 1. Changes in MAAS scores over Time as a function of Condition.

See Table 2 for the adjusted means. We found partial support for the effect of the intervention on situational awareness, for there was a significant group difference regarding SART scores (F1, 101.861 = 101.861, p = .052), but not for the SJT scores (F1, 1.519 = 1.519, p = .448). There was no group difference regarding working memory capacity (F1, 12.781 = 12.781, p = .512).

Table 2. Adjusted means for mindfulness, stress, wellbeing, situational awareness, and working memory capacity for both the waitlist-control group and the MT intervention group.

Dependent variable	Condition	Mean	Std. Error
Stress	0	2.157	.042
(PSS score at t1)	1	2.033	.057
Wellbeing	0	7.332	.070
(PWI-A score at t1)	1	7.559	.094
Mindfulness	0	3.763	.041
(MAAS scores at t1)	1	3.921	.055
Working memory capacity (OSPAN scores at t1)	0	16.406	.522
	1	15.825	.705
Self-rated situational awareness	0	19.783	.495
(SART scores at t1)	1	21.424	.669
Objective situational awareness	0	8.335	.156
(SJT scores at t1)	1	8.134	.210

Note. 0 = waitlist-control group and 1 = MT intervention group.

5   Conclusion and Discussion

In the current study, we investigated the effect of a low-dose, 10-day, self-training mindfulness program on mindfulness, stress, wellbeing, working memory capacity, and (both objective and subjective) situational awareness, which we presented to our participants as a 10-day Mindfitness self-training. The data support the MT's effectiveness in terms of increasing mindfulness and wellbeing, while reducing stress. However, contrary to our expectations, the data failed to support the MT's effectiveness on working memory capacity and only partially support the training's effectiveness on situational awareness.

In terms of the design, we were able to implement a pre-/post-test design, which has various advantages over a cross-sectional design (e.g., Hart 2007). To conclude, we established a participant pool that existed out of participants from all four armed services (i.e., army, air force, navy, and military police) instead of focusing on only one of these, which was the case in prior research (e.g., U.S. Marine Corps; Stanley et al. 2011). In doing so, we increase the generalizability, or external validity, of our findings (e.g., Barry 2005).

Given that our 10-day MT is effective with respect to mindfulness, wellbeing and stress-reduction, the Dutch military might benefit from offering (and encouraging military (wo)men to take part in) similar short, self-training MT programs. Especially since this form of MT does not require costly, time-consuming, and hard to coordinate plenary sessions, it is both economically more attractive for the organization to offer the MT program and more convenient for military (wo)men to engage in the MT. MT can serve as a preventive instead of a curative (e.g., PTSD treatment; Stanley et al. 2011) approach in terms of stress-reduction (e.g., Grossman et al. 2004; Brown et al. 2012), increased positive affect (e.g., Good et al. 2016), enhanced wellbeing (e.g., Weinstein et al. 2009) and higher military resilience (e.g., Johnson et al. 2014). As a matter of fact, prevention is better than cure.

To our opinion, these implications count for the military in other countries, and crisis management organizations in general, as well. We look forward to more research to make these organizations and professionals more robust, and to enlarge the external validity of this training. Yet, nobody can oppose to 'Train your Brain'.

Appendix 1. Intercorrelations Between Study Variables

	1	2	3	4	5	6	7	8	9	10	11	12	13	14	15	16	17	18	19	20	21	22	23
1. Gender																							
2. Age	-.130																						
3. Experience with meditation	.119	.073																					
4. Experience with mindfulness	.140	.159*	.610**																				
5. Experience with yoga	.237**	.111	.403**	.299**																			
6. Experience with Tai Chi	.053	.066	.310**	.332**	.325**																		
7. Experience (other)	-.151	-.097	.289	.403	.497*	.402																	
8. Tenure	-.150*	.914**	-.025	.110	.073	.013	-.138																
9. Military Department	-.002	-.282**	-.038	-.090	-.076	-.024	.000	-.193*															
10. Rank (low/medium/high)	.095	.437**	.108	.162*	.114	.125	-.075	.412**	-.163*														
11. t1_t0 (in days)	.050	-.154*	.053	.042	.043	.024	.304	-.151*	-.026	-.106													
12. SART (t0)	-.067	-.034	.045	-.017	-.053	.021	.219	-.007	.109	-.087	-.094												
13. SART (t1)	.015	.121	-.037	-.082	.051	-.085	-.099	.124	.153*	.062	.050	.388**											
14. OSPAN score (t0)	.079	-.061	-.028	-.061	-.039	.002	-.515*	-.069	.074	.053	.068	.198*	.247**										
15. OSPAN score (t1)	-.048	-.087	-.026	-.105	-.015	-.038	-.313	-.105	-.004	.111	-.008	.139	.121	.573**									
16. PWI-A (t0)	-.055	-.134	.078	-.037	.037	.025	-.101	-.153*	.095	.147	-.098	.302**	.344**	.158*	.125								
17. PWI-A (t1)	-.052	-.074	.034	-.068	.001	-.027	-.161	-.099	.031	.176*	-.023	.214	.339**	.194*	.218*	.780**							
18. PSS (t0)	.140	.065	.031	.000	.065	.001	.176	.079	-.054	-.084	.016	-.368**	-.356**	-.226*	-.240*	-.671**	-.511**						
19. PSS (t1)	.113	-.159*	-.004	.016	.058	-.013	.170	-.108	.071	-.207**	.073	-.289	-.426**	-.289**	-.223*	-.586**	-.615**	.703**					
20. MAAS (t0)	-.044	.134	.002	.035	-.088	-.020	-.132	.135	-.030	.096	-.074	.255**	.278**	.120	.092	.314**	.247**	-.535**	-.440**				
21. MAAS (t1)	-.047	.187*	.001	-.007	.006	.022	-.121	.189*	-.082	.155*	-.068	.279**	.336**	.148	.122	.355**	.362**	-.524**	-.540**	.792**			
22. SJT score (t1)	-.113	.182*	.012	-.013	-.037	.035	.115	.188*	.019	.211**	-.006	-.009	.122	.062	.061	.147	.115	-.083	-.134	.148	.157*		
23. MT (days) (t1)	.042	.432**	.084	-.025	.192	.133	.122	.353**	-.015	-.001	-.202	-.277*	-.021	-.195	.010	-.008	.018	.131	-.062	-.080	-.025	.068	
24. MT (minutes) (t1)	.144	.346**	.205	.129	.279*	.152	-.193	.302**	-.093	.184	-.103	-.246	.017	-.193	.106	-.097	-.022	.227	.009	.000	.002	.149	.679**

Note. The SJT score was only obtained at t1 for both the waitlist-control and the experimental condition. MT (days) and MT (minutes) refers to the self-reported practice-time of individuals in the experimental group at t1. ** = Correlation is significant at the 0.01 level (2-tailed). * = Correlation is significant at the 0.05 level (2-tailed).

References

Albers, T.: Eindelijk, mijn scriptie over uitstellen (Master's thesis) (2011)

Barry, A.E.: How attrition impacts the internal and external validity of longitudinal research. J. Sch. Health **75**(7), 267 (2005)

Bishop, S.R., Lau, M., Shapiro, S., Carlson, L., Anderson, N.D., Carmody, J., Devins, G.: Mindfulness: a proposed operational definition. Clin. Psychol. Sci. Pract. **11**, 230–241 (2004)

Brown, K.W., Ryan, R.M.: The benefits of being present: mindfulness and its role in psychological well-being. J. Pers. Soc. Psychol. **84**(4), 822–848 (2003)

Brown, K.W., Weinstein, N., Creswell, J.D.: Trait mindfulness modulates neuroendocrine and affective responses to social evaluative threat. Psychoneuroendo-crinol. **37**(12), 2037–2041 (2012)

Carson, J.W., Carson, K.M., Gil, K.M., Baucom, D.H.: Mindfulness-based relationship enhancement (MBRE) in couples. In: Mindfulness-based Treatment Approaches: Clinician's Guide to Evidence Base and Applications, pp. 309–331 (2006)

Chambers, R., Gullone, E., Allen, N.B.: Mindful emotion regulation: an integrative review. Clin. Psychol. Rev. **29**, 560–572 (2009)

Chiesa, A.: The difficulty of defining mindfulness: current thought and critical issues. Mindfulness **4**(3), 255–268 (2013)

Cohen, S., Kamarck, T., Mermelstein, R.: A global measure of perceived stress. J. Health Soc. Behav. **24**(4), 385–396 (1983)

Good, D.J., Lyddy, C.J., Glomb, T.M., Bono, J.E., Brown, K.W., Duffy, M.K., Baer, R.A., Brewer, J.A., Lazar, S.W.: Contemplating mindfulness at work: an integrative review. J. Manag. **42**(1), 114–142 (2016)

Grossman, P., Niemann, L., Schmidt, S., Walach, H.: Mindfulness-based stress reduction and health benefits: a meta-analysis. J. Psychosom. Res. **57**(1), 35–43 (2004)

Hart, M.: Design. Int. J. Childbirth Educ. **22**(1), 22 (2007)

Hülsheger, U.R., Alberts, H.J.E.M., Feinholdt, A., Lang, J.W.B.: Benefits of mindfulness at work: the role of mindfulness in emotion regulation, emotional exhaustion, and job satisfaction. J. Appl. Psychol. **98**, 310–325 (2013)

Hülsheger, U.R., Feinholdt, A., Nubold, A.: A low-dose mindfulness intervention and recovery from work: effects on psychological detachment, sleep quality, and sleep duration. Br. Psychol. Soc. **88**, 464–489 (2015)

International Wellbeing Group (IWbG).: Personal Wellbeing Index – Adult, Manual, 4th Edn. Australian Centre on Quality of Life, Deakin University, Melbourne (2006). http://www.deakin.edu.au/research/acqol/instruments/wellbeing_index.htm

Jain, S., Shapiro, S.L., Swanick, S., Roesch, S.C., Mills, P.J., Bell, I., Schwartz, G.E.: A randomized controlled trial of mindfulness meditation versus relaxation training: effects on distress, positive states of mind, rumination, and distraction. Ann. Behav. Med. **33**(1), 11–21 (2007)

Jha, A.P., Morrison, A.B., Dainer-Best, J., Parker, S., Rostrup, N., Stanley, E.A.: Minds "at attention": mindfulness training curbs attentional lapses in military cohorts. PLoS ONE **10**(2), e0116889 (2015)

Jha, A.P., Stanley, E.A., Kiyonaga, A., Wong, L., Gelfand, L.: Examining the protective effects of mindfulness training on working memory capacity and affective experience. Emotion **10**, 54–64 (2010)

Johnson, D.C., Thom, N.J., Stanley, E.A., Haase, L., Simmons, A.N., Shih, P.B., Paulus, M.P.: Modifying resilience mechanisms in at-risk individuals: a controlled study of mindfulness training in marines preparing for deployment. Am. J. Psychiatry **171**(8), 844–853 (2014)

Kabat-Zinn, J.: Full Catastrophe Living: Using the Wisdom of Your Mind to Face Stress, Pain and Illness. Dell, New York (1990)

Kabat-Zinn, J.: Some reflections on the origins of MBSR, skillful means, and the trouble with maps. Contemp. Buddhism **12**(01), 281–306 (2011)

Kristeller, J.L., Baer, R.A., Quillian-Wolever, R.: Mindfulness-based approaches to eating disorders. In: Mindfulness-based Treatment Approaches: Clinician's Guide to Evidence Base and Applications, p. 75 (2006)

Lutz, A., Slagter, H.A., Dunne, J.D., Davidson, R.J.: Attention regulation and monitoring in meditation. Trends Cogn. Sci. **12**, 163–169 (2008)

Marlatt, G.A., Witkiewitz, K.: Relapse prevention for alcohol and drug problems. In: Relapse Prevention: Maintenance Strategies in the Treatment of Addictive Behaviors, 2nd edn., pp. 1–44 (2005)

Meland, A., Fonne, V., Wagstaff, A., Pensgaard, A.M.: Mindfulness-based mental training in a high-performance combat aviation population: a one-year intervention study and two-year follow-up. Int. J. Aviat. Psychol. **25**(1), 48–61 (2015a)

Meland, A., Ishimatsu, K., Pensgaard, A.M., Wagstaff, A., Fonne, V., Garde, A.H., Harris, A.: Impact of mindfulness training on physiological measures of stress and objective measures of attention control in a military helicopter unit. Int. J. Aviat. Psychol. **25**(3–4), 191–208 (2015b)

Segal, Z.V., Teasdale, J.D., Williams, J.M., Gemar, M.C.: The mindfulness-based cognitive therapy adherence scale: inter-rater reliability, adherence to protocol and treatment distinctiveness. Clin. Psychol. Psychother. **9**(2), 131–138 (2002)

Segal, Z.V., Williams, J.M.G., Teasdale, J.D.: Mindfulness-based cognitive therapy for depression. Guilford Press, New York (2012)

Shapiro, S.L., Carlson, L.E.: The Art and Science of Mindfulness: Integrating Mindfulness into Psychology and the Helping Professions, 1st edn. America Psychological Association, Washington, D.C (2009). http://content.apa.org/books/2009-08118-000

Stanley, E.A., Schaldach, J.M., Kiyonaga, A., Jha, A.P.: Mindfulness-based mind fitness training: a case study of a high-stress predeployment military cohort. Cogn. Behav. Pract. **18**, 566–576 (2011)

Taylor, R.M.: Situational awareness rating technique (SART): the development of a tool for aircrew systems design. In: AGARD, Situational Awareness in Aerospace Operations, 17, pp. 23–53 (1990)

Trousselard, M., Steiler, D., Claverie, D., Canini, F.: Relationship between mindfulness and psychological adjustment in soldiers according to their confrontation with repeated deployments and stressors. Psychology **3**, 100–115 (2012)

Weinstein, N., Brown, K.W., Ryan, R.M.: A multi-method examination of the effects of mindfulness on stress attribution, coping, and emotional well-being. J. Res. Pers. **43**(3), 374–385 (2009)

Zimmermann, F.: Mindfulness-based practices as a resource for health and well-being. Med. Acupunct. **27**(5), 349–359 (2015)

Defense Logistics

The Military Contract Service Model of the Portuguese Armed Forces. Critical Review and Intervention Measures

Lúcio Agostinho Barreiros Santos[1](✉)
and Maria Manuela Martins Saraiva Sarmento Coelho[2]

[1] IUM Research and Development Center (CIDIUM), Military University
Institute (IUM), Lisbon, Portugal
labs0892@gmail.com
[2] Research and Development Center of the Military Academy (CINAMIL),
Military Academy (AM), Lisbon, Portugal
manuela.sarmento2@gmail.com

Abstract. This article relies on documentary and empirical data to carry out a critical analysis of the military contract model of the Portuguese Armed Forces and the incentive scheme that supports them. Theoretical and practical shortcomings, as well as measures to mitigate its negative effects are identified when analyzing the dimensions – recruitment, retention, and reintegration – and how they affect the citizens who enlist on a temporary basis. This research intersects three perspectives of analysis: macro-level (external context), meso-level (organizational context), and micro-level (individual perspective). The study used a qualitative research strategy (Bryman, 2012; Creswell, 2013) and a case study research design (Yin, 2014). The results show that there are shortcomings in the military's ability to recruit and retain contract military personnel, which indicates that the model should be revised and the incentive scheme that supports it should be updated.

Keywords: Contract service model · Incentive scheme · Recruitment
Retention · Reintegration

1 Introduction

The geostrategic, geopolitical and technological changes, as well as the strategic-military, operational, and tactical changes associated with the new transnational threats and risks, which result in new types of missions, combined with social, economic, and financial motives have justified the adoption of reformist policies regarding Portugal's military apparatus. The successive reorganizations and restructuring of the Portuguese Armed Forces (AAFF) led to the transition to a professional model that introduced new forms of military service – voluntary service (VS) and contract service (CS). The model was revised in the late 1990s [1–3] and its consolidation is still underway. Subsequently, over the last 15 years, the Political Leadership has systematically controlled the number of "authorized personnel" in VS/CS, and the situation is aggravated due to the restrictions on the recruitment and retention of military personnel due to demographic, social, and cultural reasons.

© Springer International Publishing AG, part of Springer Nature 2018
Á. Rocha and T. Guarda (Eds.): MICRADS 2018, SIST 94, pp. 195–207, 2018.
https://doi.org/10.1007/978-3-319-78605-6_16

Low employee retention rates decrease the return on the investment made in training and instruction, and often impede the proper use of human resources. Although the average term of service is between three and four years [4], whereas the total length of service allowed by law is six years [1], many citizens choose to drop out, most during the first or second year of their contract, and return to the external social and occupational environment. This "reintegration" also includes an assessment of the manner how citizens benefit after serving in the AAFF, in terms of their employability, of the military competences they obtained and/or developed, and of their certification and recognition outside the army.

This study addresses the military contract model of the Portuguese AAFF and the incentive scheme that supports it, providing a critical assessment of the influence of external (macro-level), organizational (meso-level), and individual factors (micro-level), focusing on three key dimensions – recruit, retain, and reintegrate –, which will shed light on the experiences of the citizens who serve in the army. The main problems and shortcomings will be identified across these three dimensions and measures that will be outlined to either eliminate those shortcomings or to mitigate their effects.

This article constitutes the second part of a broader investigation, which the first part is an exploratory study.[1]

The following research question was formulated to guide the study: "Are the Portuguese AAFF capable of recruiting and retaining the adequate number of contract personnel and to support their social and occupational reintegration when they go out from the military?"

The study used a qualitative research strategy [5–7] and a case study design [8]. Documentary and empirical data were obtained from interviews questionnaire (with open-ended questions) administered to a sample of officers from the three branches of the AAFF.

The article begins with a brief analysis of the legal and theoretical framework that regulates military CS, secondly the methodology is described and thirdly the main results are discussed and compared to the research questions, with special emphasis on the recruit, retain, and reintegrate dimensions. Finally, the study will draw some conclusions from the main results and suggestions will be made for future research.

2   Legal and Theoretical Framework of Military Contract Service in Portugal

2.1   Legal Framework

This study focuses on the model of military contract service of the Portuguese AAFF, which is set out in the legislation that structures and describes the form, circumstances,

---

[1] The research will conclude with a third fieldwork study to be carried out at a later date, in which the perceptions of military contract personnel will be ascertained through an inquiry on the main career anchors that lead them to join the military and, with respect to early separation, on the main reasons for the breakdown of the psychological contract with the AAFF.

conditions, limits, and procedures involved in its implementation [1, 2], and on the incentive scheme that supports it [3, 9].

The military contract service model is legally defined as "[...] the provision of voluntary military service for a length of time established in the Military Service Law, the aim of which is to meet the requirements of the Armed Forces, and may include the admission of service members into the career staff" [1, p. 6541]; the modality of recruitment that feeds this system is referred to as "normal recruitment", which aims to enlist, from a "national recruitment base", "[...] citizens who are at least 18 years of age and who wish to serve in the Armed Forces on a voluntary basis." [1, p. 6543].

In the current legal system, the military CS personnel of all categories (officers, sergeants, or enlisted) remain in the ranks for a maximum of six years; however, before serving in the military under contract, citizens can perform "voluntary service" for a period of one year, extending their term of service to a total of seven years [1, 2]. This service term means that hierarchical (vertical) career progression is substantially more limited (to three or four levels) than that of career staff, although the concept of military career, as defined in the profession's statute, is also applicable to this form of military service: "[...] a hierarchy of ranks that unfold into categories that are occupied by specialized staff and that correspond to the performance of specific positions and functions." [9, p. 3205].

With regard to career progression, in addition to the normal progression over the seven years in a given category (officers, sergeants, or enlisted), CS staff have access to internal mobility by transitioning to a higher category or by being reassigned to a different functional area within the same category, on the basis of the vacancies available and of certain requirements which must be met, and can also be admitted as career staff (at the moment, to the enlisted and sergeants categories, in the Navy, and to the officers category in the three branches of the AAFF). Long-term contract schemes have also been defined to extend the term of service [1], pending the definition of functional areas of application and the maximum length of service permitted by law; these contracts cover more educationally complex career profiles that require a longer service term to ensure an appropriate return on investment [10].

To encourage young people to join the AAFF under VS/CS, an incentive scheme was created which includes five main aspects: (1) providing support for obtaining academic qualifications; (2) providing support for obtaining professional training and certification; (3) providing financial and material support; (4) providing support for integration in the labor market; and (5) providing social support [3]. Due to financial constraints, the initial legislation underwent successive changes, which limited access to several incentives that had been originally planned, both in the number of staff covered and in the coverage provided. Thus, given that the current version of the incentive scheme has not been successful in promoting recruitment, retention, and social and occupational reintegration, partly due to successive financial restrictions, it was recently decided that the system will be revised and updated [10].

2.2   Theoretical Framework

In addition to the above description of the legal framework, the theoretical framework of this study (Fig. 1) includes a set of structuring concepts, as well as some theories that form the basis of the military CS model of the AAFF, which help to explain the circumstances, conditions, and constraints faced by the citizens who serve in the military.

The concept of model is crucial for this research, and should be understood as a simplification of the reality it seeks to represent. In fact, a model is a theoretical framework representing economic processes by a set of variables and a set of quantitative and/or qualitative relationships between them. The economic model is a simplified structure that simulates reality and considers the complex processes that underpin it [11]. A model may have various endogenous and exogenous variables, and those variables may change to create various responses [12].

The other structuring concepts of the model are recruitment (including advertising), retention (in the ranks), and reintegration (social and occupational), which when coordinated and associated with the respective mobility flows, make up a continuum that describes and explains how citizens experience their service term as contract military personnel. The "Incentive Scheme" is a (composite) variable that mediates the three dimensions through its five areas of intervention.

Military recruitment is defined in the Military Service Law as "[...] the set of operations required to recruit human resources into the Armed Forces" [1, p. 6542], and is embedded in the broader concept of "human resource recruitment" since it works from the an assumption that a "candidate attraction process" is in place for a specific function, which entails tasks such as "advertisement" and "selection" ("initial screening") of candidates and concludes with the integration (induction) of the candidates into the organization [13].

Another structuring concept is retention, which is associated, in the Portuguese military context and in this study, with the military's ability to retain a given number of staff for a suitable length of time. Thus, in 2007, a NATO working group issued a technical report defining the concept: "[...] the process of keeping adequate numbers of suitable personnel in the Military, in order to meet the needs of the organization" [14, p. D-2].

The concept of reintegration, the third major dimension of this study, is generally associated with the transition to civilian life upon separation from the military and with the entry or re-entry into the labor market, and is therefore closer to the social and occupational perspective, which implies that the Portuguese state has the responsibility, under the Military Service Law [1] and the Incentive Regulation [3], to provide support and to create the conditions to implement several of the measures provided for in the legislation.

In addition to the "recruit – retain – reintegrate" and the incentive scheme that supports it, another integral part of the theoretical framework is described in Fig. 1 where the three levels or perspectives of analysis – macro, meso and micro – have been defined to assess the military CS model, both in terms of its structure or construction (conceptual and legal framework) and its operationalization (management practices).

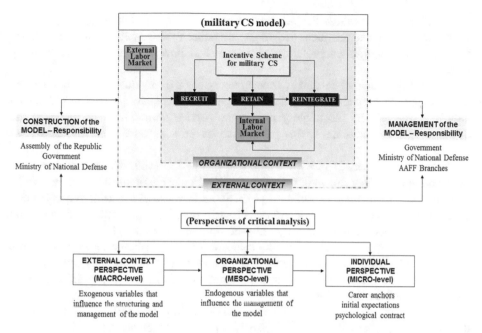

Fig. 1. Theoretical framework of the study: summarizes the theoretical framework, organized according to three levels of analysis – macro, meso, and micro. (This framework simultaneously represents the empirical study's model of analysis, where the main dimensions – recruit, retain, and reintegrate – intersect the three levels of analysis.)

The macro perspective refers to the environmental variables and their influence on the model's organization and management (operationalization) in relation to the three dimensions – recruit, retain, and reintegrate. These are exogenous variables linked to changes in the international geopolitical and geostrategic environment, the operational environment surrounding the military (new missions), the economic and financial situation (available budget), the legislative changes to the regulatory model, young people's socio-cultural context and generational profile (generation "Y"), and demographic conditions (recruitment base). The meso perspective refers to the internal context and to organizational activity (e.g. the characteristics of the military context, the requirements in terms of human resources, the career model for CS military personnel, the recruitment process, the selection profile, the geographical distribution of military units, the strategies and mechanisms with the labor market), which may either facilitate or hinder the operationalization of the three dimensions. It is influenced by the macro perspective (external environment) and influences the micro perspective (e.g. "career anchors", expectations, personal profile, psychological contract), either positively or negatively; the latter refers to individuals and the motives behind their decision to enlist in the military, as well as the reasons that lead them to serve in the military for an extended period of time or, conversely, that lead them to early dropout.

In addition to the three perspectives of analysis and the dimensions used in the study, some of the topics identified in the literature review will be presented, highlighting some theories on professional careers and how they relate to the military reality.

The existing military careers are conceptually close to organizational careers since the standard career paths are defined in a specific decree-law [9] and are managed almost exclusively by the military during the service term. They are therefore the opposite of less traditional career paths such as protean careers and boundaryless careers.

Organizational careers provide a path that can be followed, which is organized in time and space. They are linked to the organization's hierarchical structure, which controls most of the internal variables [15, 16], and are associated with vertical career paths and with access to progression [17, 18] in a specific career and with the expectation of occupational stability until the time of retirement [19].

In contrast, protean careers are the opposite of organizational careers. They are perceived as a sequential set of attitudes, behaviors, and experiences throughout a person's life, therefore the organization does not control most of the variables, which are "determined by the individual" [20–23].

Boundaryless careers, a concept introduced by Defillipi and Arthur [15], refer to the dissolution of traditional boundaries, especially within organizations, and to the creation of less structured, multidirectional career paths, and are even more flexible in terms of personal choice than protean careers [17, 24, 25].

Two other concepts relevant to this study are the "career anchors" introduced by Schein [26] and the psychological contract first addressed systematically by Rousseau [27]. These are crucial concepts to understand and categorize the motives that lead citizens to serve in the AAFF under contract (in the case of career anchors), as well as the reasons why they exit the military before the maximum service term allowed by law, which may be largely linked to the breakdown/dissolution of the psychological contract.

This study also reviewed the institutional/occupational model advanced by Moskos [28], which has proved useful in categorizing and explaining the types of motives that lead citizens to serve in the Portuguese AAFF under contract, as well as the reasons for early dropout.

3 Methodology

This study analyzes and identifies some of the main problems and shortcomings of the military CS model of the Portuguese AAFF, which corresponds to temporary military service. The investigation focuses on three key aspects: the process of recruiting contract military personnel; the factors that influence retention in the AAFF; and the process of reintegration into the external labor market after the contract with the military is completed.

The methodology used in this study includes a qualitative research strategy [5] and a case-study research design with a cross sectional time horizon [5, 8, 29]. The study also relied on inductive reasoning to construct meaning, using a research design based on discovery as described by Guerra [7].

As a result of the research strategy, the researchers adopted a constructivist onto-logical position wherein they are part of the observed reality [5, 6, 30] and a funda-mental element of the research process [7]. On the other hand, from an epistemological perspective, and in light of the methodological and ontological choices made, the position adopted by the researchers is closer to an interpretivist position [5].

As suggested by Bardin [31] and Guerra [7], the methodological procedures fol-lowed during the field research were based on the categorical content analysis of the empirical data obtained from a non-probabilistic sample of officers from the three branches of the AAFF. After a preliminary test, the final version of the open-ended questionnaire was sent by email to 210 officers. The responses were fed into a database, which was open for data entry for six months (between February and July 2017). Seventy-four responses were received, 69 of which were valid responses.

The main sociodemographic characteristics of the 69 respondents are: (1) Branch of the AAFF: Navy (n = 18; 26.1%), Army (n = 30; 43.5%) and Air Force (n = 21; 30.4%); (2) Rank: 1st Lieutenants/Captains (n = 32; 46.4%), Captain Lieutenants/Majors (n = 18; 26.1%), Frigate Captains/Lieutenant Colonels (n = 3; 4.3%) and Captains of sea-and-war/Colonels (n = 16; 23.2%); (3) Gender: Male (n = 56; 81.2%); Female (n = 13; 18.8%); (4) Professional experience: between 10 (1st Lieutenants/Captains) and 30 years (Captains of sea-and-war/Colonels).

The empirical data were processed and the information was systematized in mul-tiple input tables that refer to each topic/question. Since important content was scat-tered over several responses, the themes were then systematized in three dimensions – recruit, retain, and reintegrate – which intersect the perspectives of analysis – macro (external context), meso (organizational context), and micro (individual perspective). As suggested by Bardin [31], the keyword or theme-word was taken as the reference unit for purposes of categorization and counting (frequencies), and the sentence was used for purposes of information and contextualization.

4   Presentation and Discussion of Results

The discussion of the results of the field research focuses on the three dimensions defined above and on the three perspectives of analysis, which refer to both the structure of the legal model of military CS and to the way it is being operationalized. The results were organized into three major themes, as shown in Table 1.

The matrix in Table 1 shows that some categories are confirmed within each per-spective in relation to the "recruit" and "retain" dimensions, which indicates that they are interdependent. The "Incentive Scheme" category intersects the three dimensions of the study and is instrumental for their operationalization.

The matrix in Table 1 shows that some categories are confirmed within each perspective in relation to the "recruit" and "retain" dimensions, which indicates that they are interdependent. The "Incentive Scheme" category intersects the three dimensions of the study and is instrumental for their operationalization.

Table 1. Main factors that influence recruitment and retention of contract military personnel

	Perspectives of analysis		
	Macro-level (External context)	Meso-level (Internal context)	Micro-level (Individuals)
Recruitment	- Social representations about the Portuguese AAFF - Political action - Demographics - Competition by the external LM - Incentive Scheme	- Relations and communication with the LM – strategies and mechanisms - Characteristics of the military context/military life - Initial induction into the military - Characteristics of the Recruitment Process - Selection requirements	- Sociocultural profile - Generational profile - Vocational reasons - Geographic profile - Candidate profile - Perception about characteristics of the military context/military life
Retention	- Political action - Competition by the external LM - Incentive Scheme	- Characteristics of the military context/military life – Induction into the military – Financial and social conditions - Distance to place of residence	- Generational profile - Career Anchors - Geographic profile - Management of expectations
Reintegration	- General assessment – mostly negative (30 responses vs. 7) - Entities involved - Support model - Operationalization of support - Incentive scheme - Difficulties in improving the capacity for social and occupational reintegration		

4.1   Main Motives that Influence Recruitment in the Armed Forces

Macro Perspective (External Context).

Social Representations about the Portuguese AAFF. The progressive estrangement of citizens from national defense issues has been a growing concern of Portugal's military and political leadership. The results suggest the existence of some ambiguity and partial contradiction. The participants believe that military CS personnel, for the most part, have a generally "positive" and "favorable" opinion of the AAFF (18 responses vs. 9), however, and somewhat paradoxically, they also question their usefulness and are unaware of crucial aspects of their activity (22 responses). On the other hand, there is a markedly negative opinion (46 unfavorable responses vs. only 8 favorable) regarding the issue of the "appeal of the military career", which is described as "separate from the labor market", since the AAFF "are not a job for life". The respondents believe that limited knowledge about the activity of the AAFF may result in an "ill-defined" or "distorted" social representation, influencing both the perception about the AAFF's usefulness and generating negative opinions about the appeal of the military career. The participants also believe that this may stem largely from the difficulty that the

AAFF have in disseminating and projecting the institutional image and the jobs on offer.

Other External Factors. The remaining external factors were divided into six subcategories, and a combined analysis of the responses showed that: (1) the Portuguese AAFF are facing a shortage of personnel due to the systematic decrease of the number of citizens eligible to enter the AAFF, which stems from a reduced recruitment base (demographics); (2) the internal variables of the military organization are controlled by the political leadership, which establishes the legal model and influences its operationalization, leaving little room for maneuver by the AAFF regarding recruitment, and which has come under criticism for failing to acknowledge that there is a real recruitment problem as well as for the lack of external recognition of the Military, which has made military careers less attractive (political action); (3) the labor market is more appealing in terms of incentives and socioeconomic opportunities, which means that military employment is seen as a "last choice" (competition by the external labor market); and (4) the existence of shortcomings in the current incentive scheme, which most participants describe as "obsolete", does not promote the "appeal of the military career" and is, therefore, not very effective (incentive system) and must be revised.

Meso Perspective (Internal/Organizational Context). The organizational aspects of the "recruitment" dimension are described in five subcategories, the most important of which are: (1) the relations and communication strategies established by the MDN and the AAFF with the labor market are not articulated and are not effective in advertising the institutional job offer and the recruitment of volunteers; (2) the characteristics of the military context and military life, combined with the type of induction into the military, have a negative impact on recruitment, discouraging many candidates from joining because doing so entails a profoundly different reality from the everyday life of ordinary citizens and because the military career is perceived as "difficult to endure"; (3) the characteristics of the recruitment and selection process lead, on the one hand, to excessive "bureaucratization" and to high "response times" from application to the moment of selection, and the low quality of information available on the "conditions of military service" is seen as a shortcoming in the process; and (4) the characteristics and requirements of the selection profile, in the opinion of the participants, are both obsolete and disconnected from the current labor market, and exclude a large number of candidates.

Micro Perspective (Individuals). The micro perspective focuses on the following aspects: (1) the generational profile and the sociocultural profile of young people eligible for military CS combine to influence their ability to make a conclusive, coherent decision to join the military because they tend to take a somewhat "frivolous approach to challenges", have "trouble accepting hardship and difficulties", are generally "uninterested", and lack a "defense culture" and "civil awareness"; (2) the issue of vocation, which relates to people's perception about the characteristics of the military context and military life, serves as a first screening because many citizens self-select out of the process; (3) there is a gap in terms of "physical requirements" and "medical fitness" (visual and auditory acuity) between the average candidate profile and the "selection profile", which leads to high attrition rates; and (4) on the one hand, the geographic profile determines the main market niches of the AAFF and, on the other, it has a negative impact on recruitment in situations that entail "geographic distance" from the candidate's "place of residence" and close social circle.

4.2   Main Motives that Influence Retention in the Armed Forces

Macro Perspective (External Context). In addition to its negative impact on recruitment, (1) the external labor market also continues to attract military CS personnel by offering "better opportunities", which in turn influences the retention rate, especially in highly technical specialties; (2) the incentive scheme does not appear to be sufficiently appealing to retain a large number of staff, especially service members who endure more attrition due to multiple deployments and military exercises.

Meso Perspective (External Context). The meso perspective focuses on the following aspects: (1) the characteristics of the military context and military life, which can lead to failure to adapt and disappointment with the "physical demands", sacrifices, and restrictions of the "military condition", as well as with other requirements and restrictions that relate to the "rules" that must be followed; (2) inadequate induction and monitoring during the period of service in the AAFF, which relate to somewhat "rigid" and "excessively formal" interpersonal relationships, the most affected specialties being those in high market demand and/or that entail geographic mobility and/or operational commitments; (3) the still inadequate social conditions in terms of "living conditions" and "well-being"; and (4) the financial conditions, which are compounded by distance to the place of residence and separation from the family unit.

Micro Perspective (Individuals). The micro perspective focuses on the following aspects: (1) the main career anchors indicate, on the one hand, the importance of "security/stability" (which do not exist in the context of military CS since it does not provide "secure employment") and of "lifestyle" (given the increasing importance people attach to the social groups with which they identify), and, on the other hand, suggest that people have become estranged from the activity of the AAFF because they value "autonomy/independence" and are uninterested in "discipline" and in "obligations and rules" (generational and sociocultural profiles); and (2) the initial expectations (built around the information obtained/available) are confronted with the reality, often resulting in a significant "break with ordinary life" (management of expectations), which in turn leads to a high number of early dropouts.

4.3   Capacity to Support the Social and Occupational Reintegration of Former Service Members

The participants consider that the capacity of the Ministry of National Defense and the Portuguese AAFF to intervene in the social and professional reintegration of military CS personnel upon leaving the ranks is very limited (30 responses vs. 7) and note that: (1) the reintegration support model, which is largely based on the current incentive scheme, is "outdated"; (2) it has become less appealing, penalizing service members, and thus; (3) it must be adapted to the "mentality of the candidates" and to the "reality of the labor market"; (4) its operationalization is influenced either by "external influence" or by inadequate "articulation" between the entities in charge, both due to the lack of "advertising" or to unfairness in the "access" to that support, which clearly benefits those who live in large urban centers and coastal areas; and (5) of the five aspects of the incentive scheme, respondents find "support for reintegration in the labor market", "financial and material conditions", and "professional training and certification" to be the most relevant, although the results also suggest that there is a general lack of awareness about the incentives provided by this support system, especially among younger officers, who more directly lead with military CS personnel.

5   Conclusions

This study, which is part of a more comprehensive investigation, aimed to discuss the model of military CS of the Portuguese AAFF and the incentive scheme that supports it, identifying shortcomings and defining measures to mitigate its negative effects. The discussion focused on three key dimensions – recruit, retain, and reintegrate –, which are crucial elements of a process that will shed light on the experiences of the citizens who serve in the military. In addition, a research question was formulated to guide the investigation – "Are the Portuguese AAFF capable of recruiting and retaining adequate numbers of quality contract personnel?"

The objectives of this research were achieved and the research questions were answered using a qualitative research method and a case study research design based mainly on inductive reasoning, which situates the research mainly within the context of discovery. Through the three levels of analysis, namely macro (external context), meso (organizational context), and micro (individual perspective), the research contributed to the knowledge about the object of study in three key dimensions. With regard to data collection and analysis, a documentary review of the legal framework of military service and the reports issued by the branches of the AAFF and the MDN was carried out. Additionally, an inquiry with open-ended questions was administered to 69 respondents, and the responses were object to content analysis.

To answer the research question, the study found that the Portuguese AAFF has a specific problem that relates to recruiting and retaining CS personnel, which is now clarified. The study showed that the AAFF have a relatively limited ability to solve this problem on their own. The current incentive scheme that supports the military CS model, is instrumental for the three dimensions to function in a harmonious manner, however it has shortcomings in terms of its appeal and effectiveness and should be revised.

The research results are in line with the trends found in previous studies done by the MDN, by the branches of the AAFF, and also with the current views held by the political and military leadership. The innovation of this study is related to the sample that has one more fundamental extract, since the officers of the permanent staff are questioned and they directly work with the military of the Contract Service. In fact, they show a reduced knowledge on the incentive regime, which is crucial to the operationalization of the current model.

Thus, even if one considers that the current model is not exhausted, there is the need for swift, and coordinated measures at the national level, across the three dimensions addressed in the study, in order to reinforce the position of the AAFF in the labor market and to foster a culture of defense that reflects the positive influence of military service in the well-being of the population, for the benefit of the Portuguese citizens and of the Portuguese State.

Although the results of this research provide an important contribution that sheds light on a national concern, some aspects remain unsolved. Therefore, further research should focus on contracted military personnel, in order to determine the main "career anchors" that lead to the decision to serve in the AAFF under contract and the main motives behind the perception of "breakdown of the psychological contract" for early dropouts. Because of its implications in the recruitment process and in the process of social and occupational reintegration, it would be beneficial to deeply reflect on the relationship, communication strategies and mechanisms between the AAFF and the labor market.

References

1. Law No. 174/99 of 21 September: Approves the Military Service Law. Assembly of the Republic. Diário da República, Lisbon (1999). https://dre.pt/application/dir/pdf1sdip/1999/09/221A00/65416550.pdf
2. Decree-Law No. 289/2000 of 14 November: Approves the Regulation of the Military Service Law. Portuguese Government. Diário da República, Lisbon (2000). https://dre.pt/application/dir/pdf1sdip/2000/11/263A00/64256438.pdf
3. Decree-Law No. 320-A/2000 of 15 December: Approves the Regulation of Incentives to Service in CS (amended by Decree-Law No. 118/2004 of 21 May and by Decree-Law No. 320/2007 of 27 September). Portuguese Government. Diário da República, Lisbon (2000). https://dre.pt/application/dir/pdf1sdip/2000/12/288A01/00020011.pdf
4. Santos, L.: Reflections Arising from Military Service Under Contract in the Portuguese Armed Forces: Functional and Social and Citizenship Perspectives. Revista de Ciências Militares, III(1), 331–362, May 2015. https://cidium.ium.pt/docs/artigos/Artigo_123.pdf
5. Bryman, A.: Social Research Methods, 4th edn. Oxford University Press, Oxford (2012)
6. Creswell, J.W.: Qualitative Inquiry & Research Design: Choosing Among Five Approaches, 3rd edn. Sage Publications, Los Angeles (2013)
7. Guerra, I.C.: Pesquisa Qualitativa e Análise de Conteúdo: Sentidos e Formas de Uso. Principia, Parede (2006)
8. Yin, K.R.: Case Study Research – Design and Methods, 5th edn. Sage Publications, Los Angeles (2014)
9. Decree-Law No. 90/2015 of 29 May: Approves the Statute of the Armed Forces Military Personnel. Portuguese Government. Diário da República, Lisbon (2015). http://www.emgfa.pt/documents/cqw3zjnhvg4s.pdf
10. Decision No. 8474/2016 of 30 June: Determines the measures to be implemented in the Professionalization of Military Service, to be undertaken by the National Defense Day, Planning and Execution Committee, the Military Recruitment Planning and Coordination Committee, and the Professional Reintegration Planning and Coordination Committee. Diário da República No. 124/2016, Series II, Lisbon (2016). https://dre.pt/web/guest/home/-/dre/74813568/details/maximized?serie=II&day=2016-06-30&date=2016-06-01&dreId=74813550

11. Wash, V.: Models and theory. In: Durlauf, S., Blume L. (eds.) The New Palgrave Dictionary of Economics. 2nd edn., Palgrave-Macmillan (2008)
12. Matsuyama, K.: Structural change. In: Durlauf, S., Blume, L. (eds.) The New Palgrave Dictionary of Economics. 2nd edn., Palgrave-Macmillan (2008)
13. Caetano, V., Vala, J. (org.): Gestão de Recursos Humanos: Contextos, Processos e Técnicas. 3rd edn. Editora RH, Lisboa (2007)
14. RTO/NATO: Recruiting and Retention of Military Personnel: Technical Report of RTO (Research Task Group), HFM-107. NATO Research & Technology Organization, October 2007.   https://www.nato.int/issues/women_nato/Recruiting%20&%20Retention%20of% 20Mil%20Personnel.pdf
15. Defillippi, R.J., Arthur, M.B.: The boundaryless career: a competency-based prospective. J. Organ. Behav. **15**(4), 307–324 (1994)
16. Hall, D.T.: Careers in Organizations. Goodyear, California (1976)
17. Dutra, J.S. (org.): Gestão de Carreiras na Empresa Contemporânea. Atlas, São Paulo (2010)
18. Van Maanen, J.: Organizational Careers: Some New Perspectives. Sloan School of Management, Massachusetts Institute of Technology, Cambridge, Massachusetts (1977)
19. Martins, H.: Gestão de Carreiras na Era do Conhecimento. Qualitymark, Rio de Janeiro (2010)
20. Arthur, M.: The New Careers: Individual Action and Economic Change. Sage, London (1996)
21. Currie, G., Tempest, S., Starkey, K.: New careers for old? Organizational and individual responses to changing boundaries. Int. J. Hum. Resour. Manag. **17**, 755–774 (2006)
22. Dutra, J.: Administração de Carreiras – Uma Proposta para Repensar a Gestão de Pessoas. Atlas, São Paulo (2012)
23. Hall, D.: Careers In and Out of Organization. Sage, London (2002)
24. Arthur, M., Rousseau, D.: The Boundaryless Career: A New Employment Principle for a New Organizational Era. Oxford University Press, New York (1996)
25. Peiperl, M., Arthur, M., Goffee, R., Morris, T. (eds.): Career Frontiers: New Conception of Working Lives. Oxford University Press, Oxford (2000)
26. Schein, E.: Career Anchors: Discovering Your Real Values. Pfeiffer & Company, University Associates, San Diego (1990)
27. Rousseau, D.: Psychological Contracts in Organizations: Understanding Written and Unwritten Agreements. Sage, Thousand Oaks (1995)
28. Moskos, C.: From institution to occupation: trends in military organizations. Armed Forces Soc. **4**, 41–50 (1977)
29. Saunders, M., Lewis, P., Thornhill, A.: Research Methods for Business Students, 5th edn. Pearson Education Limited, Essex (2009)
30. Denzin, N., Lincoln, Y. (eds.): Collecting and Interpreting Qualitative Materials, 4th edn. Sage Publications, Los Angeles (2013)
31. Bardin, L.: L'Analyse de Contenu. Presses Universitaires de France, Paris (1977)

Strategy, Geopolitics and Oceanopolitics

Evolution of the International Regime for Oceans Under the Hobessian Image View

Luis Piedra[1], Teresa Guarda[2,3,4(✉)], and Ramiro Armijos[1]

[1] Armada del Ecuador, Guayaquil, Ecuador
{luispiedraaguirre, r_armijos}@hotmail.com
[2] Universidad de las Fuerzas Armadas-ESPE, Sangolquí, Quito, Ecuador
tguarda@gmail.com
[3] Universidad Estatal Península de Santa Elena – UPSE, La Libertad, Ecuador
[4] Algoritmi Centre, Minho University, Guimarães, Portugal

Abstract. The evolution of the international regime for the oceans use, was materialized in the United Nations Convention on the Law of the Sea (UNCLOS), which regulates sovereignty and rights over maritime spaces and its resources. This paper analyses the historical stages established by Keohane and Nye (1988), under the conceptual view proposed by Hobbes in Chap. 13 of Leviathan, regarding the natural condition of men and the anarchic state of nature similar to war. According to Hobbes, a government or a common power would avoid anarchy and conflict. However, after the UNCLOS was established, disputes arose in the process of defining maritime zones and boundaries between neighboring coastal states, expecting to obtain more resources and spaces at sea. Thus, conflict may be diminished in the presence of a common power represented by an international regime or authority, but not completely eliminated. That is man's nature.

Keywords: Hobbesian image · Maritime conflicts · UNCLOS
International relations

1 Introduction

The Hobbesian image of international relations is one of the characteristics that has been suggested and recognized by many theorists in the international political system, which establishes that anarchy in the international system and in relations among states would be similar to an anarchic state of nature explained by Thomas Hobbes in his book Leviathan.

Hobbes (1651) in Chap. 13 of the aforementioned work, describes the natural life of men, as supposedly was carried out before a central power with a set of laws governs the behavior of members of society. According to Hobbes, men in the state of nature would have lived in a permanent war of all against all, that is, they would have lived in a situation of permanent hostility and threats [1]. Of course, this situation to be applied in international relations and in the issue that brings us together -conflicts in the evolution of the International Regime related to the oceans- needs to replace men by nation-states.

© Springer International Publishing AG, part of Springer Nature 2018
Á. Rocha and T. Guarda (Eds.): MICRADS 2018, SIST 94, pp. 211–221, 2018.
https://doi.org/10.1007/978-3-319-78605-6_17

In this context, the absence of a central authority that governs states, which act according to their own interests, would be one of the aspects that produce a situation of anarchy, which in our case would be at sea. On the other hand, for the subject in question, it should be noted that during the historical evolution of the International Regime relative to the oceans, the United Nations Convention on the Law of the Sea (hereinafter UNCLOS), the situation of conflicts and disputes was a constant in the relations between states, whose realistic motivation has always been to protect and satisfy their own interests. Thus, conflicts over the delimitation of maritime spaces and especially over the control of the natural resources contained in it, increasingly took on greater relevance, as the advance of technology allowed to discover new resources and modern methods to extract them.

Szekely in his study of the law of the sea confirmed "History testifies that most of the conflicts that have taken place, in the contemporary era, have been due to the attempts of some state to increase the spatial scope of its sovereignty, obviously at the expense of another, or the lack of agreement between neighboring states regarding the criteria to be used for fixing the dividing line of their respective sovereignties" [2].

So it could be said, in the pure style of the Hobbesian image, that evolution to define the international regime in the oceans, that is, a common authority that regulates, among other things, the delimitation of maritime spaces between states and the sovereignty of the resources contained in the seas, has been resembled to a situation of "war of all against all".

In this sense, this work proposes the following purpose: To verify if some aspects of the Hobbesian anarchic image are identified in the historical process for establishing an authority for the use of oceans, and determine similarities between Hobbes' state of anarchic nature and the evolution process for developing an international regime to the oceans.

For this purpose, the evolution of the international regime relating to the oceans, UNCLOS, is briefly explained, considering as a basis the study carried out by Keohane and Nye (1988) with respect to international regimes in the maritime policy system. Next, we go over the most important aspects written by Hobbes in Chap. 13 of the Leviathan, referring to the natural condition of the human race. Afterwards, we try to establish similarities between the state of nature of Hobbes and the conflicts in the evolutionary process of the aforementioned international regime. Finally, the respective conclusions are proposed.

2   Evolution of the International Regime Relate to the Oceans

Aristotle in his book Republic, referring to the sea said "there is no doubt that, taking into account the security and abundance necessary to the State, it is very convenient to the city and the rest of the territory to prefer a shore by the sea" [3]. In that sense, oceans since antiquity have always had strategic importance. In the first place, in the economic aspect, by its inexhaustible source of both living and non-living resources, and because it facilitates trade through maritime transport; secondly, in the political area, related to security and defense; and thirdly, as a consequence of the two previous ones, a legal importance, to regulate its use in the face of conflict and anarchy.

In the sixteenth century there was a first attempt to regulate the authority in the oceans and their resources, through the great debates of Hugo Grotius, who defended the concept of Mare Liberilum or freedom of the seas -free navigation-, and John Selden who defended the concept of Mare Clausum or property of the seas, postulating the latter that: the sea belongs to who can appropriate it and defend it [4]. In 1882, civilized nations recognized that the open sea must be internationalized. By then, territorial waters were in the vicinity of the coast, at a distance of three miles [5].

Keohane and Nye in their book Power and Interdependence [6], indicate three main stages to explain the International Regime referred to the sea, in the system of maritime policies, as indicated in the following (Table 1).

Table 1. International regimes in the maritime policy system 1920–1975 (Source: [6]).

Period	Years	Status of the regime	Action at the beginning of the period
1	1920–1945	Maritime freedom regime	Great Britain reaffirms its leadership after the First World War
2	1946–1966	Quasi strong regime	Declaration of Truman 1945 and expansion of jurisdictions in Latin America
3	1967–1975	Quasi weak regime	Speech by Pardo, United Nations in 1967

The first period comes from the nineteenth century, in which the regime of freedom of the seas was associated with the interests and power of the main maritime country, England. Thus, the first period that Keohane and Nye indicate is initiated in 1920, and marked by the mentioned maritime hegemony. In 1930, during the Conference of the League of Nations in The Hague, where small states had voice and vote, they began to question the three-mile limit and made efforts to reach other extensions. They initiated a series of disputes; twenty countries that represented the main powers supported the territorial limit of three miles, except the Union of Soviet Socialist Republics (USSR) that supported twenty and Italy six miles. However, efforts made by countries such as Ecuador, Mexico and Iran to extend jurisdiction were not recognized by maritime powers [6].

A conflicting situation arose, due to the great interests of the powers to continue with the regime of freedom of the seas, and thus have larger maritime extensions to be exploited. A series of claims and tensions were present in the international scenario by the rights of coastal nations claimed over resources existing in distant waters, and the prospects of obtaining resources at the bottom of the sea, before the increase in the presence of maritime powers eager to exploit them. All this favored turning the oceans into a scenario of instability and confrontation. After the First World War, until 1945, the principle of freedom of the seas began to decline. In the absence of international norms governing the use of the sea as a source of wealth, the world entered then into a new era of unilateral measures of self-protection adopted by the respective states [7].

In the second moment proposed by Keohane and Nye from 1945, there was a change in the international regime of the oceans, with the declaration of President

Truman, with the new technologies of fishing and oil drilling in the marine platform, The United States of America unilaterally established jurisdiction over all the natural resources of its continental shelf (oil, gas, minerals, fisheries, etc.), until a depth of 200 m. In this situation, other countries such as Egypt, Ethiopia, Saudi Arabia, Libya, Venezuela and some Eastern European countries claimed a territorial sea of 12 miles, all of which clearly departed from the traditional three-mile limits [8].

At the same time, on the west coast of South America, countries such as Ecuador, Peru and Chile, where there is very little continental shelf, argued that the depth criterion had no value for them and claimed jurisdiction in terms of distance, which was concretized with the Santiago Declaration of 1952, establishing 200 miles here for its territorial sea. This brought several diplomatic conflicts, before the seizure of North American tuna vessels, found within the jurisdictional waters of Ecuador and Peru [6].

Situations that occurred in less developed countries, with hope of preventing access by large fishing fleets and preserving the existing fish resources within their adjacent seas, for which warships and coast guard boats were required to defend jurisdictional spaces.

The United States and Great Britain guided their efforts to protect the weakened regime of the freedom of the seas, during the two conferences on maritime law held in Genoa in 1958 and 1960, where no agreement was reached on the limit of territorial waters. Several countries tried to transcend the freedom of the seas to the territoriality of the seas, pretending to submit to their jurisdiction, vast areas of what had traditionally belonged to the high seas [6]. Conflicts continued, for example in the North Sea oil found on the high seas was interest of Britain, Denmark and Germany, which sparked confrontations because of the sovereignty of the continental shelf, due to the rich resources of this mineral [8].

The third moment of Keohane and Nye started in 1967, when Ambassador Pardo's speech impacted the international community, regarding the enormous benefits of the seabed and focused attention on ocean resources and distributional issues, as well as the conservation of the seabed. Since then, it has been considered that the efficient management of the oceans can achieve great benefits to the states; however, it has been thought at the same time that the benefits of a state turn out to be the losses of other states [6].

During this time, dangers were numerous in the presence of nuclear submarines with the possibility of exploring the seabed and designed with missile systems; supertankers that transport oil from the Middle East to the ports of the world leaving traces of oil spills; and the increase in tensions among nations over the demands for maritime spaces and resources; what generated in the oceans an anarchic environment, with a multitude of demands, counterclaims and sovereignty disputes [8].

Less developed countries, fearful that the global goods of the seas would be exploited only by the countries with the most technology capabilities, insisted on a greater extension of the national jurisdiction and to strengthen the international regulatory body. Thus, a series of controversies have arisen, China maintained that the freedom of the seas was maintained by both powers (United States and Great Britain), to exercise hegemony and expansionism in the oceans and the plundering of the maritime resources of other countries [6].

On the other hand, Canada and Australia allied with the United States and England during the cold war, changed their political point of view in favor of their coastal interests. Thus, in 1970, Canada asserted the right to regulate navigation in an area extending up to 100 miles from its coasts in order to protect water against Arctic contamination [8].

Finally, in December 1973, the Third International Conference on the Law of the Sea was convened, which after eleven sessions, over the course of nine years, adopted in 1982 in Montego Bay, Jamaica, the Convention of the Nations United on the Law of the Sea, better known as the constitution of the seas, which came into force in 1994. Thus, after several years of conflicts, controversies and discussions, UNCLOS adopted, among other things, the following maritime spaces for the coastal states: territorial sea with an area of 12 nautical miles with full sovereignty; exclusive economic zone with an area of 200 nautical miles with sovereignty to explore, exploit, conserve and manage natural resources; continental shelf, which includes the bed and subsoil of the underwater areas up to the outer edge of the continental margin or up to a distance of 200 nautical miles, with the possibility of extending up to 350 nautical miles; and the high seas, beyond 200 nautical miles for global use; and likewise, the way to delimit with the neighboring coastal countries was established, by means of the equidistant line [9].

3   Review of the State of Nature of Hobbes Concepts for the Analysis

According to the above, taking into account the conflict and anarchy of the process to agree on a common authority to govern the activities and conduct of the states at sea, it has been considered a review of the book written by Hobbes, Leviathan, exclusively Chap. 13 [1]: "Of the Natural Condition of Mankind, as Concerning Their Felicity and Misery".

For which, the following paragraphs and main texts have been extracted as core issues, in order to subsequently find similarities with the evolution of the International Regime explained above.

"Men are equal by nature", Hobbes begins his description of the state of nature by referring to a basic natural equality between all men, and based on this fact he urges the inescapable conflict between them. This is indicated:

> Nature hath made men so equal in the faculties of the body and mind, as that, though there be found one man sometimes manifestly stronger in body or of quicker mind than another, yet when all is reckoned together the difference between man and man is not so considerable (…) For, as to the strength of body, the weakest has strength enough to kill the strongest, either by secret machination or by confederacy with others that are in the same danger with himself (…)
>
> (…). For such is the nature of men that, howsoever they may acknowledge many others to be more witty or more eloquent or more learned, yet they will hardly believe there be many so wise as themselves, for they see their own wit at hand and other men's at a distance. But this proveth rather that men are in that point equal than unequal (…).

Based on this equality of capacity, Hobbes continues with "From Equity comes distrust", and mentions that:

> From this equality of ability ariseth equality of hope in the attaining of our ends. And therefore, if any two men desire the same thing which nevertheless they cannot both enjoy, they become enemies; and, in the way to their end, which is principally their own conservation and sometimes their delectation only, endeavour to destroy or subdue one another. And from hence it comes to pass that, where an invader hath no more to fear than another man's single power.

In other words, as Miranda also reasons, "equality of capacity leads to equal expectations of reaching the same ends, and this is the source of conflicts between men, because when two men want the same thing, and that thing cannot be enjoyed by both together, they become enemies" (1984, p. 72) [10].

And Hobbes continues saying that "Of the distrust, the war", that is, before this situation of mutual distrust, men try to dominate through force or by cunning, until they no longer have a threat, this is understood it is for their own conservation and interests. And it is necessary for the preservation of a man, to increase his dominion over other men.

Hobbes distinguishes three main causes of discord: competition, diffidence and glory. And it indicates:

> The first maketh man invade for gain; the second, for safety; and the third, for reputation. The first use violence, to make themselves masters of other men's persons, wives, children, and cattle; the second, to defend them; the third, for trifles, as a word, a smile, a different opinion, and any other sign of undervalue, either direct in their persons or by reflection in their kindred, their friends, their nation, their profession, or their name.

That is, the first cause drives men to use violence to obtain a benefit; the second, the defense to obtain security; and the third, resorts to force to achieve reputation.

Next Hobbes says:

> Hereby it is manifest that, during the time men live without a common power to keep them all in awe, they are in that condition which is called war, and such a war as is of every man against every man.

Consequently, for Hobbes a common power is necessary to control and regulate the activities of men, including, frightening to avoid war.

He also mentions:

> For 'war' consisteth not in battle only or the act of fighting, but in a tract of time wherein the will to contend by battle is sufficiently known (…) the nature of war consisteth not in actual fighting but in the known disposition thereto during all the time there is no assurance to the contrary. All other time is 'peace.'
> (…)Whatsoever therefore is consequent to a time or war where every man is enemy to every man, the same is consequent to the time wherein men live without other security than what their own strength and their own invention shall furnish them withal.

It is understood that these moments of war of all against all, are not exclusively referred to the real moments of struggle, but also, at times when there is a willingness to go to the war, there will be a war in pot. Your safety is through your own resources.

He continues:

(...)Let him therefore consider with himself, when taking a journey, he arms himself and seeks to go well accompanied; when going to sleep, he locks his doors; when even in his house, he locks his chests; and this when he knows there be laws and public officers armed to revenge all injuries shall be done him; what opinion he has of his fellow-subjects when he rides armed; of his fellow-citizens, when he locks his doors; and of his children and servants, when he locks his chests. Does he not there as much accuse mankind by his actions as I do by my words? (...)

Here, it is reiterated with respect to the distrust existing among men, which motivates them to implement assurances, such as armed walking and placing enclosures or locks on their property.

And later on it continues referring to the common power that must be had to avoid war:

(...)Howsoever, it may be perceived what manner of life there would be where there were no common power to fear, by the manner of life which men that have formerly lived under a peaceful government use to degenerate into, in a civil war.(...)
(...)To this war of every man against every man this also is consequent, that nothing can be unjust. The notions of right and wrong, justice and injustice, have there no place. Where there is no common power, there is no law; where no law, no injustice.(...)

In the aforementioned, Hobbes insists on the need for a common power, which imposes laws, so that there is justice.

(...) It is consequent also to the same condition that there be no propriety, no dominion, no 'mine' and 'thine' distinct, but only that to be every man's that he can get, and for so long as he can keep it. And thus much for the ill condition which man by mere nature is actually placed in (...)

Here, Hobbes reiterates about the natural human condition of war, where everything is anarchy.

Finally: The passions that incline men to peace are fear of death, desire of such things as are necessary to commodious living, and a hope by their industry to obtain them. And reason suggesteth convenient articles of peace, upon which men may be drawn to agreement.

At this point, it is highlighted that peace is motivated by the desire to achieve well-being, and can be achieved through standards reached through agreements or agreements between men.

From the paragraphs reviewed and in order to summarize the most important aspects, the following are two condensed central ideas:

- First idea: In the natural state of the man of equality among them, the weakest would have the possibilities of equaling the strongest, through alliances with others or using their ingenuity; thus the two consider that in equal capacities they would be in the same conditions to attack or be attacked. Furthermore, given this equality comes the distrust between them, therefore, when both have the same goals and interests for something, such as their survival, need for resources or simply satisfy even trivial issues, both become enemies. Mistrust is evident when men must be armed and must place their assets secure.

- Second idea: If there is no common power or authority to regulate the activities among them, including obliging or intimidating the rules, the situation would always be one of anarchy and constant conflicts. The war situation of all against all, it is understood that not only refers to the real moments of struggle, but also to the moments when you have the disposition to go to war or to the will to face the enemy. When there is no common power or authority, there are no regulations, in other words there are no laws to comply with, therefore, there is no justice to apply. In the situation of war, what would affect to reach peace would be the fear of dying, the desire to maintain a comfortable life, for which, mutual peace agreements should be established.

4   Some Similarities Among the Conflicts in the Evolution of the International Regime for Oceans and the Hobbes' State of Nature View

Before moving forward with the similarities, it is worth mentioning the analogy expressed by Miranda regarding the actions of the countries in international relations and the state of nature of Hobbes: "The search for preservation, security, and in general, the own interests, remain the primary and primary causes of the actions of states in the international arena, in the same way that the search for the preservation of one's own life, security and one's own interests were the main causes of actions of individuals in the state of nature described by Hobbes" [10]. Thus, below it is proposed an analysis of the similarities found between the core parts of Chap. 13 of Hobbes' work, the Leviathan, summarized in the two last ideas and the evolution of this international regime referring to the oceans, taking into account the three periods initially explained:

The first period of the evolution of the international regime considered by Keohane and Nye, from 1920 to 1945, was characterized by the particular interest of the maritime powers, whose hegemony was maintained by Great Britain, in continuing with the regime of freedom of the seas and thus maintain access to a larger maritime area to freely exploit the resources of the sea. During this period, there were tensions and claims in the coastal nations, for the rights that they also have to the existing resources in distant waters, mostly exploited exclusively by the great maritime powers.

Similarities: The first idea previously indicated, referring to that natural state of man of equality between them, which leads to mistrust and become enemies if they have the same interests, is directly related to the aspects of this first period analyzed, in the sense that the natural attitude of the great powers was to take advantage of their condition of hegemonic states to satisfy their own interests, and in the same way, the natural attitude of the less developed countries was also to protect their interests, which provoked conflicts and claims of the latter, so that their rights were respected. The marked distrust between them is observed, that when coinciding in their same ends and interests -added to other geopolitical issues-, therefore some of them became potential enemies.

In the second period, from 1946 to 1966, where the beginning of the decline of the principle of freedom of the seas that had been sustained by the maritime powers, the absence of international standards to regulate the uses of the sea, promotes states to adopt unilateral measures of self-protection. Less developed countries such as Ecuador, Peru and Chile unilaterally proclaim 200 miles of territorial sea. There are arrests of North American tuna vessels in the waters of Ecuador and Peru. In short, the least developed countries sought to prevent access by foreign fishing fleets off their coasts, in order to prevent and control the depletion of fishing in their adjacent seas.

Similarities: Like the previous case, this period shows analogies with the first idea established, in the sense that the equality that becomes distrust and enemies when you have the same ends, continues to appear, before the coinciding interests for natural resources from sea. Additionally, it is observed in this period that in a similar way to what was raised by Hobbes regarding the possibilities of equaling the strongest, through alliances with others or using their ingenuity, the less developed countries sought pacts or alliances between similar to face the power of the strongest countries, this is the case of the South American countries of the south pacific coast (Ecuador, Peru, Chile).

In the third period, from 1967 to 1975, the principle of freedom of the seas is completely questioned. For this time, the dialogues and conferences that concluded with the establishment of the regulatory authority of the seas, UNCLOS were in process. However, tensions and conflicts between states over the sovereignty of oceanic spaces and their resources continued, despite the fact that certain norms and regulations were already being applied. It was known that the benefits that a state would obtain resulted from the loss of these same benefits in another state.

Similarities: This period is more similar to the second idea described, in the sense that in the absence of a common power or authority to regulate the activities between them, the situation remains conflicting and anarchy. That was exactly happening with the coastal states, which did not reach agreements, especially regarding the extension of the sea, where the resources can be explored and exploited with total sovereignty, the environment was conflictive.

In addition, in the second idea that war is also considered during the moments when there is a willingness to face the enemy, an analogy is observed with the three periods, because although there were several conflicts that did not degenerate into wars declared, with the exception of some disputes such as the cold war between Iceland and Great Britain for fishing rights in the North Atlantic -between the 1950s and 1970s-, it is understood that all countries are predisposed to go to war when it comes to defending their sovereignty, especially to protect what they consider their rights and interests in the exploitation of the resources of the sea, as these undoubtedly serve for their survival and development.

Likewise, an analogy is found with the second idea, in reference to what Hobbes said: "where there is no common power, the law does not exist; where there is no law, there is no justice," that is, in the absence of a common authority or regulations, there would be no justice to apply. This was exactly what was happening in the case of the oceans, there was no law or justice, until the authority was established with UNCLOS and rules and regulations were implemented.

It is worth mentioning that is indicated by Armitege [11], regarding the fact that initially "Hobbes was not associated with international relations, and that he began to be considered a theoretician of international anarchy once a consensus emerged on the fact that the scope of international relations was certainly anarchic" (2006, p. 34). In this sense, the several similarities found in this analysis corroborate the direct relationship of Hobbes' anarchic state, with the evolution of the International Regime referred to the oceans, within the anarchic situation of the international system.

Also, Miranda in his analysis of Hobbes and international anarchy, concludes by saying that: "the Hobbesian description of the anarchic state of nature as a state of war, is a suggestive image that can be profitably used in the study of international relations contemporary, from the moment in which the main guide of the States when designing their international policies is the search of what safeguards their interests, in view of which, they are all even willing to resort to war, forgetting any moral consideration" [10].

Therefore, once UNCLOS was established, with a series of regulations designed to avoid disputes and conflicts between states, paradoxically, disputes and confrontations arose again in the process of defining maritime zones and boundaries between coastal states, motivated by the aspiration of states in reaching agreements to obtain greater maritime territory and greater resources from the sea.

5   Conclusions

During the three periods considered to explain the evolution of the international regime referring to the oceans, UNCLOS, there are several similarities with the state of nature of Hobbes. Among them, from 1920 to 1945, when there was a general acceptance of the principle of freedom of the seas, the natural state of man of distrust and enemy (if they have the same interests), resembles the attitude of the great powers in taking advantage of its hegemonic condition to satisfy its ends, and with the attitude of some other countries, which coincided in their interests for the resources of the sea, and became potential rivals. From 1946 to 1966, in which there is a decline in the principle of freedom of the seas, the idea that the weakest would have the possibility of equaling the strongest through alliances, is related to the pacts or agreements of the less developed countries to face to the powers regarding their interest at sea, likewise, that distrust among men that causes it to arm itself, resembles the acquisition of ships to maintain their security. From 1967–1975, when the principle of freedom of the seas is questioned and new norms appear, the idea that the situation is anarchic in the absence of a common regulatory authority, is related to what happened between the coastal states, which without rules to delimit and exploit the resources of the sea, the situation was conflictive.

All these similarities and others found confirm that the Hobbesian anarchic image has been present in the historical process that has elapsed to establish authority in the oceans. However, established the UNCLOS, paradoxically, there arose again disputes and confrontations between the states when defining their limits and maritime zones, in the pretension to reach greater spaces of the sea. What finally leads to rethink about the state of nature of Hobbes and the power or common government that could avoid anarchy and conflict, because in the analyzed case, after many disputes an authority was established, but in the path of apply the new rules, new conflicts appeared. This leads us to a final reflection, based on the Hobbesian image: conflict may be diminished in the presence of a common power, but not completely eliminated, that is the nature of man!

References

1. Hobbes, T.: Leviathan. Epublibre (1651)
2. Szekely, A.: Derecho del Mar. Universidad Nacional Autónoma de México, México (1991)
3. Aristóteles: República (384–382 a.c.)
4. Salom, N.: Dos colosos frente al mar. Fundación Cultural Javeriana de Artes Gráficas-Javegraf, Bogotá (2003)
5. Henning, R.: Körholz: Introducción a la Geopolítica tercera edición edn. Editor Pleamar, Buenos Aires (1977)
6. Keohane, R., Nye, J.: Poder e Interdependencia. La política mundial en transición. Grupo Editor Latinoamericano, Buenos Aires, Argentina (1988)
7. García, E.: La Doctrina de las 200 millas y el Derecho del Mar. Revista de la Facultad de Derecho de la PUCP, pp. 12–27 (1974)
8. Organización de las Naciones Unidas: The United Nations Convention on the Law of the Sea. A historical perspective. http://www.un.org/Depts/los/convention_agreements/convention_historical_perspective.htm
9. Organización de las Naciones Unidas: Convención de las Naciones Unidas sobre el Derecho del Mar., Jamaica (1982)
10. Miranda, C.: Hobbes y la anarquía internacional. Revista de Ciencias Políticas 6(2), 71–84 (1984)
11. Armitage, D.: Hobbes y los fundamentos del pensamiento internacional moderno. Derechos y Libertades 15, 17–46 (2006)

The Validity and Influence of Clausewitz's Trinity on the Development of Naval Power

Carlos Guzmán[1(✉)], Óscar Barrionuevo[1], and Teresa Guarda[2,3,4]

[1] Armada del Ecuador, Guayquil, Ecuador
cguzman@armada.mil.ec, oscarbarrionuevovaca@gmail.com
[2] Universidad de las Fuerzas Armadas-ESPE, Sangolqui, Quito, Ecuador
tguarda@gmail.com
[3] Universidad Estatal Península de Santa Elena – UPSE, La Libertad, Ecuador
[4] Algoritmi Centre, Minho University, Guimarães, Portugal

Abstract. General Carl Von Clausewitz, based his work "On War" on the lessons obtained from the Napoleonic Wars; defining in this work the denominated "Trinity of Clausewitz": Town, Government and Armed Forces. The purpose of this paper is to analyze if this theory is still valid and to demonstrate whether the conceptual applicability in the process of planning and development of the naval power of a State. The problem is approached from the context of the Grand Strategy and the Maritime Strategy, based on the Trinitarian theory of Clausewitz. The aim is to visualize the relationship that exists between this conception and the theories that sustain the development of naval power and the design of forces. The argumentation of this research was developed through the analytical-deductive method, describing Clausewitz's theory about war, and then interpreting what concerns the development or planning of forces, thus having the elements to correlate to determine the validity and incidence of "The Trinity of Clausewitz" in the development of Naval Power.

Keywords: Naval power · Trinity of Clausewitz · Design of forces
Maritime Strategy

1 Introduction

Karl von Clausewitz, was a Prussian general, born in Burg-Germany in 1780, who was noted for the profound study of the war phenomenon, materializing much of his legacy in his work "On War", which was published by his widow in 1832. In this work, which was written after what happened in the Napoleonic wars, it exposes the most transcendental phrase about the war, defining it as "The continuation of politics by other means" [1].

It also highlights the strategic theory known worldwide as "The Trinity of Clausewitz", which in synthesis links three elements: the People, the Government and the Armed Forces, for what he called "political act" or "political enterprise of high flight", the war [1]. When configuring this thought probably it was proposed that this theory, besides being complete, considering that the war not only involved physical forces but also moral forces have a permanent value, here is a key aspect to be analyzed in this paper, the validity of the theory called "The Trinity of Clausewitz".

© Springer International Publishing AG, part of Springer Nature 2018
Á. Rocha and T. Guarda (Eds.): MICRADS 2018, SIST 94, pp. 222–234, 2018.
https://doi.org/10.1007/978-3-319-78605-6_18

On the other hand, naval power ensures, defends and protects the maritime interests of a nation. A Navy, at present, must develop the strategic capacities that satisfy the achievement of the objectives imposed by the State; therefore, this naval potential must be configured through a process involving, inter alia, national interests and objectives, national and military strategy, and maritime operational concepts; this discipline is known as planning and development of forces.

The Trinitarian theory of Clausewitz is the fundamental basis of the analysis to be carried out, in such a way, to visualize the relationship that exists with the development of naval potential.

Under this context, the following question arises: Clausewitz's Trinity is a theory still in force and consequently its conceptual applicability affects the process of planning and development of the naval power of a State? The purpose of this paper is precisely answer the question.

The field of research and analysis is limited to the scope of Strategy and National Policy, knowing that the Strategy relates ends and means, and the Policy uses the instruments of power to safeguard or achieve their interests, being the naval power the defender of national interests in the maritime field. The Trinity of Clausewitz is immersed in phenomenological abstraction and is based on a circular thought. The development or planning of forces tends to materialize the theory, using a logical process that facilitates determining the means that military power should possess. Polemology and sociology are theoretical fields that will also be analyzed, but in a superficial way.

The scope and coverage of the research is defined by the understanding of the concept known as The Trinity of Clausewitz, and what involves the development of forces, all framed within the spectrum of the Grand Strategy and the Maritime Strategy.

The argumentation of this research will use the analytic-deductive method, describing Clausewitz's theory about war, and then interpreting what concerns the development or planning of forces, thus having the elements to correlate to determine the validity and incidence of "Clausewitz's Trinity" in the development of Naval Power. The paper will culminate exposing the conclusions obtained from the study.

2   The Trinity of Clausewitz

General Karl Von Clausewitz, as a scholar of war, widely explored this phenomenon, based on other theorists and the experiences lived in his time, developing what he thought was the nature of war. Generated thus, some concepts about war, among which are relevant to bring up: "War is a political act", "War is a high-flying political enterprise", and "War is the continuation of politics by other media".

As a starting point, what the author of the work "On War", about what is now known as "Clausewitz's Trinity" is exposed:

> "...War is not, therefore, not only a true chameleon, due to the fact that in each specific case it changes its character, but it also constitutes a singular trinity, if it is considered as a whole, in relation to the tendencies that predominate in it. This trinity is composed of both hatred, enmity and primal violence of its essence, elements that should be considered as a blind natural impulse, as by the game of chance and probabilities, which make it an activity devoid of emotions, and by

the subordinate nature of the political instrument, which induces it to belong to the scope of mere understanding. The first of these three aspects is of special interest to the people; the second, the commander in chief and his army, and the third, only the government..." [1].

From the conceptualization of the "Clausewitz Trinity" three specific elements are identified, which according to one's own understanding, indicate that the people, the government and their Armed Forces coexist, which interrelate to make war, so much so that if an element will be missing, there would be no such war antagonism.

In a first approximation, what Clausewitz described as an "Inseparable Trinity" is represented in Fig. 1.

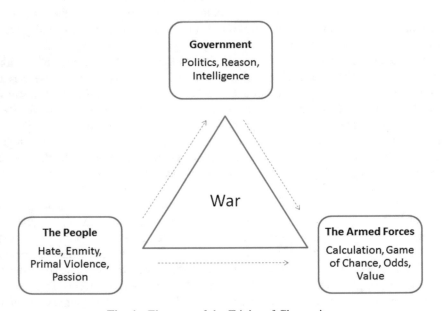

Fig. 1. Elements of the Trinity of Clausewitz.

It is understood that of the three elements, the one that produces the first ignition is the society, the people, the one that expresses the will of man for survival, passion and hatred, against those who oppose his will or affect his well-being or interest; then this will shows it to the Armed Forces in terms of what they want from them for their security, protection, revenge, etc., and to the government in the same way, but the latter, which is the rational and intelligent element Clausewitz postulate, process and decide whether to use or not, his call by the Prussian general, "political instrument", military power.

The government is responsible for translating the will of the people, and transmitting them in terms of political objectives to be achieved by military power. The latter, is who represents the value and talent to calculate and play with the odds and chance, and who ultimately depends on the achievement of the objectives that the political element set. The success or failure of military action to achieve the desired effects will also depend on the character of the commander and the military force available.

The first link between the Government and the Armed Forces is established with the establishment of its phrase: "War is the continuation of politics by other means", and refers to the means by which force is applied, which are available to the Military power. Clausewitz subordinates war to politics in terms of fixed objectives, and it is this phrase that conceives the dependence relation of war to politics, where its main instrument for war is the military potential of its Armed Forces.

It follows as an axiomatic assumption that to understand war it is necessary to accept that the most significant in it is its social and political nature, thus evidencing the relationship between the people and government elements. But Clausewitz (1832) in his work "From the War", on this relationship states: "... Here we can only treat politics as a representative of the interests of the community!" [1].

The last interrelation of the Trinitarian theory of Clausewitz, takes place between the town and the Armed Forces, in such a way, that the vigor of the population determines, to a great extent, the efficiency of the personnel and the material of those forces.

When we analyze the "the people" component, it can be pointed out that its thought about war is closely related to sociology, since it states that "... the war action is the responsibility of social life, being in essence a conflict of great interests..." [1], being this way, this theory also is related to the Polemology, which studies the war from the sociological point of view; since the creator of the concept, the French Gastón Bouthoul, maintained that the war is a social phenomenon.

Of the three pieces of the theory under study, its author says that the people are the main component, and it is because it considers that in all the constituents of "the trinity", there are actors that belong to society, i.e. soldiers and politicians. They are also part of town.

Analyzing the elements Government and Armed Forces, it is considered that it is the ends and means, respectively. Clausewitz (1832) noted that: "The political goal is the objective, war is the means to achieve it and the media can not be considered isolated from its purpose," which notes that the military means available to the Armed Forces and the Political objective of the war are interdependent. When talking about ends and means, we are talking about strategy, which in simple and particular terms is defined as the art of adapting ends to means or vice versa.

The elements that make up the trinity under study, I believe are at the highest level of driving, that is, the National Strategy or also called Grand Strategy, which manifests the following definition:

> "... strategy is the art of controlling and using the resources of a nation - or coalition of nations - including its armed forces, towards the end that their vital interests are effectively promoted and secured against real, potential or simply presumed enemies. The strategy at its highest level - sometimes called grand strategy - is that which integrates the policies and armaments of a nation for which the recourse of war is considered unnecessary or is undertaken with the maximum chance of victory ..." [2].

In this area, the strategy is aimed at activities of a collective nature, to prevent or facilitate human actions that generate significant changes in a social conglomerate. It relates philosophical, political conceptions and the objectives that emanate from it, making them reality mainly through the military component.

After going deeper into the Trinitarian theory of Clausewitz, this thought can be represented as presented in Fig. 2.

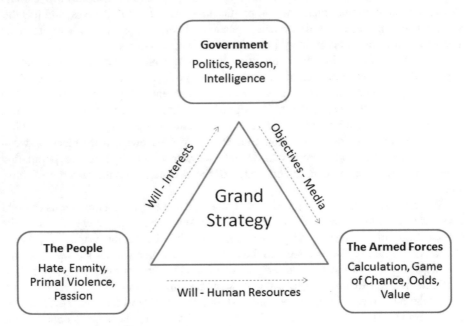

Fig. 2. Relation and scope of the elements of the Trinity of Clausewitz.

3   Planning and Development of Forces

The field of Force Planning is the other variable to be analyzed, in such a way of subsequently comparing and interpreting the relationship that exists with the Clausewitz theory. The definition of force planning that is considered the most appropriate for purposes of the study, says: "It is the process of assessing the security needs of a nation, establishing the military requirements arising from those needs and selecting, with the restrictions of resources, military forces to meet such requirements" [3].

It is understood that the Armed Forces must have the naval, air and army means that satisfy the achievement of the objectives that the policy determined based on the needs of national security.

To assent that this process is based on the art of strategy, John Lewis Gaddis is quoted "By strategy I mean simply the process by which the ends are related to the media, intentions to the capabilities and objectives to means" [4].

It is highlighted, then, that the definition of force planning speaks of national security, national objectives (referring to ends in terms of strategy) and means, thus identifying the components of the highest level of leadership, the National Strategy, and when dealing with the national objectives to be obtained by the armed forces, we speak of the subordinated Defense Strategy and Military Strategy.

From the referenced document of Bartlett et al., it is extracted that other ingredients are added to those already mentioned, in order to expose an integral approach on the strategy and the planning of forces, these are: the security environment, the risks or threats and limited resources. The so-called "key variables" and their relationship are represented graphically, according to the author of this conception, called "Bartlett's Model" [3] (Fig. 3).

Fig. 3. Bartlett model for force planning (Source: [5]).

In a simple way you can differentiate five steps in the planning of forces, namely: (1) Specification of specific objectives or interests to promote; (2) Evaluation of the threat in terms of capabilities, intentions, vulnerabilities and opportunities; (3) Formulation of the strategy, choosing one or a combination of several of the multiple possibilities that exist; (4) Allocation of resources in terms of financial resources and means to materialize the chosen strategy; and, (5) Reconciling objectives with media [5].

The process of force development is dynamic and continuous, since it feeds back, by re-appreciating the key variables, in such a way that the variation of one of them leads to the revision of the others, determining their impact in dependence on the eventual importance that is attached to each of the elements. Then, methodologies are evidenced that according to the modification of the ends, the modification of the means, the modification of strategy or the re-evaluation of the risks, are focused in accordance with the emphasis that is given to the key variable. Thus, we have approaches: descending, ascending, by scenarios, fiscal, by existing capacity, by threats and vulnerabilities, among others, Bartlett et al. [3].

The Chilean Navy Frigate Captain Mauricio Arenas [6] represented, as the force planning process is, the graphic representation I have allowed myself to modify, with the purpose of simplifying the steps, which I consider pertinent to the field to which the present study is circumscribed.

Figure 4, when referring to national interests, it is considered that they are the ones that define the basic and fundamental needs of the nation, representing the will of the population. The national objectives are those specific achievements that the nation seeks to materialize, in order to promote, support or defend their national interests [7].

Fig. 4. Simplified force planning process (Adapted from [7]).

The national objectives are those that the State establishes as essential, for the preservation of interests that are considered vital or important for the community. These objectives usually say about the welfare of the national community, security, balance and prosperity.

4 Validity and Incidence of the Clausewitz Trinity in the Development of Naval Power

Clausewitz philosophized about the war and defined the elements that compose it, finally shaping his Trinitarian theory, but it must be emphasized that his thought was developed in the first decades of the nineteenth century, based mainly on what was raised during the Napoleonic wars, that took place mostly, in the first fifteen years of the same century. I consider that to determine the validity of his theory, it is necessary to transfer what was understood by war in those times, which in synthesis was a warlike confrontation between regular military forces of two or more States; to what could be described as war in the current era.

In order to analyze the validity of Clausewitz's theory, the following are the criteria of some writers that reject their applicability at present. For example, Delmas [8] states that "In the current international system, there may also be situations of rupture of the trinity, by loss of state control in the relationship government - people, and government - army"; and, Van Creveld [9] referring to the wars that are not fought against another regular army, says that "... the Clausewitzian world figure is obsolete or incorrect, since it does not allow to understand the modern war and especially the conflicts of low intensity between and with non-state actors..." These types of war, where an element does not exist, are known as "non-Trinitarian wars", as stated by Fernando Thauby [10]. In this context, it is cited:

> "... Scarcely are relations of Clausewitz's vision of war, with the postmodern definitions of war as unrestricted; but if you intend to apply this thought to unrestricted war, the error is not Clausewitz, but those who do not know the nature of conflicts and have not understood its philosophy and historical context, being able to point out that Clausewitz's theory has not lost validity, the problem is that it has taken a more ethereal, which requires rethinking Clausewitz paradigm to address new forms of violence and not deny the basis of war and modern Western strategy, but to make it more flexible in its understanding and application in the military or academic environment" [11].

We must note that "The Trinity of Clausewitz" is a theory that covers a broad spectrum, which depends on the approach towards which the person who uses the postulate and the understanding of what war is for himself, conceiving each conflict as a particular situation, in which the three aspects (People, Government and Armed Forces) are amalgamated in different tonalities.

Another reasoning, states that "Clausewitz's Trinity" is only present in democratic states, where the "people" factor influences and controls government decisions, demanding to be informed timely, truthful and in full, reserving the right to change their opinion and withdraw your support, depending on the state of your emotions; For its part, the "government" factor would be obtaining greater representation with the support of society, but it must take care that its decisions are unpopular; and, the factor "armed forces" through the fulfillment of military operations is evaluated by public opinion, with pressure from the policy. The preeminence of the "people" factor stands out in this type of State [11].

In contrast, it can be deduced that in the non-democratic states, the "people" factor does not have an active participation in the decision making of the "government" factor, and it seems that one of the elements, the main one for Clausewitz, disappears. Thus, it is considered that the detachment of the element "people", in the formulation of the objectives that the policy establishes, is temporary, because when the policy is executed through the application of force, the consequences of the actions fall significantly on war, is the population of the State, which is the majority of society.

Therefore, in non-democratic states, the need to maintain the support of public opinion, creating dependence on it, is also present for the success of the other two factors, revealing the Trinitarian principle. On this (that the aspect "people" apparently is not present, Clausewitz [1] maintained that "... even when a conflict began without that emotional basis, its development would necessarily influence this plane".

It is noted that his theory had two times, planning and execution, and that in non-democratic states, the population or society may not influence the process of preparing the strategy, but finally, during the development of the actions of application of military power, highlight its presence, product of what your emotional state makes you solve.

An example that allows us to visualize how "the Trinity of Clausewitz" works in the current era, is the terrorist act against the twin towers, on September 11, 2001 in New York, United States. The effects after the attack led to the war against terrorism, its ingredients place the conflict in an environment of low intensity and asymmetric characteristics. This war against an Islamist group, had its first outbreak of emotions, synthesized in hatred, passion and impulses of violence, in the American population, which manifested its will to the government and supported the decisions that it took; the state policy transformed these needs into political objectives of the war, and its Armed Forces committed all their effort to combat the asymmetric threat and achieve the strategic objectives; the Armed Institution received the moral support of its people.

From the time of Clausewitz to the present times, the actors that threaten the interests or national objectives of a State have varied the order of appearance and priority, among which irregular armed groups are added to the regular forces of another country. Be these revolutionaries, guerrillas, insurgents, etc., which give rise to revolutionary war, asymmetric, etc. If it is a question of combating the so-called "new threats" such as: the smuggling of arms or merchandise, illegal migration, drug trafficking, the illegal exploitation of living or non-living resources and other illicit activities that undermine state interests, the armed forces they would operate against the actors that materialize these activities.

So we have, that the military power of a country, if the government decides as a representative of the will of the population, could be used to combat, everything that threatens national objectives. Therefore, for purposes of the study that is being carried out, it is proposed to understand as a war any expression of conflict or combat to irregular forces or groups and the "new threats", in which the regular armed forces belonging to a State, take participation as an "Instrument of Politics" (qualified by Clausewitz), in defense, protection or scope of the national objectives set, for the good and development of the society to which they are owed.

As stated earlier, Clausewitz, when writing his work, intended to develop a theory about war that would have permanent validity. It was not about giving formulas to win war, but tools for the analysis of each particular war, helping to develop the thought of the leaders and commanders who decide and make war. This is accentuated by his phrase: "… the first act of judgment, the most important and decisive one incumbent upon a statesman and the general in chief […] is to know the war he is undertaking" [1].

In short, the validity of "The Trinity of Clausewitz" depends on the approach for which it is particularized, being the reason for this analysis the conceptual application of trinitarian thinking in the field of strategy, noting its impact on the development of naval power.

After analyzing the validity of the Clausewitz trinity, and noticing its flexible nature to adapt to the variety that the war typology contains, then, more than a transfer in time of the Trinitarian concept, the observation of the correlation will be made that exists between the component elements of the two concepts previously described,

"The Trinity of Clausewitz" and the Planning of Naval Forces. It is understood, as stated in previous sections, that the two concepts are immersed in the field of strategy, this being the common spectrum that allows correlation.

Society appreciates the real or ideal value that it gives to an object, such as: living and non-living resources or sovereignty. When that intrinsic value is treasured, it becomes interest. It is the people who determine the set of national interests, which are the simple translation of the will of the population of a nation.

National objectives are the transformation of national interests, which are expressed by adding a verb, according to what you want to do with national interests, for example: protect living and non-living resources at sea or ensure territorial sovereignty; This task corresponds to the policy, thus obtaining what you want to do. To this is added the way the government plans to do it, which is nothing more than the determination of the national strategy. The responsibility for these steps corresponds to the National Government.

The political objectives related to defense will be reached by military power, through the establishment of military strategy. If it is the naval component, it will be done through the Maritime Strategy derivative. The naval media are the ones that materialize the strategic capacities, through the operational concepts that determine the types of naval operations to execute for the fulfillment of the tasks. This part of the process corresponds to the Armed Forces jointly and to the National Navy in relation to the naval component.

To reaffirm this correlation between the Trinitarian postulate of Clausewitz and what defines the development of naval power, Pertusius expressed: "The Navy is in the seas what the army on earth and together with the people and the government constitutes the amalgam of the national will in the maritime and the fluvial" [12] (Fig. 5).

The Argentine Rear Admiral Roberto Pertusio, carried out a study called "Design of a possible Navy for the Argentine Republic for the next 25 years" in which he carried out an analysis of the incidence of Clausewitz's theory in the design of naval power. An interpretation of your study will then be made.

The incidence, to personal consideration is approached from the theoretical point of view, according to what Ballesteros says [13] "The strategy is an art, in which the theoretical knowledge, the experience and the intelligence of the person have great importance. He develops it."

As mentioned above, the Clausewitz Trinity is always present whatever the conflict in question and one or more components of military power intervene. The current times impose important changes in all the activities and organizations of a State, and the Armed Forces are not free to have to assimilate such changes. The external and internal order has presented different threats to those that guided in the past, the design of the military instrument. The latter has depended to strengthen its power of the country's economic capacity, but the military power is the one that in turn supports national policy and development. The threats materialize in a broad spectrum and take advantage of the vulnerabilities presented by state institutions [12].

The design and development of a naval force is directly related to what the constituents in a democratic state, the population, demands from their navy, being the naval power the safeguard of their interests in the maritime field. These needs must be in accordance with the naval means available, in such a way that the objectives to be

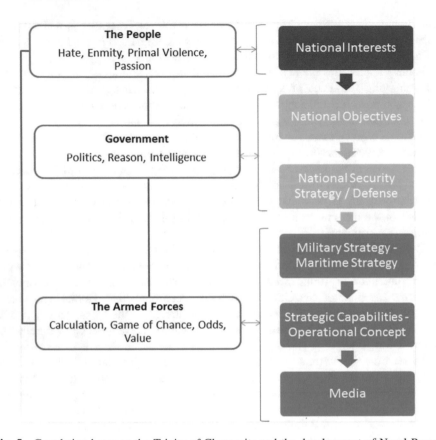

Fig. 5. Correlation between the Trinity of Clausewitz and the development of Naval Power.

obtained by the naval component are achievable. The National Government is the mediator of the two parties.

People and government must perceive that they have adequate and balanced naval power, capable of facing, with possibilities of success, the threats that could be presented to them at sea. This context is related to what was expressed by Liddell Hart: "The military objective should be guided by the political objective, but according to a basic condition, politics should not demand from the military what is impossible to fulfill" [14].

The people show their will to their representatives in terms of interests (maritime field); the democratic government through politics and strategy agrees with the naval power to achieve the objectives that correspond to it, the Navy plans the "how to do it" through the maritime strategy, which plans the development of the strategic capabilities of the naval power that leads to produce the desired effects. The main condition is the availability of naval means, which corresponds to the National Government to provide, justifying these requirements to society.

The above, is in other words 'The Trinity of Clausewitz', which shows that the theory under study has a positive impact on the development of naval power, and that through its contextual applicability can be promoted to generate the programs and projects that strengthen the naval power, according to the multiple demands, of the current and future security environment.

Finally, the strategic conception of Clausewitz, tells the members of the Ecuadorian Navy, that it is important to strengthen the maritime consciousness of the Ecuadorian population, making them notice the wealth that comes from the sea and the need to ensure its protection and control. Of these awareness-raising actions, the national interests referred to the maritime environment, would improve their position in the scale of priorities. The government would positively modify its ocean political vision and consequently the naval power would develop in a way, if not optimal, satisfactory.

5   Conclusions

Immersed in the field of phenomenological abstraction, the strategic theory conceptualized by Karl von Clausewitz, allows us to understand that war is of a political and social nature, that in it, besides physical forces, moral forces intervene, and where they coexist forming a trinity, the people, the government and its armed forces, instituting that war as a whole, is a conflict of social life of great interests.

The validity of the Trinitarian theory of Clausewitz, which is based on flexibility and adaptability to the varied typology of war, allows correlating with the process of force planning the component factors of the trinity, highlighting the interdependence of the concepts in the design of the Naval Power.

The contextual application of the Trinity of Clausewitz, which is configured in the spectrum of strategic thinking of the rulers and military power of a democratic regime affects the design of a balanced naval power and coherent with current and future demands.

The maritime awareness of the population of a State, as the main identifying element of the importance of the maritime interests of the nation, fosters the government's ocean political vision in order to have a naval power with adequate strategic capabilities to guarantee the use of the sea for the benefit of the country.

References

1. Von Clausewitz, C.: On War, ed. and trans. (1832) Michael Howard and Peter Paret. Princeton University Press, Princeton (1975)
2. Earle, E.: Makers of Strategy. Princeton University Press, Princeton (1971)
3. Bartlett, H., Holman, P., Somes, T.: The Art of Strategy an Force Planning. Naval War College Press, Newport (2004)
4. Gaddis, J.: Strategies of Containment. Oxford University Press, Oxford (1982)
5. Lloyd, R.: Fundamentals of Force Planning: Defense Planning Cases, vol. 2. Naval War College Press, Newport (1990)
6. Collins, J.: Defense planning steps. In: Fundamentals of Force Planning Faculty, pp. 143–148 (1991)
7. Arenas, M.: Planificación de Fuerzas y La Estrategia Marítima. Revista de Marina (2013)
8. Collins, J.: Grand Strategy, vol. 2–3. Naval Institute Press, Annapolis (1974)

9. Delmas, P.: El Brillante Porvenir de la Guerra. Andrés Bello, Santiago de Chile (1996)
10. Creveld, M.: La Transformación de la Guerra. Ejército Argentino (1991)
11. Thauby, F.: Escenarios Bélicos Futuros - La "extraña trinidad de Clausewitz". Revista de Marina de Chile (1997)
12. De Pablo, M.: La Guerra Irrestricta Un nuevo modo de hacer la Guerra? Memorial del Ejército (2013)
13. Pertusio, R.: Diseño de una Armada posible para la República Argentina para los próximos 25 años. Centro Naval (2005)
14. Ballesteros, M.: Fundamentos de la Estrategia para el siglo XXI. Las Estrategias de Seguridad y Defensa. Monografías del CentroSuperior de Estudios de la Defensa Nacional, vol. 24 (2003)
15. Hart, L.: Strategy, the Indirect Approach. Faber & Faber, London (1954)

A Maritime Vision for Geopolitics

Galo Ricardo Andrade[1]([✉]) and Datzania Villao[2]

[1] Armada del Ecuador, Guayquil, Ecuador
gandradedaza@hotmail.com
[2] Universidad Estatal Península de Santa Elena – UPSE, La Libertad, Ecuador
datzaniavillao@gmail.com

Abstract. The current research, analyses the importance of a maritime vision of geopolitics for the South American coastal countries. A bibliographical research was conducted which included national, regional and international authors, that served to evidence the way geopolitics was originally conceived, its evolution and the need for changes in its approach. It was analyzed, the threats that States would face in the future, such as the lack of food, energy and water, as well as the existence of maritime interests that could supply them. The importance of the maritime vision of geopolitics focuses on the conjugation of elements mentioned before and allows determining that given the future conditions, this geopolitical approach is fundamental for the States, in which the maritime consciousness is one of the essential elements for its construction. In order to reach the different levels of maritime awareness, it is necessary a system involving all actors linked to the maritime sphere, to coordinate and plan all actions leading to achieve that maritime interests are appropriately sized and constituted in the essence of a national development with international vision.

Keywords: Geopolitics · Maritime interests · Future

1 Introduction

Faced with the great challenges posed by the future, States must structure their geopolitics adequately. In addition, a maritime vision is not only necessary but essential for the South American coastal States.

Different authors have proposed several ways to define Geopolitics, but in social sciences, there is no unanimity of criteria of the definition. However, it is clear the need to analyze and rethink the perspectives used, taking into account the changes taken place in the world.

In this sense, several studies determine that, in the future of humanity, the lack of food, water and energy are factors that will be presented with intensity. These, together with the increase of the population in the world, make a complex scenario, against which States must explore various alternatives to plan their Geopolitics.

These alternatives must be based on basic aspects of geopolitics, such as time and space. Over time the only thing that can be controlled is the determination of the appropriate moments to implement state actions based on the analysis of the space factor, which, from the perspective set out in this work, must consider the sea as a

© Springer International Publishing AG, part of Springer Nature 2018
Á. Rocha and T. Guarda (Eds.): MICRADS 2018, SIST 94, pp. 235–243, 2018.
https://doi.org/10.1007/978-3-319-78605-6_19

substantial element, given the interests that keeps in its interior and the lack of basic elements for the human being as a constant factor in the future.

This article starts describing the origin of geopolitics to determine its essence as a previous step to the visualization of the approach that different authors have given over time. In this part, it is established the link between maritime vision and geopolitics. Then, the study continues with the analysis of maritime interests, which make the sea, the sustenance of humanity and necessary to shape geopolitics with maritime vision, which will be developed as a final part of the paper.

The current work aims to determine the importance of the maritime vision of geopolitics for countries of the region and the influence of the maritime awareness in its construction.

2   The Geopolitics

2.1   Geopolitics Origin

Different publications and studies about geopolitics show how difficult it is to find an exact definition, which makes it is analyzed and described from several perspectives. First, the origin of geopolitics will be explained in a general way, as a preamble to the analysis of some of the main elements that different authors have included in their way of seeing this discipline.

In its origins, geopolitics was conceived as a useful tool to analyze States and their international relations in function of the physical space. This conception that could be considered the essence of geopolitics, allows making a first approach to the importance of the maritime field in the geopolitical context of coastal States.

Sven Holdar (cited by O'Loughling, 1994), the Swedish Rudolf Kjellén was the one who initially defined geopolitics in 1899, in the Swedish geographic magazine Ymer, influenced by Friedrich Ratzel. In this first approach, Kjellén does not define the geopolitical term, but indicates that he prefers its use as anthropography, when analyzing the political implications of the borders [13]. His studies deepened the subject and were later published as Staten som Lifsform (The State as a way of life) (1916: 94), where he visualized a State with characteristics as dynamic as those of living beings. Finally, the compilation of his writings translated into German was published as Grundriß zu einem System der Politik (Scheme of a Policy System) in 1920.

Explained in a general way, it can be determined that Kjellén proposes a system to analyse the State through five categories of analysis: geopolitical, ethnopolitical, econopolitical, sociopolitical and political management regime. According to Sven Holdar, cited by O'Loughling (1994) the names in their original language are: "Geopolitik, Etnopolitik, Ekonomipolik, Sociopolitik, Regemenspolitik" [13].

These five categories, of which the last three correspond to cultural views of the State, are subdivided into fourteen, from which, three are related to geopolitics: topopolitics, morphopolitics, physiopolitics. However, for Sven Holdar, cited by O'Loughling (1994) in their original language are: "topopolitics, morphopolitics and physiopolitics" (1916: 94). With these three approaches, States should be analysed according to their position with respect to other States and needs. This analysis included both the ideal form that a State should have from the physical point of view as well as its extension and characteristics [13].

According to Holdar, this idea originally visualized by Kjellén (1916) as a form of state analysis, was misrepresented and taken as a basis by Karl Haushofer to develop his concept of geopolitics from the perspective of vital space for German development [13]. The rest of the story is better known, and the use of the concept by the Nazis in World War II caused it to be discredited. At the time of the Cold War, mainly in the 1980s, this discipline reappears with important contributions from France and the United Kingdom, and was called critical geopolitics

Moncayo [12] points out that the end of the Cold War brought with it a sense of stability that generated the tendency to consider the end of geopolitics. The time and the emergence of different types of threats showed the wrongness of this perspective and the need to rethink all the schemes in function of a new world dynamics. Geopolitics is not exempt from this revision, an approach with maritime vision could be fundamental for coastal countries, especially South American countries.

2.2 Important Aspects About Definitions and a Necessary Revision of Geopolitic Approach

This work does not intend to establish a geopolitical definition, but to determine important aspects in some of the existing definitions, and the most adequate perspective to link geopolitics with the sea. In this sense, Moncayo [12] determines that there are common points in the conceptualization of both classical and postmodern writers.

The general points of the classic writers presented by Moncayo [12], posit that there is an approach focused on the relationship between geography, as a space, and its influence on the conduct of international politics, as a planning element for the security and development of States. On the other hand, critical or postmodernist writers observe geopolitics in a less orderly and more subjective way, oriented towards the construction of a world politics from the understanding of the existence of elements such as the hegemonic states and the world conflict. However, for the two groups of authors, to a greater or lesser extent, the understanding and analysis of the geographical space in a historical, current and/or future environment constitute the common denominator of their definitions.

It is evident that at the present time space, in itself, must be dimensioned beyond the traditional conceptions of land, sea and air, reaching even to consider by many authors, cyberspace as a dimension of modern analysis. Nevertheless, the following section aim to determine the reason why, for some countries like the South American riverside, the maritime perspective is important for their geopolitics.

One of the questions that arise when analysing this topic is: Why has not the sea been given the importance it should, despite having so many resources, interests and influence? Possibly one of the answers is found in the words of Bernal [3], who during his historical research on the Pacific Ocean, determined that this sea has been taken into account little in the Mexican historical development and expresses "history is the study of man's happening and in the sea there are no men. The sea itself is anti-historical, because it is immutable in its essence and form" (p. 28). From this perspective, the sea is seen only as an element "through which" is carried; marketed, transported and different civilizations are communicated. That means, it prevails the conception of the sea as a space in which there are countless maritime routes, so known by marines and so important for the "use of the sea", a substantial element of the maritime strategy.

Deepening in this aspect, the surface of that great space called planet Earth, is composed by 149 million km^2, which means that only 29.22% of the terrestrial surface, (Weihaupt [14]: 47); and the remaining 70.78% corresponds to maritime spaces. From this traditional approach, the sea is a large space that must be crossed to get from one place on land to another, to such an extent that even the importance of maritime routes is related to the cargo and the type of transport they carry the ships that materialize them. With this vision Mackinder named as the pivotal area and then Heartland, the center "heart" of a great "World Island" formed by Europe, Asia and Africa, where the control of the world centered in the domain of the Heartland by the concentration of resources in this "world island" [13].

Although it seems contradictory, even authors such as Mahan, who always emphasized the importance of maritime power to counteract the expansion of the Heartland [13] analyzed in detail, focus the value of the sea on its ability to project land power, rather than in their same maritime interests. The study of maritime interests at the time the author pointed out the last, should be deepened in order to confirm the validity of this assertion.

This reflection is useful to establish the necessary revision and deep analysis that the change of time requires. This correct vision for its time could lose momentum now and especially in the future. As it was indicated before, one of the components of geopolitics is time, and there is no doubt that there is a world in which the only constant is the vertiginous change of the different variables. Therefore, the way of appreciating geopolitics, especially for the South American States should be revised. The analysis of the new circumstances in which humanity develops, especially establishing the nature of the main threats that States will face now and in the future, is the main element to raise this new vision of geopolitics in which the sea is a transcendental element.

3   The Sea in the Future of Humanity

3.1   The Threats in the Future of Humanity

Some prospective studies determine that the essential resources for humanity related to energy, food and water will be increasingly scarce on the planet. At the same time, it is clear that the demand for these resources will gradually increase with the population increase as well and, consequently, this need will be increasingly determinant in geopolitics.

From the perspective of polemology, study of war as a social phenomenon aimed at understanding the conflicts that trigger it and how to prevent them, the condition described above is ideal for the emergence of conflicts, so undoubtedly the analysis of these factors must be permanent in the institutes and strategic education institutions of each country.

The States of the region are not far from this reality because they belong to an international community that must anticipate responses to future challenges, responses to new conditions generated by the emergence of new ways of attacking security in general. This situation, that gives the current evidence, will surely be the trend of the coming years.

Some prospective studies even consider how hard for the States is to act in front of these new conditions in an independent way. The study of the Millennium Project Latin America 2030 research program, Delphi Study and Scenarios, prepared by Alegría et al. (2012), with the collaboration of 800 specialists from 70 countries, concluded that there should be common responses to common problems [1]. Under the same approach, the conclusions of the "Global Risk Report 2017: executive summary, of the World Economic Forum, highlighted the key challenges that the world is currently facing. One of them is to protect and strengthen the systems of global cooperation in the face of environmental risks, among them the lack of water.

Another challenge facing humanity is the production of food. For Friedrich (2014), the current production satisfies in general terms the demand of the population. However, due to the population increase, in the next 30 years this production will have to be doubled, situation that faces an obvious fact "There is no more land available, and agricultural land is constantly lost for other uses [9]. This perspective is important because it brings together two crucial elements for the future of humanity, the lack of food and the need for energy. The latter is even influencing the former by using cultivation fields for the generation of forms of energy such as biofuels that "have increased by 500% between 2001 and 2011 [9].

From the energy point of view it cannot be ignored that the population, according to Amell Arrieta (2014), will increase to eight billion people by 2030, which means that energy demand will surpass 35% of the energy required in 2005, which represents a challenge for the States that will have to respond to this demand [2].

It can be determined that there are constants in the near future of humanity such as the increase in population, the difficulty of satisfying their energy and water needs, as well as the saturation of the earth to extract the necessary resources that meet the needs of the population. It is precisely due to this situation that urges to think that the sea could be constituted for the coastal States in the sustenance of their future, especially for the South Americans. Therefore, it is not only possible but necessary to act and face in an integrated manner to face any threat to their interests, in order to improve their security conditions [5].

3.2   The Sea, Sustenance of Humanity

If the above perspective is taken into account, it becomes evident that the sea in the future could increase its value as a source of resources. Therefore, countries faced with the pressure to meet the energy, food and water needs of their citizens, will not only need to increase the protection of their aquatic spaces and specifically their maritime interests, but also to determine those spaces in which their fleets can extract resources safely, beyond their borders. That is why, it should not be ruled out that large fishing or extractive fleets in the future sail with the protection of their navies.

In this scenario, to think of an increase in disputes over the resources of the sea is not foolish. Therefore, it is necessary to raise awareness about the importance of the sea for the future of all coastal States, where see the ocean not only as a space through which maritime routes circulate, but with a vision that inserts maritime interests in state geopolitics.

3.3   State Protection of Maritime Interests

The importance of the sea lies in maritime interests, for which there are different definitions of its meaning and scope. Brousset [4] points out that maritime interests are the expression of the Nation's collective desire to use the "maritime environment" and take advantage of its resources, through the development of activities in the political, economic, legal, scientific, cultural and other fields, in order to contribute to a per- manent national welfare and certain aspects of security, and strengthen the capacity of the State (1998: 18). On the other hand, Codina defines them as the "set of benefits of a political, economic, social and military nature obtained by a nation from all activities related to the use of the sea" [6].

The maritime interests are constituted by practically all the activities that take place in the sea that are beneficial for the State. In a proposal of curricular insertion of the subject of maritime awareness for basic education in Ecuador, carried out by the Marine Environmental Education Program PEAMCO of the Navy, the following constituent elements of these maritime interests were determined [8]:

- Geo-maritime field.
- Merchant navy.
- Scientific and technological infrastructure.
- International organizations, treaties and agreements.
- Mineral resources.
- Fishing resources.
- Tourism.

MARITIME CONSCIOUSNESS. (Capital letters are part of the original document and are preserved with the intention of maintaining its essence, expressing the importance of maritime awareness).

These elements that make up the maritime interests are interesting because they give a real perspective of everything that is at stake when it comes to the sea. However, two components must be highlighted: the maritime consciousness and the geo-maritime field.

The maritime consciousness influences political will, and according to Liger (2012), it is the way in which citizens perceive the need for optimal and rational use of the sea [10]. In this sense, without maritime awareness, the proposed maritime vision would not exist. However, more than an element of maritime interests, maritime awareness must be seen as an essential condition for achieving the state's objectives in the maritime field, and must be forged in the citizen since childhood, so the national education system also is part of the actors linked to the process.

For its part, the geo-maritime field is understood as the set of maritime features that distinguish a country, derived from its condition, geographical environment and how it has been used by the human being. Moloeznik posits that "it comes given the geographical situation, the physical configuration and the merchant marine (commercial and sports fishing)" [11]. In other words, it is related to the geographic space of management of the coastal State and it is important to visualize the real scope of maritime interests, which can go beyond the maritime limits that structure its maritime territory according to its maritime power and capacity to project its influence.

If at this point of the analysis the elements considered so far are combined, there is a future scenario characterized by the lack of basic elements such as water, food and energy. At the same time it is important because as a large maritime space with essential interests for the States are constructed, maritime vision of geopolitics becomes a current necessity for the future of the coastal countries. Specially, the South American perspective should consider a regional maritime strategy in the future, an aspect that could be the object of future research in the centers and institutes of strategic and maritime studies.

4 Shaping Geopolitics with a Maritime Vision

The construction of this maritime vision of geopolitics still has a long way to go, or better expressed a long journey to navigate. Although it seems only a question of form, this assessment serves to understand that the key to shape the maritime vision of geopolitics lies in the change of mentality, in transforming national thinking and orienting it towards the sea.

Most of the countries of the region, to a greater or lesser extent, have carried out work in this regard through their respective navies. At this point, it is essential to note that the mere fact that the navies are the institutions that promote these processes shows the deficient maritime vision of state geopolitics. In other words, the navy should be a component, possibly the main component, of a comprehensive system that involves all the state actors merged and directed by the guidelines established by the state authority.

The Brazilian case is interesting because of the positive results obtained. Under the concept of Blue Amazon has managed to reach important levels of maritime awareness that, among other aspects, is evident in the strengthening of its Armed Forces, to guarantee the free use of the sea. According to Diégues (2010), the Brazilian perspective of security and defense in the South Atlantic considers that the activities and interests of Brazil in this sector contribute in such a way to the economic and social development of the country that it is necessary to consider all the internal and external factors that may affect them [7]. This demonstrates the essence of the maritime vision of geopolitics to focus on the national interest in the use of the sea.

However, to put in practice this experience, which is still in the process of consolidation, the Blue Amazon, that bio-diverse area with a large amount of energy resources that constitutes the great Brazilian heritage, should be evidenced from different perspectives that gradually matured in both understanding as in acceptance.

From the observation of this case and the analysis of the elements linked in this maritime vision, levels can be established in the construction process of this state maritime vision:

- The first level corresponds to the understanding of its great dimension. No having too much essence is important because it stops looking at the sea as a limit and it begins to consider it as a space of opportunities and projection. At this level, however, the idea focuses only on the great extent of this space, and the action of the actors involved in maritime development is not yet fully coordinated.
- A second level partially leaves aside the point of view of the dimensions of the sea and is focused from the perspective of the resources it possesses, the exports that circulate through its communications lines, the existing gas and oil reserves and discover, the minerals contained in its waters, the economic aspects, infrastructure, and everything that is understood as maritime interests, including its global projection capacity. Here all sectors that act in the maritime development of a country are clearly identified, which are coordinated by the state authority as well.
- The third point is the real understanding of the importance of maritime interests and the application of a maritime vision of geopolitics that analyses and prioritizes their proper management. To be at this level, a State will have clearly configured a system that integrates all the actors according to a maritime vision of its geopolitics and manages its actions. Finally, in order to achieve these levels, a long-term project is necessary, which structures the actions to obtain measurable achievements over time and which can be permanently re-provided. That is, to establish a system which primary function is to build this maritime vision. Until reaching the corresponding level of maritime awareness, the navies are called to act in favour of this national project, under a basic premise that points out that in the sea there are not only more square kilometres of extension but a great geopolitical space vital to guarantee the future of the citizens of a country and a region.

5   Conclusions

To sum up, it can be pointed out that the construction of this maritime vision of geopolitics requires changes in paradigms and consequently begins with changes in the way of thinking.

Due to maritime vision for geopolitics is very important for the countries, it is necessary public policies oriented in this sense, that involve all the actors that directly and indirectly influence and act in the maritime field - without neglecting the national education system, which is fundamental for shaping the thinking of the new generations - in such a way that, with a broad vision, we can understand the need for a geopolitics that, given the future threats that States will face, prioritizes maritime interests as the essence of its national development and international projection.

References

1. Alegría, R., Agreda, V., Artaza, M., Boix, J., Casanueva, H., Cordeiro, J., De Hoyos, A., Durán, E., Gutiérrez, M., Mojica, F., Olavarrieta, C., Ortega, F., Sosa, Y., Vitale, J., Zugasti, I.: Latinoamérica 2030: Estudio Delphi y Escenarios. Yurí Berríos, Santiago de Chile. mousecolors,cl. (2012). http://studylib.es/doc/2280543/lectura-de-referencia-latinoam%C3% A9rica-2030.pdf
2. Arrieta, A.A.: Panorama y futuro energético mundial. Revista Ingeniería y Sociedad (7), 26–29 (2014). http://aprendeenlinea.udea.edu.co/revistas/index.php/ingeso/article/view/18986/ 16207
3. Bernal, R.: El Gran Océano. Fondo de Cultura Económica, México D.F. (2012)
4. Brousset, J.: Los intereses marítimos del Perú. Una visión resumida. Revista Agenda Internacional 5(11), 15–22 (1998). http://revistas.pucp.edu.pe/index.php/agendainternacional/ issue/view/721
5. Bustos, M., Rodríguez, P.: La disuasión convencional, conceptos y vigencia. MAGO Editores, Santiago de Chile (2004)
6. Codina, R.: Visión de la Armada de Chile. Sobre los intereses marítimos y su contribución al desarrollo nacional. Revista de Marina 120(876), 1–11 (2003). http://revistamarina.cl/ revistas/2003/5/codina.pdf
7. Diégues, F.: O Atlantico sul na perspectiva brasileira de seguranza e defesa. Revista Marítima Brasileira 130, 23–32 (2010)
8. Ecuador. Programa de Manejo de Recursos Costeros: Propuesta de inserción curricular de la temática de conciencia marítima, manejo costero integrado y educación ambiental marino costera (2008). http://simce.ambiente.gob.ec/sites/default/files/documentos/geovanna/Libros %20Azules%20para%20el%20Programa%20de%20Educaci%C3%B3n%20Ambiental% 20Marino%20Costero%20PEAMCO.pdf
9. Friedrich, T.: La seguridad alimentaria: retos actuales. Revista Cubana de Ciencia Agrícola 48(4), 319–322 (2014). http://www.redalyc.org/pdf/1930/193033033001.pdf
10. Liger, J.: De la Geopolítica a la Oceanopolítica. Participación de la Fuerza Naval como componente del poder naval, en apoyo a la Oceanopolítica del Ecuador. Academia de Guerra Naval, Guayaquil (2012)
11. Moloeznik, M.: Hacia un marco teórico y analítico del poder naval. Contribución doctrinaria al desarrollo de la Armada de México. México y la Cuenca del Pacífico 12(35), 81–109 (2009). http://www.mexicoylacuencadelpacifico.cucsh.udg.mx/index.php/mc/issue/view/35
12. Moncayo, P.: Geopolítica Espacio y Poder. Universidad de Fuerzas Armadas ESPE, Sangolquí (2016)
13. O'Loughlin, J.: Dictionary of Geopolitics. [Diccionario de Geopolítica]. Greenwood, Westport (1994)
14. Weihaupt, J.: Exploración de los Océanos. Compañía editorial Continental, S.A., México, D. F. (1984)
15. World Economic Forum: Informe Riesgos Globales 2017: Resumen ejecutivo (2017). http:// deres.org.uy/wp-ontent/uploads/GRR17_Executive_Summary_Spanish.pdf

Ethical Evaluation of International Regimes and Challenges for Scientific-Technological Cooperation in the Global South

Ramiro Armijos[1(✉)], Teresa Guarda[2,3,4(✉)], and Luis Piedra[1]

[1] Armada del Ecuador, Guayaquil, Ecuador
r_armijos@hotmail.com, luispiedraaguirre@hotmail.com
[2] Universidad de las Fuerzas Armadas-ESPE, Sangolquí, Quito, Ecuador
tguarda@gmail.com
[3] Universidad Estatal Península de Santa Elena – UPSE, La Libertad, Ecuador
[4] Algoritmi Centre, Minho University, Guimarães, Portugal

Abstract. The current system of international regimes has made cooperation between States possible, appealing to rational self-interest and the compatibility of complementary interests, fundamentally in the field of world economy. However, these multilateral institutions have suffered serious ethical questions because the benefits they have produced for the global south have been marginal. As a result, these countries have sought to create their own institutional spaces to cooperate, especially in the field of science, technology and innovation (STI). This article attempts to recover the conceptual proposals of Professor Robert O. Keohane about the ethical value of international cooperation within the system of current regimes, extrapolating into the domains of international cooperation in STI to identify the biggest challenges for the global south; and, thus, try to understand the motivations and efforts that the countries of the South have made to achieve fairer institutional cooperation spaces based on South-South cooperation (SSC), under the aegis of the UN.

Keywords: International regimes · Utilitarianism · Theory of rights
South-South cooperation · Science and technology

1 Introduction

Through the emergence of the system of international regimes from the second post-war, and under the hegemonic influence of the winners, multilateral mechanisms and political-institutional spaces for international cooperation were established, which essentially appealed to the rational self-interest of the States in the achievement of their foreign policy objectives, especially related to issues of the world economy: financial, energy, commercial, monetary, among others. However, ethical questions have arisen about the practical results of the functioning of these institutions in terms of the distribution of their global benefits, especially towards the most vulnerable actors in the international system, many of whom belong to the geopolitical south.

Concomitantly, cooperation in science, technology and innovation (STI), as an area of international cooperation, has come to be used in recent decades as a fundamental

© Springer International Publishing AG, part of Springer Nature 2018
Á. Rocha and T. Guarda (Eds.): MICRADS 2018, SIST 94, pp. 244–253, 2018.
https://doi.org/10.1007/978-3-319-78605-6_20

instrument for the development of States, so that the global South, through South-South cooperation (SSC), has sought to establish more favorable institutional spaces for their development aspirations based on cooperation in STI.

This paper attempts to recover the conceptual proposals made by Professor Robert O. Keohane about the ethical value of cooperation within the current system of international regimes, essentially relating it to the distribution of its benefits. Professor Keohane put two theories face to face: the morality of states and cosmopolitanism, leaning for the evaluation of the latter in the conceptual positions of utilitarianism and Rawls' theory of rights. Keohane's ethical assessment of international institutions will serve as a trigger to understand: how the ethical and practical deficiencies of international regimes, extrapolated to the domain of STI cooperation in the global south, have fostered the search for institutional spaces of their own in the Southern countries, although these have been built, mostly, under the aegis of the UN.

2 Basic Ethical Considerations About the Origin and Functioning of International Regimes and Cooperation

Keohane (1988) in one of his emblematic works: "After Hegemony: Cooperation and Discord in the World Political Economy", he conducted an analysis, under a liberal institutionalist approach, of the international regimes created by the central powers after the Second World War. Thus, in the final proposal of his book, includes an assessment of the ethical value of cooperation that States have carried out within these regimes by appealing to two completely opposed theories, which are: the morality of States and cosmopolitan morality, whose fundamental difference is given by the establishment of the subject of the moral valuation that would be the State or the person, respectively [1].

In the first theory, the main concern revolves around state sovereignty, that is, the ability of the State to make autonomous decisions regarding its actions, which, for the specific case of the regimes, would be applied to the voluntary action of the States to cooperate with each other; however, this theory has no consideration of the consequences of distributive justice within cooperation. That is, under this approach a system of values would be assumed for the group of States that are part of the international system - within the realistic conception of rational state action - which would revolve around sovereignty and its autonomous action. Under this premise, those who support this theory would agree with international regimes and the institutional spaces they have generated for cooperation [1]. These international forums, for the case of an economic nature, would provide interaction channels to facilitate that the States, through a voluntary adhesion, manage the achievement of the objectives of their foreign policy. On the other hand, cosmopolitan morality would judge the individual in the moral evaluation of his actions and in the consequences -total or global- within international cooperation, since "… he denies that state borders have moral significance …" [1]. In fact, the author does not take a position on either of the two theories - although he does show some affinity for the second - emphasizing that the cosmopolitan approach always requires a more strict study of the global consequences of cooperation [1], and which state autonomy would go to a secondary level.

3   Cosmopolitan Utilitarianism and Rawls' Theory of Rights

Focusing the evaluation on international regimes and cooperation under a cosmopolitan moral approach, Keohane (1988) relies on the theoretical arguments of Professor John Rawls (1971), who established a theoretical alternative to utilitarianism, which he called *the theory of rights* [2].

Keohane (1988) mentions that Rawls, did not extend this theoretical proposal to International Relations, but that Charles Beitz (1975) does take up this challenge. Based on Beitz's arguments the author questions that if a person X "… would approve international regimes and the cooperation they imply without knowing their nationality or their position in the structure of their society" [1–3]. That is, applying, hypothetically, a *veil of ignorance* (Rawls 1971) to those who designed and implemented the international system of regimes.

On utilitarianism, one of the main criticisms is that social utility could be confused with a pragmatically entrepreneurial and administrative vision, since it seeks to maximize benefits to as many citizens as possible (Keohane 1988), at the expense of others who could be sacrificed [1].

On the other hand, with the theory of justice with equity, when considering a veil of ignorance, this does not allow a premeditated calculation of the final benefits [2], but the results are left at random to social evolution. In this way, benefits could not be calculated in advance - that could guide in a biased way the design of the institutions - and any inequality in positions and rewards would be oriented to the common benefit with free and equal access to those who make up the social group [4].

4   International Regimes, Cooperation and Their Ethical-Moral Deficiencies

Interdependence refers not only to an interconnection between States, but to a linkage that to a greater or lesser extent implies shared costs or benefits, even if these are sometimes unequal [1, 5]. The increase in global interdependence, especially since the 70s, has made cooperation between states more necessary under the stimulus of complementary interests [1].

In this regard, as the industrialized world experienced a boom in post- second World War cooperation, initiated with a Marshall Plan aimed primarily at the former European allies, this became the juncture for the birth of Foreign Aid for Development that later would evolve to what we know today as International Cooperation for Development [6]. This scenario took place under a hegemonic context of the USA, a country that promoted the creation of supranational institutions, which would provide political spaces for cooperation especially among later industrialized countries, to the extent that American hegemony deteriorated. Institutions survived by allowing post-hegemonic cooperation schemes helping to realize common interests of world politics [1]. On the other hand, the historical dynamics in the appearance of these institutional spaces were preceded during the first half of the 20th century by the emergence of the theory of rational choice, which produced a theoretical and methodological revolution in the social sciences, establishing that the intentional action is always guided by interests, be they of any kind [7]. Also, during those years, realistic currents of International Relations are consolidated, within which States act as unitary and rational actors motivated by self-interest.

This, however, would not inhibit them from cooperating with each other to obtain the best possible results; to which the game theory, through the prisoner's dilemma, would provide with some logical reasoning. Thus, the international regimes bequeathed during the post-war period would have become institutional political spaces within which States could find channels of dialogue and convergence to promote cooperation, motivated by the rationality of complementary interests and under a principle of reciprocity. However, from a cosmopolitan point of view these would have serious moral limitations, both in their origin and in their consequence, when applied to developing countries [1].

In order that results produced by the current international system were accepted on a global scale, under what Rawls (1999) calls a firm morality, all countries, including those in the process of development, must have, at least, participated in its design [4].

5 Challenges of International Cooperation in Science, Technology and Innovation

One of the challenges of cooperation in science, technology and innovation and, in its most basic version, technical cooperation (TC), has been primarily the difference in capabilities that may exist between cooperating countries.

With respect to cooperation in STI, Troyjo (2003) considers that it is essential that exist a certain equivalence, or at least minimum requirements in the capacities of the cooperators, in order to establish and achieve common goals in terms of knowledge transfer and technological innovations that bring economic benefits [8]. However, this does not always happen in an optimally way, since each State has a specific level of development that has been produced, in large part, by sustained economic growth, which, according to Schmokler (1966), is strongly determined by the national STI capabilities for invention and innovation [9].

On the other hand, TC in its native concept is carried out under an asymmetric scheme of donor-recipient that also requires an equalization of the technical-scientific capacity of cooperants, so that an adequate transfer of knowledge becomes possible [8].

According to Malacalza (2016), some authors have a more drastic vision and consider that STI cooperation is feasible only when it occurs horizontally and symmetrically [10]. From this affirmation, it can be deduced, that the cooperation in STI could hardly be concretized between an industrialized country and another one in the process of development; however, it must be taken into account that the symmetries and asymmetries could be heterogeneous depending on a specific STI thematic. For example, a subject such as information and communication technologies (ICT) encompasses a wide variety of topics and subtopics; therefore, two cooperating countries could present symmetries in certain topics or subtopics and asymmetries in others, regardless of whether cooperation takes place between developed or developing countries, or in a combination of these, that is, in a north-south axis.

On the other hand, an alternative that contributes to a viable cooperation in STI and TC to countries with less scientific-technological development is what is known as triangular cooperation. This consists, precisely, in the triangulation of cooperation in STI from an industrialized country, which includes the provision of financial resources, through a third country in the global south with a smaller breach in STI development, compared to the final recipient of cooperation [6]. In this way, the technological gap between direct cooperants and even the cultural distance could be reduced therefore facilitating cooperation in STI.

Although at present it could be considered utopian to speak of a total scientific-technological autarky, there are countries, especially from the global north, that decide to execute projects in STI individually since they have certain self-sufficiency in research and innovation of specific STI themes; they also do it for security, strategic or commercial considerations. However, when this is not the case, one of the essential drivers of cooperation in STI comes from the complementarity of capabilities; that is, it is assumed that a country could not achieve certain research and innovation objectives individually, either because of its complexity or its costs. This has led to the creation of multinational research networks, some of which are linked to large-scale research projects called "big science" [11] which is very common to observe in cooperation STI north-north. These research projects on a global scale, involving: participation of multinational research teams; significant investments in infrastructure and equipment; complex and long-term objectives, and others. One of these cases would be the project created for the deciphering of the human genome, which lasted more than a decade and integrated research institutes and universities from: USA, United Kingdom, Japan, France, Germany, Canada and China. The main theme of the project was ICT, with the specialized topics of Bioinformatics and Supercomputing.

Likewise, the impact that the STI has had on the security and development of the States has been so significant that it has transformed the international diplomatic agenda, to the point of linking STI cooperation with foreign policy. As a result, during the last decade there has been talk of the emergence of science diplomacy, as a reference to the fusion of these two very different elements: science and diplomacy [12]. However, science diplomacy, as a practice, is not a recent phenomenon of international politics, one of its most important historical evidences being the conformation of the European Community, in which it has been considered key because of the political nexus it helped to build among Europeans [13].

On the other hand, it is considered that in the last decades there has been a serious widening of the STI capacity gap between developed and developing countries, being this especially critical for certain countries in the global South (Latin America, Africa and Asia), which present important limitations in their capacities in STI. In this sense, diplomatic efforts have been unsuccessful, even in the context of multilateral organizations, since a specific commitment from the north would involve: expanding their concessions in terms of technology transfer and intellectual property, which would undermine the interests of their competitiveness [14].

The emerging economies of the BRICS (Brazil, Russia, India, China and South Africa) deserve special treatment, whose market size and STI capabilities would be very attractive for northern countries. This is corroborated by Flink and Schereiterer (2010) in their studies on the science diplomacy of: USA, United Kingdom, France, Switzerland,

Germany and Japan, for whom the BRICS would be among the priority target countries. According to this study, the most used bilateral diplomatic tools would be the International Science and Technology Agreements (ISTA), which integrate projects related to economic cooperation for development in the southern hemisphere [13].

But the asymmetries would not be the only challenge to face. Gaillard (1994) tells us that there has also been a dominance of the north, in reference to the power relations that would be reproduced within the cooperation in STI, which has resulted in the imposition of conditionality's on its southern counterparts that, in tacit or expressed, would be combined with foreign policy interests and with the northern research agendas [15]. With this agree Oregioni and López (2014), indicating that the framework agreements of donors would establish research agendas that do not respond to the realities and social and development needs priority for developing countries [16].

6 South-South Cooperation and the Search of Institutional Spaces for STI Development

The countries of the global South, among them Latin America, have imperatively required to improve their STI capabilities as a central strategy to face the challenges of their socio-economic development; and in this effort, international cooperation becomes an essential instrument. However, as indicated above, the conditionality's and asymmetries that could arise within a bilateral STI cooperation scheme between north and south have become part of the complex obstacles that have to be overcome.

On the other hand, appealing to the option of multilateral schemes has meant confronting international regimes that have been created to promote the interests of the central powers. In this regard, Keohane (1988) admits that the motivations of origin of these institutional spaces have been ethically questionable, although he affirms that their existence has produced benefits for world politics in general and, also, some benefit for developing countries [1]. That is to say, it supposes in some way, a maximin criterion for the countries of the South, within a set of interests that takes place in a scenario of high uncertainty like that of the international system [2]. Keohane (1988) considers it preferable to have an institutional space that channels the cooperation that could be perfected and not an international state that approaches what Hobbes defined as the "war of all against all" [1].

These limitations that the international institutions have presented in terms of a distribution of benefits, product of the moral shortcomings of their results rather than their origin, which should consider the special conditions of the global south, has motivated the construction of alternative cooperation systems, such as the South-South Cooperation (SSC), which has been mounted on the same institutional bases of the post-war period. Ayllón (2013), argues that the SSC should be considered a break in the aid industry, which has traditionally come from the global north combined, in large part, with ideological and hegemonic motivations, and that has been transformed into a counter-hegemonic subsystem based on principles of: equity, horizontality, respect for domestic policy issues, non-conditionality, mutual benefit, and others [6].

The SSC had its origins in an idea of the south, which emerged from the beginning of the Cold War, converging on a transcendental milestone for south-south relations in the Bandung Conference of 1955. The emerging geopolitical links between countries of the Global South would support, among other political actions, in technical cooperation (exchange of experts, granting of scholarships). Thus, due the spirit of Bandung, it was visualized the need to create a financial fund to boost TC, which would evolve with the creation of the Special United Nations Fund for Economic Development (SUNFED) [6].

During the 60s and 70s, this awareness of the South continued to grow, consolidating within the UN with the creation of the G77, deepening the debate on the need for a New World Economic Order to reduce gaps and economic asymmetries between North and South. In 1974, the unit for technical cooperation among developing countries was created within the United Nations Development Program; and, in 1978, during the Buenos Aires Action Plan, a total of 80 countries in the global south established a division of agendas between the TC (knowledge and expertise transfers) and economic cooperation (finance, trade and investment). Subsequently, in 1983, the Pérez Guerrero Fund was created for the TC within the framework of the G77, although it was born with scope and budget limitations. Also, between the late 1980s and the mid-1990s, some countries in the region created the first bureaucratic structures in accordance with this TC agenda.

These first efforts to develop institutional spaces that facilitate TC among countries of the global south did not have, in the later, milestones to be highlighted; but until 2005, when the Doha-Qatar Action Plan established by the G77 plus China created under the auspices of UNESCO, the International Center for Science, Technology and Innovation for South-South Cooperation (ISTIC). This center was established in Kuala Lumpur-Malaysia and has had as its main objective: to facilitate a development approach as part of the national STI policies of the southern countries; build STI capabilities by advising on these policies; share experiences and best practices; and, implement a network of centers of excellence among developing countries for problem solving and exchange of researchers, students, scientists, among others [17].

In this regard, it should be considered that only developing countries that have a minimum capacity of STI could make this type of cooperation feasible. Malacalza (2016), tells us that cooperation in STI, in the framework of SSC, occurs rather by exception since few southern countries really have the minimum STI capacities required for a cooperation [10]. Evidently within this "exceptional" group would be the BRICS and other countries of the south with relatively important STI developments - in the region: Mexico, Chile and Argentina. According to Ribeiro and Baiardi (2014), BRICS would also have, among them, STI asymmetric capacities, but there are many sectors in which they would be highly competitive, which would make cooperation viable [18]. They mention as an example the case of Brazil, in that it could lead research and innovation in topics related to plant and animal production [18].

We could say then that the spirit of Bandung has resurfaced, converting the SSC, within the specific field of the STI, into a techno-political response that has sought to tune in with the new geopolitical circumstances of multipolarity and progressively achieve the overcoming of the gaps scientific-technological between the different countries of the global south, creating these new institutional spaces on the traditional multilateral structures of the Organization of the United Nations. In this way, the expectation of being able to build institutional systems that are not solely motivated by the self-interest of the States has been generated; they also show an ethical consistency that combines with proposals such as those of Rawls (1971) and its principle of difference or justice, this despite the asymmetries of power.

Of course, before the emergence of the SSC the criticisms of the North have not been made wait, they speak of a supposed idyllic vision of the SSC, whose philosophy and forms of execution would place it in a position of Moral Superiority with respect to North-South Cooperation, as a consequence of the developmental character of that one [6].

On the other hand, a major challenge of South-South STI cooperation would be to avoid inheriting some of the institutional weaknesses that the SSC has shown up to now, related to a possible unilateral interpretation of the cooperators, especially donors, on the qualitative and quantitative aspects that would define the quality of cooperation. In this regard, Ayllón (2013) offers us an analysis in his book: South-South and Triangular Cooperation: Subversion or adaptation of international cooperation? whose title would suggest that the SSC would have the following paths to choose: the subvert the international cooperation scheme and transform it radically in favor of the global south, or, simply adapt it to the traditional schemes of cooperation -with its vices and shortcomings-; but in the field of south-south relations. The author makes a critique of the SSC referring precisely to the unilateral interpretation made by some cooperators in the south on the quantification of the same, without having a standardized form for evaluation.

The foregoing could lead to the establishment of whimsical STI cooperation typologies and, therefore, to a questioning of the true spirit of cooperation due to the deviations that other foreign policy interests of certain countries of the South could cause. In this way, despite the fact that the flaws of origin that we criticized from traditional international regimes would be corrected, the results in the distribution of benefits would continue to be ethically dubious.

For this, the application and measurement of these processes should have pre-established standards which consolidate an institutionalization of cooperation in STI, whose institutional value would not be given by the headquarters and material facilities they provide, but by a set of practices and expectations - fairer- that are generated around its operation [1].

An adequate institutionalization of south-south STI cooperation should preserve a spirit of Bandung that not only respects the sovereignty of the States, but also focuses on the priority needs of the research and innovation agendas of the less developed countries, and aim towards the creation and/or strengthening of STI capabilities that contribute to the real solution of the socio-economic problems of the global south.

7   Conclusions

In short, the results of traditional international economic regimes, even if they are morally deficient in their motivations of origin, are feasible alternatives to achieve an interaction between States, including cooperation in STI; this does not inhibit the need to review the principles on which they were built. The practical limitations of this institutional system, which was created by the central powers, has forced the global south to look for alternative institutional spaces, such as the CSS, on which the TC and cooperation in STI have been developing, although incipiently. The improvements of this institutional system will have to be gradual and cannot rule out a natural self-interest of the States. It is preferable that the global south seek to influence politically to modify them, in order to create greater spaces to achieve cooperation in STI that contributes to truly improve global distributive justice.

We leave open the question: Would there be the possibility that the asymmetries in STI cooperation, which also occur between countries of the global south, reproduce the same vices that we question the northern countries? With the aggravating circumstance that: it would be ethically questionable to play with the expectations of the most vulnerable countries and not achieve a better distribution of benefits according to what John Rawls proposed in his theory of justice.

References

1. Keohane, R.: Después de la Hegemonía. Cooperación y Discordia en la Política Económica Mundial. Grupo Editor Latinoamericano. Colección Estudios Internacionales (1988)
2. Rawls, J.: Teoría de la Justicia. The Belknap Press of Harvard University Press, Cambridge (1971)
3. Beitz, C.: Justice and International Relations. Philos. Public Aff. **4**(4), 360–389 (1975)
4. Rawls, J.: Justicia como equidad. In: Justicia como equidad. Revista Española de Control Externo. Tecnos, Madrid, pp. 129–58 (1999). Cap II del libro "Justicia como equidad"
5. Pelfini, A.: Posibilidades de la integración latinoamericana en condiciones de escasa interdependencia. Revista Signos Universitarios **53**, 69–82 (2017)
6. Ayllón, B.: La Cooperación Sur-Sur y Triangular. Subversión o adaptación de la cooperación internacional? Primera edn. IAEN, Quito (2013)
7. de la Rosa, G.V.: La Teoría de la Elección Racional en las ciencias sociales. Sociológica **23** (67), 221–236 (2008)
8. Troyjo, M.: Tecnología & Diplomacia. Desafios da Cooperacao Internacional no Campo Científico-Tecnológico. Aduaneiras, Sao Paulo (2003)
9. Schmookler, J.: Invention and Economic Growth. 1a ed. (1966)
10. Malacalza, B.: International Cooperation in Science and Technology: Concepts, Politics and Dynamics in the case of Argentine-Brazilian Nuclear Cooperation. Instituto de Relaciones Internacionales: Contexto Internacional (2016)
11. Wagner, C., Yesril, A., Hassel, S.: International Cooperation in Research and Development: An Update to an Inventory of U.S. Government Spending. Reporte, Science and Technology Policy Institute, Santa Monica (2001)
12. Turekian, V.C., Macindoe, S., Copeland, D., Davis, L.S., Patman, R.G., Pozza, M.: The Emergence of Science Diplomacy. In: Science Diplomacy. New Day or a False Dawn? pp. 3–24. University of Otago, Otago (2015)
13. Flink, T., Schereiterer, U.: Science diplomacy at the intersection of S&T policies and foreign affairs: toward a typology of national approaches. Sci. Public Policy **37**(9), 665–677 (2010)
14. Calestous, J., Cosmas, G., Allison, D., Audette, B.: Forging New Technology Alliances: The Role of South-South Cooperation. Cooperation South (2005)
15. Gaillard, J.: North-South research partnership: is collaboration possible between unequal partners? Knowledge and policy. Int. J. Knowl. Transf. Utilization **7**(2), 31–63 (1994)

16. Oregioni, M., López, M.: Cooperación Internacional en ciencia y tecnología. La voz de los investigadores. Rev. iberoam. cienc. tecnol. soc. **8**(22) (2014). http://www.scielo.org.ar/scielo.php?pid=S1850-00132014000100004&script=sci_arttext
17. ISTIC: International Science, Technology and Innovation Center for South-South Cooperation. https://istic-unesco.org/index.php/features/module-positions. Accessed 17 July 2017
18. Ribeiro, M., Baiardi, A.: Contexto Internacional **36**(2), 585–621 (2014)

Maritime Security and Safety

Comprehensive Security in Artisanal Fisheries: An Indedudible Commitment of the State and the Citizens

Emanuel Bohórquez[1](✉), Teresa Guarda[1,2,3], José Villao[1],
Linda Núñez[1], William Caiche[1], Roberto Lucas[1], and Silvia Renteria[4]

[1] Universidad Estatal Península de Santa Elena – UPSE, La Libertad, Ecuador
ema_bohorquez@hotmail.com, jvillaov@hotmail.com,
lnunez_ing@hotmail.com, robertauro2l@hotmail.com,
tguarda@gmail.com, caichewilliam@yahoo.com
[2] Universidad de las Fuerzas Armadas-ESPE, Sangolquí, Quito, Ecuador
[3] Algoritmi Centre, Minho University, Guimarães, Portugal
[4] Tecnológico Espíritu Santo, Guayaquil, Ecuador
sprenteria@tes.edu.ec

Abstract. This research focuses on integral security, being a subject of global interest that covers all the contexts and that depends mainly on two elements, as is the state and the society, for that reason, one investigates the current situation of integral security of the artisanal fishermen of the fishing ports of the canton of Salinas. For this study, a quantitative approach and descriptive research was used, concluding the existence of a low integral security that the artisanal fishermen have, specifically the human and social insecurity due to the piracy and the lack of social security originated by the lack of policies and support from state and local public institutions, and in the absence of ethical values and behaviors of society.

Keywords: Comprehensive security · Artisan fishing · State · Society
Artisanal fishermen

1 Introduction

Integral security is a topic of great interest to countries at a global level, increasingly important in the context of a society free from fears and needs, involving the individual and the state itself.

Human security report views violence as the key measure of human security, which does not mean that government and development issues do not matter; on the contrary, the causes of political and criminal violence are mainly a result of the lack of government capacity present in most poor countries [1] and the media and criminal logic that are also evolving, provoking new security demands, increasing the number of components to be taken into account, so that physical security can now be guaranteed; with globalization, new spaces emerge that make the action of the state scarce as far as social obligations are concerned [2].

Á. Rocha and T. Guarda (Eds.): MICRADS 2018, SIST 94, pp. 257–265, 2018.
https://doi.org/10.1007/978-3-319-78605-6_21

The problem of research results from the low integral security that artisanal fishermen have in carrying out their daily tasks, which involves not only human security but also social, legal, technical and economic security; aiming at identifying the current situation regarding the integral security of artisanal fishermen and proposing an integral security model for artisanal fishing.

This research is structured as follows: the second section presents the principles underlying integral security and artisanal fishing; the third section describes the methodology used. In the fourth section, the results obtained are presented using descriptive statistics. Finally, in the fifth section, the conclusions are presented with proposals aimed at promoting the increase of integral security.

2 Comprehensive Security and Artisanal Fishing

2.1 Comprehensive Security

The contributions of Gairín & Díaz under a globalizing conception indicate how this security must consider the human, social, legal, technical, and economic aspects of all risks that may harm the people participating in an organization [3].

It should be emphasized that maintaining the constitutional order is one of the main functions of the state, but this mission presents a great problem, due to a security established on the idea of protecting individuals from risks of all kinds, revealing the inability of the organizations intended in this regard, due to the continuous emergence of new threats, resulting in the mobilization of different levels of intervention and contribution [4].

In terms of human security, it is worth noting the United Nations Development Program's argument, first referred to in the Human Development Report for 1994, that human security originates from a combination of "freedom from fear" and "freedom of necessity". Noting that human security is composed of seven elements [5]:

– Economic Security: In the face of the threat of poverty.
– Food Security: Facing the threat of hunger.
– Health Security: In the face of the threat of injury and disease.
– Environmental Security: In the face of the threat of pollution, deterioration of environmental and depletion of resources.
– Personal Security: Faced with a threat that includes various forms of violence.
– Communal Security: In the face of the threat to the integrity of cultures.
– Political Security: In the face of the threat of political repression.

Mack describes human security as a condition or state characterized by the freedom of dominant threats about people's rights, about their tranquility and even about their lives [1].

Bauman says that the main contexts to satisfy the security need of this heterogeneous panorama are the following: the power to keep the property of the material elements achieved without fear; as well as the certainty that the individuals who make up the community know how to differentiate according to their customs and culture what is right and wrong in the daily decisions that must be taken, liberating the culture from fear; and finally, that when they do rightly they feel protected, and that no extreme danger threatens their integrity, that of their family and their belongings. These conditions allow the heart of political life to remain in a deep and insatiable yearning for security [6].

Thus, In a world where the protection of individuals becomes inescapable and the social is increasingly prevalent, it becomes imperative that approaches and security strategies become increasingly efficient, ensuring that citizens can think about realizing their life projects without being exposed to being victims of any crime [2].

In the Constitution of Ecuador, art. 393, states: "The State shall guarantee human security through integrated policies and actions to ensure peaceful coexistence of persons, promote a culture of peace and prevent forms of violence and discrimination, and commit offenses and offenses. Policies will be entrusted to specialized agencies at different levels of government." [7].

With regard to social security, it has been recognized as a human right by the world community in the Universal Declaration of Human Rights (1944) and is present in the International Covenant on Economic, Social and Cultural Rights (1976) [8].

In this sense, the Constitution of Ecuador in art. 34 disclaim: "The direct social security and unrestricted directness of all persons, e is the main responsibility of the State." Social security will be governed by principles of solidarity, observability, universality, equity, efficiency, subsidiarity, sufficiency, transparency and participation, for attention to individual and collective needs [7]; in addition, art. 373 mentions that: "The peasant social insurance, which is part of the Ecuadorian Social Security Institute, will be a special regime of compulsory universal insurance to protect the rural population and those engaged in artisanal fishing; it will be financed with the solidarity contribution of the insured persons and employers of the national social security system, with the differentiated contribution of the heads or heads of the protected families and with the fiscal allocations that guarantee their strengthening and development. The insurance will offer health benefits and protection against the contingencies of disability, disability, old age and death."

With regard to technical security, the contributions which referring to occupational health as safety and health at work, determined as a discipline dealing with the prevention of injuries and illnesses caused by working conditions, and the protection and promotion of workers health. It aims to improve the working conditions and environment, as well as health at work, which involves the promotion and maintenance of the physical, mental and social well-being of workers, in all occupations [9].

In the context of economic security, in the Constitution of Ecuador, art. 283, mentioned that "The economic system is social and solidarity; recognizes the human being as subject and end; it tends to a dynamic and balanced relationship between society, state and market, in harmony with nature; and aims to ensure the production and reproduction of the material and immaterial conditions that enable good living." [7].

So and consequently, communities and experts managed to determine the next hierarchy of places that deserve priority attention to make the urban environment safer [10], and that in our opinion would influence society at large:

- Transport system: whereabouts, pedestrian walkways, cycle routes, parking.
- Commercial areas: non-exclusive, but integrated with housing centers.

- Industrial areas: also designed to better accommodate employees and customers.
- Parks: Make them more integrated into the activities of the community.
- Residential areas: they must transmit a sense of security for all.
- Universities and colleges: they must attract the whole community and expand their schedules.

In addition, the real experts in citizen security are the same citizens. The other entities are collaborators to make that security effective; being, for some, security an institutional issue, the legal system that has a specific society; for others rather; security is a particular institution, the public force, the police or its equivalents. In one case or another it is defined that it is human societies that have given origin to the ordering, expressed in laws, organization of the State and a moral acceptable for all its members [11].

The public force is another expression of this social order, and it may be understood that the members of the public force are the same citizens whom the rest of society has delegated to them who exercise the power to control those individuals who try to disturb the organization accepted by the majority; it is clear that the exercise given to the members of the public force implies the use of violent degrees and means in which the same society gives some acceptance [12].

It should be emphasized that the State cannot be conceived as only the national government and national security as the office of the central authorities, because it is in essence the responsibility of the regional and local governments to monitor the fundamental work of Security and citizen coexistence [13].

While the police, judicial and defense functions remain to protect people, what is relevant today is to organize people to establish a society that starts from the reality of groupings and forms of solidarity, fulfilling the state an activity of "instituting of the social", no longer as the power of ordering of a conglomerate of individuals, but as a coordinating force for a multiplicity of functional and autonomous public services oriented to the general interest [14].

Likewise, in this scenario of continuous evolution, it should be have in mind that the social factor becomes relevant for the State, making the culture and the habits of the citizens, essential factors. In this sense, the State must assume its protector and executor role, multicultural society where citizens demand recognition of their difference, and at the same time their protection is insufficient, so it is necessary that this conglomerate not only be demanded, but also become a security offered to through the prophylaxis of their behavior [2].

With regard to security models, the best model will be one that encourages the strengthening of democracy and interaction with citizens; and in which the public force is committed to educating and protecting, given its direct contact with the community [12].

No doubt that the social factor determines the efficiency and effectiveness of the public security system. For Francesc Torralba in order to achieve coexistence, the city requires its inhabitants to assume an ethical category, adopting a certain lifestyle that cultivates certain virtues [15].

2.2   Artisanal Fishing

Benavides Rodriguez (2014) mentions that fishing is a primary activity that is based on three elements: the fisherman, the boat, and the fishing gear. Considering fishing as an economic activity, it incorporates a market for those who sell, facilities to land and preserve production [16].

González Laxe (2008) defines artisanal fishing in the following way: artisanal fishing vessels are those that are devoid of trawl gear and frozen equipment on board with a power number equal to or less than 200 HP; boats that do not meet these parameters are considered industrial fishing vessels [17].

The INEC figures in the 2010 Census indicate that the province of Santa Elena has 301,168 inhabitants, distributed in three cantons: Salinas, Santa Elena and La Libertad, its population represents 6% of region 5 and 2% of the total country. 64% live in rural areas and only 36% live in urban areas; y contributes 5% of the regional CAS. Santa Elena's contribution to GNP is of the order of 4% regional and 1% nationally [18].

The fact of having a majority of the rural population determines the nature of the economic activities that are mainly dedicated to agricultural production and fishing, 24% of the EAP, followed by activities related to commerce, 18%, manufacturing with 10%, construction of 9% of PEA, and education with 4%, 35% of activities consists of a series of occupations [18].

Based on the sectorial structure and contribution generated for GNP in 2007, the Santa Elena fishing sector contributes 6% of GDP, [18] this contribution is mainly due to the production of the 2 main fishing ports of the province, both located in the canton of Salinas, Santa Rosa and Anconcito.

The Salinas fishing ports bring together some 4,000 artisanal fishermen who carry out their work in a park of boats, mainly motorized, that reach approximately 1300 units and about 33 mother boats [19].

3   Study Case

The main objective of this work is to identify the current situation of artisanal fishermen in terms of integral security.

In this sense, the population of the study is 4500 fishermen, from the fishing ports of Anconcito and Santa Rosa; The study sample corresponds to 323 artisanal fishermen, for whom a margin of error of 5% and a confidence level of 95% was considered.

The survey distributed by the fishermen is composed of 22 direct closed questions and polytomics, being structured in two distinct parts: one with the sociodemographic data and another one with the questions related to integral security.

It is shown in Fig. 1 that 84% of the human security of artisanal fishers indicate that it is a dangerous activity, mainly due to piracy that threatens fishermen's lives, and in most cases deprives them of the engines of the boats and of the fish caught.

Figure 2 shows that in terms of social security of artisanal fishermen, 59% mention that the State should help fishermen to obtain health insurance, and 41% do not consider it a priority for daily work.

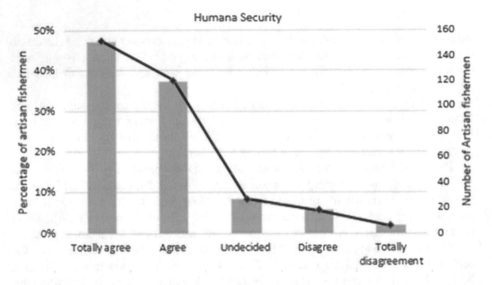

Fig. 1. Human security.

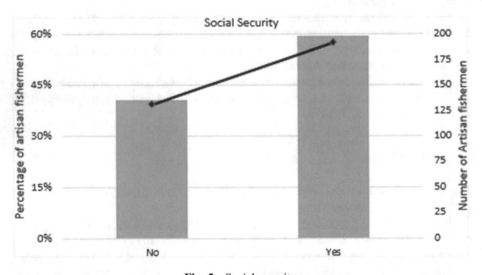

Fig. 2. Social security.

In the case of Fig. 3, with regard to legal certainty for artisanal fishers, 48% receive little or no government support and 9% indicate that they receive much state support.

In terms of technical safety of artisanal fishers, 44% consider the environmental and labor conditions provided by the state, and 42% consider it Bad and lousy, clarifying that the fishing port of Anconcito has been repotenced by the state, contrary of Santa Rosa fishing port (Fig. 4).

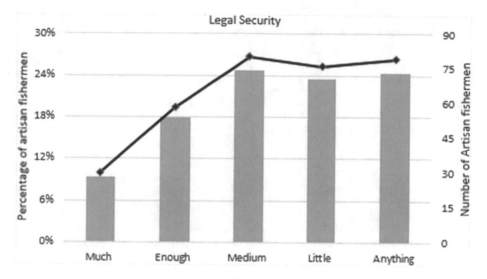

Fig. 3. Legal security.

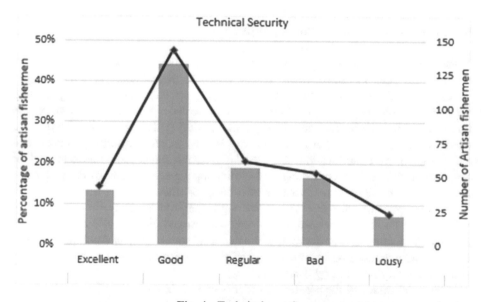

Fig. 4. Technical security.

Figure 5 refers to the economic security of artisanal fishermen, where 50% are satisfied with what they are caught in the fishery, and 42% indicate that they are between undecided and very unsatisfied; According to fishermen, satisfaction is mainly given in the seasons of abundance and tends to fall in the seasons of scarcity.

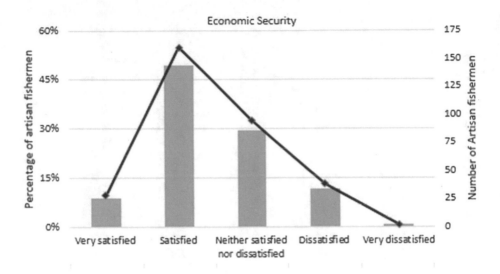

Fig. 5. Economic security.

4 Conclusions and Recommendations

In the present investigation, the current situation regarding integral security in fishing activities has been identified. Being able to conclude the existence of a low integral security that artisanal fishermen have when carrying out their daily tasks, highlighting the human insecurity that originates when the sea whistles lash out at the lives of the fishermen taking the engines and fishing captured during the day; social security that refers to the fact that artisanal fishermen have health insurance against the threats that are exposed in their daily work; and the legal and technical insecurity due to the lack of sufficient support in terms of fishing controls and having an adequate working physical environment for the good performance of fishing activities.

It is essential that the State establish, implement and supervise integral security policies, for the benefit of artisanal fishermen. Armed Forces of Ecuador, being responsible in times of war to preserve the maritime sovereignty of the country and in times of peace is responsible for controlling illicit activities in maritime territory, must ensure the safety required by fishermen.

The maintenance of the constituent order is one of the main functions of the State, but this mission contemplates a wide range of problems, because a security established on the idea of protecting individuals against risks of all kinds, means maximizing the capacities of the organizations consigned for this purpose, since new threats are frequently arising, causing the congregation of the different levels of intervention and contribution.

The present investigation has as limitation, the geographical factor of the population under study, which limits the possibility of generalizing the results in other contexts. Future research could identify the situation of integral security that artisanal fishermen have and that make their daily tasks difficult. Likewise, we hope that the proposals to increase integral security will be analyzed and implemented by the governments of the different countries and implemented and promoted by the local public institutions.

Acknowledgment. The collaboration of artisanal fishermen from the fishing ports of Anconcito and Santa Rosa of the Province of Santa Elena is appreciated.

References

1. Mack, A.: El concepto de Seguridad Humana, Papeles de cuestiones internacionales, pp. 11–18 (2005)
2. Jiménez Vega, R.: Las nuevas demandas de seguridad. Revista Opera **17**, 127–147 (2015)
3. Gairín, J., Díaz, A.: Diagnóstico de la seguridad integral en centros educativos: El cuestionario de autoevaluación EDURISC. Revista del Fórum Europeo de Administradores de la Educación **1**, I–VIII (2012)
4. Chevallier, J.: El Estado posmoderno. Universidad Externado de Colombia, Colombia (2011)
5. PNUD: Informe sobre desarrollo humano, Washington D.C. (1994)
6. Bauman, Z.: En busca de la política. Fondo de Cultura Económica, Buenos Aires (2001)
7. Asamblea Nacional Constituyente de Ecuador: Constitución del Ecuador. Ecuador (2007)
8. Bertranou, F., et al.: Encrucijadas en la seguridad social argentina: reformas, cobertura y desafíos para. CEPAL y Oficina Internacional del Trabajo, Buenos Aires (2011)
9. Henao, F.: Seguridad y salud en el trabajo. ECOE Ediciones, Colombia (2016)
10. Wekerle, G., Whitzman, C.: Safe Cities: Guidelines for Planning, Design, and Management. Van Nostrand Reinhold, New York (1995)
11. Aradau, C.: Security and the democratic scene: desecuritization and emancipation. J. Int. Relat. Dev. **7**(4), 388–413 (2004)
12. Triana, R.: Modelos de seguridad. Centro de Estudios Estratégicos sobre Seguridad y Defensa Nacional, Bogotá (2014)
13. Martin, G., Ceballos, M.: Bogotá: Anatomía de una Transformación - Políticas de seguridad ciudadana. Pontificia Universidad Javeriana, Bogotá (2004)
14. Rosanvall, P.: La Legitimidad Democrática: Imparcialidad, reflexividad, proximidad. Manantial, Buenos Aires (2009)
15. Torralba, F.: Cien valores para una vida plena: la persona y su acción en el mundo. Milenio, Lleida (2003)
16. Benavides Rodriguez, A.: El sector pesquero de Santa Elena: Análisis de las estrategias de la comercialización, Ciencias pedagógicas e innovación UPSE (2014)
17. González Laxe, F.: La actividad pesquera mundial. Netbiblo S.A., España (2008)
18. MCPEC: Agendas para la transformación territorial provincia de Santa Elena, Ministerio de Coordinación de la producción, empleo y competitividad, Quito (2011)
19. VMAP: Situación actual de la pesca en Ecuador SRP-MAGAP. MAGAP, Quito (2014)

Offshore Wind Farms – Support or Threat to the Defence of Polish Sea Areas?

Benedykt Hac[1] and Kazimierz Szefler[2(✉)]

[1] Department of Operational Oceanography,
Maritime Institute in Gdansk, Gdansk, Poland
benhac@im.gda.pl
[2] Maritime Institute in Gdansk, Gdansk, Poland
kaszef@im.gda.pl

Abstract. Coastal sea areas, both the water column and the sea bottom, are very attractive and important for the power industry, fisheries, tourism etc., but also for the State armed services due to the strategic importance of open sea waters in the National Defence Doctrine. Free access to this area is very important to the Command of the Polish Army, which sees the build-up of the sea bottom with industrial objects (offshore wind farms) as a significant impediment to the fulfilment of defence tasks. At the same time, there is a high interest of the industry to locate wind farms in the sea resulting in significant tension between the industrial investors and the armed forces. The Authors think that it is possible to link the different concepts of using these sea areas by giving the status of a multifunctional installation to selected elements of offshore wind farm structures. Analysis indicates that it is possible not only to resolve the problems but also to generate new solutions of a hitherto unachievable quality.

Keywords: Offshore planning · Offshore wind farms · Opto-electronic head
Surveillance radar · Acoustic barrier

1 Introduction

Polish sea areas offer one of the highest potentials for offshore wind farms in the Baltic Sea Region, both in terms of economic effectiveness and the wind conditions in the Polish exclusive economic zone. However, spatial conflicts with other uses of sea space (navigation routes, pipelines, mining, fisheries, military requirements) significantly reduce the actual possibilities of using the Polish sea space for offshore wind.

One of the ways of solving the conflicts is to give a multifunctional status to some forms of sea area usage. Offshore wind farms are an excellent example since their locations allow not only effective power production but may be also a basis for a development of a military surveillance and control system covering areas adjoining the territorial waters and the national border of the country.

© Springer International Publishing AG, part of Springer Nature 2018
Á. Rocha and T. Guarda (Eds.): MICRADS 2018, SIST 94, pp. 266–274, 2018.
https://doi.org/10.1007/978-3-319-78605-6_22

2 Present Status of the Offshore the Wind Farms

2.1 Potential of the Marine Power Production in the Polish Exclusive Economic Zone (EEZ)

It is assessed that the theoretical potential installed power of the Offshore Wind Farms (OWFs) in the Polish EEZ is about 35 GW and the theoretical economical potential of these areas is about 100 billion €. This translates into a theoretical number of 40 thousand jobs during the construction, and 24 thousand permanent jobs related to operation and maintenance of the OWFs.

In reality, in spite of solving the spatial conflicts, the real technical potential will not exceed 20 GW installed power, with the economic potential close to 60 billion €. The real number of new jobs will be 23.5 thousand during the construction, and 14 thousand permanent jobs during the operation of the OWFs.

2.2 Maritime Spatial Planning in the Polish EEZ

For every coastal state, maritime spatial planning of its internal sea waters, territorial sea and EEZ (Fig. 1) is an important element allowing realisation of an effective maritime economy.

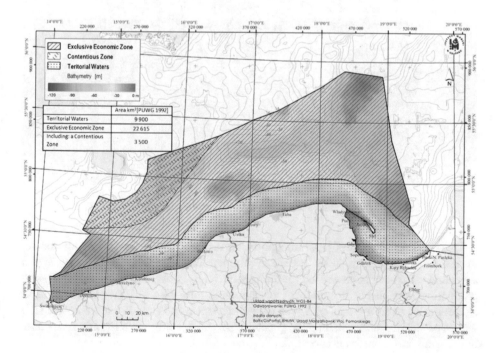

Fig. 1. The polish sea areas

Poland is currently carrying out spatial planning of its sea areas, and it is assumed that all the present and future maritime activities and investments, activities aiming at nature protection and at defending the territory of the State will be taken into account in a comprehensive and coherent maritime spatial plan. The main aims of this plan are to distinguish the important (not only economically) sea areas and solve present and future conflicts concerning the use of available space in order to allow effective and diversified investments important for the national economy.

At present, from the point of view of the investors planning to develop offshore wind energy, building OWFs is economically justified in waters less than 60 m deep. Sea areas with such water depths are shown in Fig. 2.

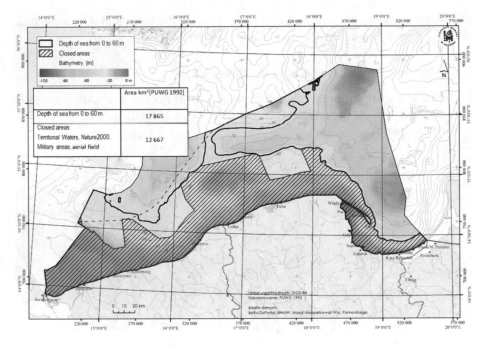

Fig. 2. Areas excluded from areas useful for the construction of OWF

After taking account of legal limitations, in result of which OWFs are not allowed to be located in the internal sea waters and the territorial sea (also in nature protection areas and some special areas used for military exercises), investments in offshore power production have been "forced out" of the most attractive sea areas. Water depth exceeding 40 m in the area of a planned investment requires using more costly techniques of building and installing the generator towers, resulting in a significant increase in the wind farm construction cost. All this markedly limits the sea area appropriate for long-term investments.

Conflict analysis suggests that locating economically justified wind farms in these areas is impossible without giving the multifunction status to the selected areas. In result of agreements between the present and future users of the sea, a compromise was reached, which allows marking off the significant sea areas for future OWFs. Figure 3 shows the existing, and some of the planned uses of Polish sea areas and the preliminary division of plots designated for OWFs.

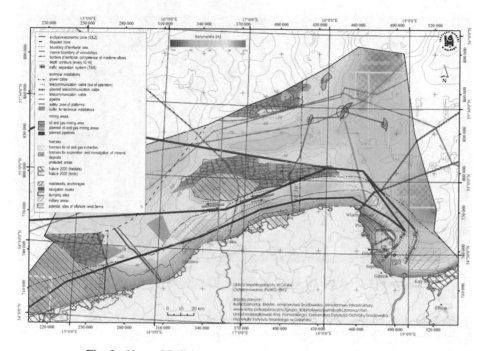

Fig. 3. Uses of Polish sea areas and plots designated for OWFs

3   Use of Offshore Wind Farms for Supporting Military Systems for Recognising Tactical Situation at Sea

The present political and military situation in the Baltic Sea Region is stable, and there is no need to build special military observation installations on permanent structures located at the faraway offshore sites.

However, knowing the distribution of the planned OWFs which are to be located in an area between 22 and 70 km from the coast, it can be observed that the external towers of the wind farms form a continuous line of permanent structures distant from each other by no more than 45 km. This means that by using selected towers (Fig. 4) as a basis for radio, hydro and optoelectronic recognition systems it may be easy to develop a permanent barrier, located rather far in the sea, crossing of which unobserved would be nearly impossible. This concerns airborne vessels, including drones, and vessels sailing on the sea surface and underwater.

One of the main characteristics of wind farm towers is their very solid structure, resistant to extreme wave and wind action, with stable foundations in the sea bottom. Energy transmission and steering systems require laying power transfer and generator control cables. Usually for controlling each of the offshore generators there are used

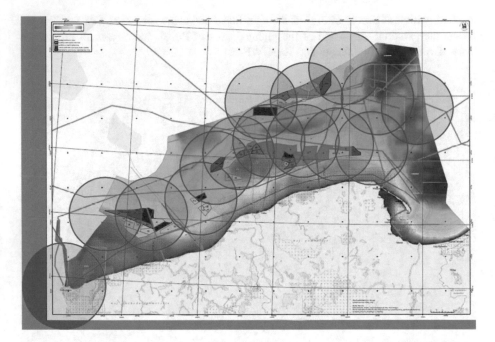

Fig. 4. The general concept of the arrangement of electronic surveillance devices on OWFs

fibre-optic cables with very high data transmission capacities. Besides their normal function of transmitting data and steering signals to/from the coast, they can be easily used also for transmitting data from the radar, hydro-acoustic or optoelectronic systems. All these features of the OWFs are very desirable in case of technical observation stations, which could be placed on the towers.

The starting point would be an agreement between the sea space users, i.e. the General Staff of the Polish Army and the private investors building the wind farms. The State, which is the sole owner and administrator of the sea areas, can determine appropriate requirements to be fulfilled when applying for permits for OWFs, which will ensure that needs of national defence are properly taken into account.

The best way to solve the conflicts which may occur when planning industrial, military and other uses of sea space is to prepare maritime spatial plans for all the Polish sea areas. The plan should account for the needs of:

- institutions responsible for the defence of Polish territory;
- companies planning to build offshore wind farms;
- companies mining minerals, gas and oil;
- shipping companies (passenger and cargo);
- fisheries;
- tourism and recreational navigation.

4 Devices Used in Electronic Surveillance Systems

It is possible to build a modern system supporting the protection (defence) of the marine border of Poland.

This system could consist of:

- a system for radar control of sea surface;
- a system for radar control of the situation in the air;
- an optoelectronic observation system (television cameras, infrared cameras, laser range finders, etc.);
- an underwater electromagnetic barrier system, detecting vessels and submerged objects crossing the barrier (determination of the signature of the vessels based on a measurement of the electromagnetic field);
- a hydroacoustic system for detecting objects on the water surface and below the water surface, divers, etc. (acoustic signature of the object for detecting and automatic identification of objects).

4.1 Elements of the Radar System

The planned devices for technical surveillance should have the following technical properties:

system for radar observation of sea surface operating in the S or X range (wave length $\lambda = 10$ cm or $\lambda = 3$ cm). Detecting objects on water surface within 24 nautical miles from the point at which it is installed;

system for radar observation of situation in the air. Depending on the adopted concept and tasks determined for the system, it may consist of:

- *active short range radar* with active antenna based on semiconductor transmission modules cooled by a liquid, which can digitally form received beams and can perform digital synthesis, coding and filtering of signals. Provided with a subsystem for drone and helicopter detection (in that hovering helicopters);
- *active long-range radar* with a long-range radiolocation station operating in the metre band (VHF). This radar allows detection of airborne objects using stealth techniques (with the small effective surface of radiolocation reflection;
- *passive radar* operating in PCL-PET technique, i.e. multistatic passive radiolocation. Operates basing on two methods of detection of objects:
  - PCL – Passive Coherent Location – using electromagnetic radiation of local emitters,
  - PET – Passive Emitter Tracking – with detection of electromagnetic radiation generated by installations on board the detected objects (airborne, on land and floating).

System for optoelectronic observation based on a head which integrates passive optoelectronic sensors operating in wavebands from visible to thermal radiation as means for detection, identification, observation and after integration with the platform for determining the coordinates of distant objects in the visible light and infra-red spectrum [1].

Integrated optoelectronic systems are characterized by passive detection – identifi-cation – thermal-video determination without disclosing own position. They can be used for airspace control, allow a safe for eyes laser measurement of distance within the range of 100 m to 30 km. They provide information about the distance and bearing of the detected object, position the object and calculate its coordinates. The main element of the system is the optoelectronic head.

4.2   Elements of the Optoelectronic Head

"The optoelectronic head is furnished with the following:

- *thermal camera* serves the purpose of surveillance of remote air targets. The system provides for target detection and observation independent of lighting conditions both during the day and at night and allows detecting thermal signals emitted by objects in the detector field of view due to high temperature sensitivity.
- *LLLCCD television camera* serves to observe remote targets within the range of visible light in the conditions of minimum lighting. It is furnished with an advanced zoom lens. Thanks to the application of a modern converter, the camera may work under the minimum lighting conditions.
- *laser range finder* consists of a laser transmitter where a strong optical impulse is generated and a laser receiver which is able to detect extremely low levels of laser radiation.

The main computer system consisting of digital display and control displays images coming from the television camera or thermal camera on an LCD screen. It provides for a controlled drive in azimuth and elevation. In this system, the digital display serves as an integrated display to present data from other channels, such as, e.g., radar tracking sta-tions, IFF target identification systems, or laser range-finders. The picture from the tele-vision camera or from the thermal camera is superimposed on the microcomputer-generated image with the target parameters from the radar station, IFF system and laser range-finder (azimuth, elevation, altitude, range, 'friend or foe')" [1].

4.3   Elements of the Underwater Surveillance Systems

The underwater electromagnetic barrier system and system for hydroacoustic detection of objects on the water surface and under water, divers etc. can operate separately or in an integrated system for detecting moving underwater objects. Such an integrated, multi-sensor system is designed for monitoring and counteracting under-water sabotage, detecting "alien" submarines of any size entering Polish sea areas.

The proposed solution is based on linking a passive magnetic barrier with an active acoustic barrier. Magnetic barriers react to environmental changes of the magnetic field caused by movement of objects generating weak magnetic fields. Each crossing of the protection line formed by the magnetic and hydroacoustic barriers is processed and visualized on a map of the protected area. The process is realized by simultaneous recording and analysis of signals from sensors detecting anomalies of various physical fields of potential objects in the protected underwater zone, ensuring an appropriate probability of detection. Additionally, a DDS sonar is used in such systems for detecting divers.

5 Existing Possibilities for Using the Offshore Wind Farms for Building Technically Advanced Detection and Surveillance Systems

The process leading to the appearance of the first OWFs began in 2015 when licenses for environmental investigations in the area of the proposed wind farms were issued. The projects to build four OWFs, the construction of which will begin in 2019/2020 (Bałtyk III and Bałtyk II of Polenergia, and Baltica and Baltica II of PGE EO) could be a basis/starting point for building a full system. Figure 5 shows the possibilities and scope of surveillance of the Polish EEZ in the area of the first two OWFs.

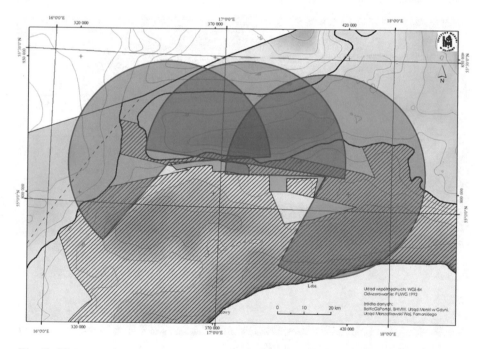

Fig. 5. The general concept of the arrangement of electronic surveillance devices on OWF

The concept of building the system in small steps is justified by the gradual development of wind power production in the Polish EEZ.

It also seems that technical progress cannot be stopped. Therefore, a way for most effective solving of conflicts between participants in the licensing processes should be developed. Linking of several projects – as described above – ensuring both power supply safety of the country and protection of its sea borders seems an example of an optimum solution.

6 Conclusions

Construction of offshore wind farms (artificial islands) in the Polish EEZ forms a good basis for development of a modern technical surveillance system at sea located 40 to 100 km from the coast, thereby introducing a completely new quality into the defence capacities of Poland. Information about the position of vessels in the observed area (both in navigation routes and outside the routes) is priceless for the SAR services. It enables speedy, well-coordinated salvage actions since real-time information allows optimising and improving the effectiveness of salvage operations.

Reference

1. Wojskowe Zakłady Uzbrojenia S.A. Grudziądz, pp. 1–2. www.wzu.pl, Glowica_ optoelektroniczna.pdf

Naval and Military Engineering

Horseshoe Vortex Suppression with a Strake

Jun Pei Lee[1], Jiahn-Horng Chen[2]($\boxtimes$), and Ching-Yeh Hsin[2]

[1] CSBC, 3 Jhonggang Road, Siaogang District, Kaohsiung 81234, Taiwan
[2] National Taiwan Ocean University, 2 Pei-Ning Road, Keelung 20224, Taiwan
jhcntou@gmail.com

Abstract. We conducted computationally a parametric study to investigate the horseshoe vortex suppression due to a boundary-layer flow past a wing of finite span by a leading-edge strake. Both boundary-layer flows over a flat and curved wall were explored. In total, 48 cases were studied for various strake geometries. The Spalart-Allmaras model (1-equation model) is employed for the turbulence effect. The computational results show that different ratios lead to different flow development. Some of them can effectively suppress the horseshoe vortex. The detailed flow fields near the leading edge of the wing and the wake development are also investigated for different cases.

Keywords: Strake · Horseshoe vortex · Junction vortex

1 Introduction

This paper investigates the control of horseshoe vortex caused by a turbulent boundary layer past a wall-mounted wing. Horseshoe vortex has great influences on the overall flow; unfortunately, most of them are undesired. Due to the nature of secondary flow, horseshoe vortex deteriorates the uniformity of downstream wakes and creates a local high shear stress region; thus a greater drag might be induced. Since the flow has become less uniform, a higher noise level is also expected. Several different methods or approaches were proposed to control or even mitigate these disadvantages.

For bridge piers in rivers, civil engineers may place some collars or riprap at the juncture regions to prevent scouring, but their major purpose is to protect the bridge foundations rather than control the flow field [1]. To get rid of the horseshoe vortex, various methods have been proposed. For some examples, Kairouz and Rahai [2] placed a ribbed surface in front of the wing-body junction. Kang et al. [3] devised a cavity upstream of a circular cylinder. Hassan and Hua [4] installed another smaller circular cylinder ahead of the main circular cylinder. Barberis et al. [5] placed a fillet-like object in front of the wing. Liu et al. [6] installed two bafflers on the two sides of the primary wing to destroy the vortex core of the horseshoe vortex. Ölçmen and Simpson [7, 8] conducted a series of experiments of turbulent flow past wings with different nose shapes. They observed that a blunter nose led to a stronger vortex.

© Springer International Publishing AG, part of Springer Nature 2018
Á. Rocha and T. Guarda (Eds.): MICRADS 2018, SIST 94, pp. 277–283, 2018.
https://doi.org/10.1007/978-3-319-78605-6_23

Devenport *et al.* [9] studied the flow past a wing-body junction with a continuous radius fillet. Their experimental results did not display much improvement of the flow field. Since the horseshoe vortices originate in the upstream flow separation, it is more effective to control the upstream flow development. Both Devenport *et al.* [10] and Van Oudheusden *et al.* [11] conducted experiments to study wing-body junction flows with leading edge fairings or strakes. Experimental results indicated that the wake uniformity had been much improved, and the separation ahead of the nose region was not found.

2 Formulation

2.1 Governing Equations

The flow is governed by the Reynolds-averaged Navier-Stokes equations

$$\frac{\partial \bar{u}_i}{\partial x_i} = 0, \tag{1}$$

$$\rho \bar{u}_j \frac{\partial \bar{u}_i}{\partial x_j} = -\frac{\partial \bar{p}}{\partial x_i} + \mu \frac{\partial^2 \bar{u}_i}{\partial x_j \partial x_j} - \frac{\partial}{\partial x_j} \left(\rho \overline{u_i' u_j'} \right), \tag{2}$$

where u_i is the velocity vector, p the pressure, ρ the fluid density, and μ the dynamic viscosity of the fluid. The variables with overbars denote the time-averaged quantities. In the present study, the coordinates $x_1 = x$, $x_2 = y$, and $x_3 = z$ denote the streamwise, spanwise, and lateral directions, respectively. The origin of the coordinate is at the leading edge of the body on the plate. The last term in Eq. (2) is the Reynolds stress tensor. There are many turbulence models available for it. The Spalart-Allmaras model (one-equation model) [12] was employed here due to its better performance in terms of computational accuracy and robustness shown in our previous study [13].

2.2 Numerical Procedure

The commercial CFD software ANSYS Fluent 17.0 was used. It employs the finite volume method for discretizing the governing equations, and the SIMPLEC algorithm was chosen for the steady state velocity-pressure iterations.

Polyhedral grids are generated for computations. They can reduce the grid skewness and improve the convergence quality. The total gird number ranges from 5 to 6 million. If y^+ denotes the (scaled) coordinate direction normal to a solid wall and P denotes the center point of the cell closest to the wall, we have in our mesh $y_P^+ \approx 50$ which satisfies the condition to capture the junction vortex [14]. Moreover, the present grid can ensure a grid-independent solution [15].

3   Results and Discussion

The present study focuses on the effect of strakes of different sizes and shapes on the development of horseshoe vortices. Two different kinds of strakes with various geometric ratios are mounted in the juncture region. The wing in the study is composed of a 3:2 ellipse nose and a NACA 0020 tail which join together at the location of their maximum thickness [10]. The upstream inlet condition is available in [16]. The height of the boundary thickness is about a third of the wing maximum thickness and the location is at $x/T = -18.24$. The Reynolds number based on the boundary layer momentum thickness is 6700, and the Reynolds number based on the chord length and free-stream velocity is 5×10^5. We varied the strake shapes systematically and conducted computations for 48 cases. In the first part of the results, the wing is mounted on a flat plate, and a two dimensional planar boundary layer is specified upstream. In the second part, the wing is mounted on a curved surface of a circular cylinder. The radius of the cylinder is $R = 0.69C$.

Figure 1 shows the schematic of the strake; h and l are the height and the length of the strake, respectively. From the top view, the cross section of the strake is a semi-ellipse, and the center of each semi-ellipse is located at the center of the nose region. The radius along the z-axis of each semi ellipse remains $T/2$, and the other radius is dependent on the profile on the symmetric plane. By defining the geometry this way, the configuration of the strake is solely determined by the length, the height, and the profile. This particular configuration was proposed by [11]. Two profiles are investigated here; they are the linear and elliptical profiles. The length of the strake ranges from $0.25T$ to $1.00T$, and four different length-height ratios are specified. For wing mounted on a curved surface, only the strake of linear profile is considered.

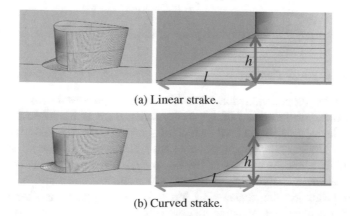

(a) Linear strake.

(b) Curved strake.

Fig. 1. Schematic of the strake.

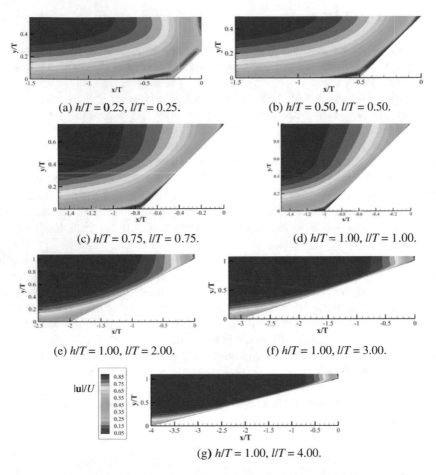

(a) $h/T = 0.25$, $l/T = 0.25$.

(b) $h/T = 0.50$, $l/T = 0.50$.

(c) $h/T = 0.75$, $l/T = 0.75$.

(d) $h/T = 1.00$, $l/T = 1.00$.

(e) $h/T = 1.00$, $l/T = 2.00$.

(f) $h/T = 1.00$, $l/T = 3.00$.

(g) $h/T = 1.00$, $l/T = 4.00$.

Fig. 2. Velocity distribution on the plane of symmetry (strake with linear profile).

3.1   Wing on a Flat Plate

For the present wing-plate junction, available experimental result shows that there is a horseshoe vortex on the plane of symmetry when there is no strake [13]. Figure 2 shows the velocity distribution on the plane of symmetry due to different linear strakes. Obviously, a larger strake produces more significant effect. Figure 3 shows closer views of certain cases. As the size of the linear strake grows, the size and the strength of vortex core decrease. If the length-height ratio is above 2, the vortex core disappears. Unfortunately, around the strake-wing junction, there appears another vortex. This additional one is due to the non-smooth connection. However, it is not so strong as the original one. As the length-height ratio is raised, it grows in size and strength.

Figure 4 shows the results of strakes with elliptical profile. There is no additional vortex around the junction between the wing and the strake because the connection is now smoothly faired. Similar to the flow with linear strakes, the vortex can be suppressed completely as the strake is sufficiently large.

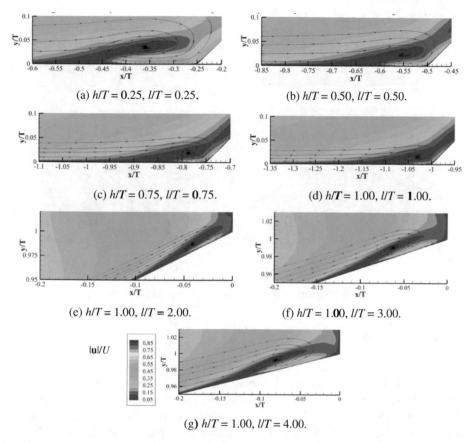

(a) $h/T = 0.25$, $l/T = 0.25$.

(b) $h/T = 0.50$, $l/T = 0.50$.

(c) $h/T = 0.75$, $l/T = 0.75$.

(d) $h/T = 1.00$, $l/T = 1.00$.

(e) $h/T = 1.00$, $l/T = 2.00$.

(f) $h/T = 1.00$, $l/T = 3.00$.

(g) $h/T = 1.00$, $l/T = 4.00$.

Fig. 3. Velocity distribution on the plane of symmetry (strake with linear profile, closer views).

We also studied the wake distribution. For strakes with a linear profile, the wake uniformity is much improved when the strake is large enough but is worsened if the strake is too large. For strakes of an elliptical profile, the trend is similar, but the wake distribution remains favorable even for the largest strake. This is due to the fact that the juncture is now smoothly faired.

3.2 Wing on a Curved Plate

The computations for the linear strakes show that the velocity distribution on the plane of symmetry is almost identical to the results in the previous section. If the strake is chosen carefully, the horseshoe vortex can be entirely diminished. As to the wake, its uniformity can also be much improved if the strake is sufficiently large.

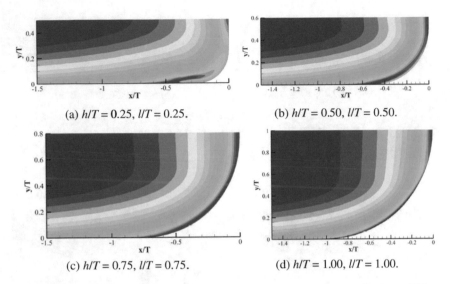

(a) $h/T = 0.25$, $l/T = 0.25$.

(b) $h/T = 0.50$, $l/T = 0.50$.

(c) $h/T = 0.75$, $l/T = 0.75$.

(d) $h/T = 1.00$, $l/T = 1.00$.

Fig. 4. Velocity distribution on the plane of symmetry (strake with elliptic profile).

4 Conclusions

In this study, we have investigated the effect of a strake on the horseshoe vortex development due to a wing-plate junction. Both linear and elliptical strakes were studied. The computational results show that a larger strake can effectively reduce the size and strength of the horseshoe vortex. Elliptical strakes are better because the linear strakes induce an additional vortex due to non-smooth junction to the wing. Furthermore, the wake due to an elliptical strake is also more uniform than that due to a linear one. As to the wing on a curved plate, the results are similar to those on a flat plate.

References

1. Zarrati, A.R., Nazariha, M., Mashahir, M.B.: Reduction of local scour in the vicinity of bridge pier groups using collars and riprap. J. Hydraul. Eng. **132**(2), 154–162 (2006)
2. Kairouz, K.A., Rahai, H.R.: Turbulent junction flow with an upstream ribbed surface. Int. J. Heat Fluid Flow **26**(5), 771–779 (2005)
3. Kang, K.J., Kim, T., Song, S.J.: Strengths of horseshoe vortices around a circular cylinder with an upstream cavity. J. Mech. Sci. Technol. **23**(7), 1773–1778 (2009)
4. Hassan, J., Hua, Z.: Juncture flows around cylinders of unequal finite-height and diameter in tandem arrangement. J. Vis. **18**(2), 343–348 (2015)
5. Barberis, D., Molton, P., Malaterre, T.: Control of 3D turbulent boundary layer separation caused by a wing-body junction. Exp. Therm. Fluid Sci. **16**(1–2), 54–63 (1998)
6. Liu, Z., Xiong, Y., Wang, Z., Wong, S.: Numerical simulation and experimental study of the new method of horseshoe vortex control. J. Hydrodyn. **22**(4), 572–581 (2010)
7. Ölçmen, S.M., Simpson, R.L.: Influence of wing shapes on surface pressure fluctuations at wing-body junctions. AIAA J. **32**(1), 6–15 (1994)

8. Ölçmen, S.M., Simpson, R.L.: Influence of passive flow-control devices on the pressure fluctuations at wing-body junction flows. J. Fluids Eng. **129**(8), 1030–1037 (2007)

9. Devenport, W.J., Agarwal, N.K., Dewitz, M.B., Simpson, R.L., Poddar, K.: Effects of a fillet on the flow past a wing-body junction. AIAA J. **28**(12), 2017–2024 (1990)

10. Devenport, W.J., Simpson, R.L., Dewitz, M.B., Agarwal, N.K.: Effects of a leading-edge fillet on the flow past an appendage-body junction. AIAA J. **30**(9), 2177–2183 (1992)

11. Van Oudheusden, B.W., Steenaert, C.B., Boermans, L.M.M.: Attachment-line approach for design of a wing-body leading-edge fairing. J. Aircr. **41**(2), 238–246 (2004)

12. Spalart, P.R., Allmaras, S.R.: A one-equation turbulence model for aerodynamic flows. In: 30th Aerospace Sciences Meeting and Exhibit, Reno, Nevada (1992)

13. Lee, J.P., Chen, J.-H., Hsin, C.-Y.: Study of junction flow structures with different turbulence models. J. Mar. Sci. Technol. **25**(2), 178–185 (2017)

14. Jones, D.A., Clarke, D.B.: Simulation of a wing-body junction experiment using the Fluent code. DSTO-TR-1731, DSTO Platforms Sciences Laboratory, Fishermans Bend, Victoria, Australia (2005)

15. Lee, J.P.: Numerical study of junction flow and its control. Master thesis, National Taiwan Ocean University, Keelung, Taiwan (2017)

16. Devenport W.J., Simpson, R.L.: An experimental investigation of the flow past an idealized wing-body junction: final report. Virginia Technical report, VPI-AOE-172 (1990)

Hybrid Joint Between Steel Deck and Fiberglass Superstructure

Franklin J. Domínguez Ruiz[1(✉)] and Luis M. Carral Couce[2]

[1] Escuela Superior Politécnica Del Litoral, Guayaquil, Ecuador
jdominguez@tecnavin.com
[2] Universidade da Coruña, La Coruña, Spain
lcarral@udc.es

Abstract. In semi-displacement vessels, the reinforced fiberglass composite superstructure, or FRP, is one of the best construction options due to its low weight, shape, and surface finish. A problem with the FRP composite material lies in the joint with the steel hull due to the lack of adherence in the joint between the deck and the superstructure's FRP panels. This research presents a solution for hybrid joints between the deck and the FRP superstructure using a steel tubular structure and laminates with Isoftalic Resin. The design shear stress for the lamination of this hybrid joint has been considered to be 4.55 MPa, according to the recommendation of Lloyd's Register. The analysis initially makes an estimate of the reactions at the hybrid joint based on a 36.80-m ship and then performs a critical layer analysis with the finite elements method. This is followed by an analysis of possible hybrid joints to find the best option for a construction that fulfills design stress. The final hybrid joint presents better results and consists of ASTM B53 steel tubes, Sch40, and uses vertical tubes 2 in. in diameter and 60 cm in height as well as two longitudinal tubes of 1 in. in diameter at 30 cm and 60 cm from the deck, respectively.

Keywords: Hybrid joint · Composite panel · Composite superstructure

1 Introduction

Fiberglass reinforced panels are used in commercial and naval ships for areas where there is no need for high structural resistance, such as interior division panels, stairs, interior decks, etc. Currently, hybrid joints between steel decks and fiberglass superstructures on semi-displacement vessels have not been widely used even though fiberglass-panel superstructures with a balsa core provide a weight reduction of approximately 40% compared to a naval steel superstructure [3].

The development of this type of construction has led various researchers to make proposals regarding hybrid joints.

Salih and Patil [8] mentioned the benefit of using bolts in a hybrid joint, provided it is flexible joint. In their research, Ritter [4] and Hentinen [7] showed the advantages and disadvantages of using hybrid joints with adhesives and described applications on military ships.

© Springer International Publishing AG, part of Springer Nature 2018
Á. Rocha and T. Guarda (Eds.): MICRADS 2018, SIST 94, pp. 284–295, 2018.
https://doi.org/10.1007/978-3-319-78605-6_24

Weitzenböck [5] analyzed hybrid joints using adhesives and bolts, where the bolts were used as an alternative method in case of failure of the adhesive.

In their study, Babazadeh and Khedmati [1] analyzed the effects of the main parameters of a hybrid sandwich joint with adhesive, applying tension loads. The analyzed parameters included the length of the hybrid joint, the thickness of the adhesive, and the thickness of the upper and lower laminates. They concluded that the performance of the bond hybridized with adhesive depends on the geometry of the joint.

In a similar study, Kotsidis et al. [2] used modification in the joint; where the inner end of the steel plate had a bend of length L and an angle θ, concluding that the optimum angle of the bend is 15°, and that the bend contributed to the reduction of flexion stress.

The construction of ships with a mixed superstructure, i.e. with a superstructure above the main deck of with tubular steel and fiberglass panels and an upper superstructure composed only of reinforced fiberglass panels, first began in Ecuador in 2011 [3].

The hybrid joints proposed here consist of a tubular structure welded to the deck. The vertical structure comprises steel tubes 2 in. in diameter, Sch40, with a horizontal stiffening structure composed of a 3 × ¼-in. flat bar at the base welded to the deck and 1-in. tubes spaced 30 cm apart (welder tubes). The number of welder tubes depends on the finite elements analysis.

1.1 Introduction to Finite Elements Analysis

The following assumptions have been made for the analysis of the models using the finite elements method:

- The weight of the superstructure is 9.09 tons.
- The forces to be applied in the proposed hybrid model have been obtained considering the reactions of the ship under critical conditions, namely 30° of heel and 15° of trim. In each case, the design pressures given by Lloyd's Register [6] have been applied.
- There is no contact between the steel tube and the fiberglass panels; however, surface contact has been applied between the fiber laminate of the tubes and the steel tubes.
- The steel structure and the fiber structure do not use the same nodes.
- The edges of the fiberglass panel are simply supported and the steel structure moves in the direction of the applied forces.
- The mesh size is between 1.5 mm and 9 mm, with quadrilateral and triangular elements.
- In the finite element stress results, the occurrence of "hot spots" has been taken into consideration in order to avoid erroneous conclusions.
- ASTM B53 steel is used as the modeling material for the pipes. The following is used for the composite panels: a balsa core of 144 kg/m^3, fiberglass of 450 g/m^3 (Mat 450), bidirectional fiberglass of 400 g/m^3 and 800 g/m^3 (WR 400 and WR 800).

1.2   Behavior of the Hybrid Joint

Critical Zones of the Laminate: To study the hybrid joint between the steel tubular structure and the reinforced fiberglass, a local analysis was carried out on a panel of dimensions 1 m × 1 m with a vertical 2-in. tube and a horizontal (welder) ¾-in. tube. The FRP panel used was of the sandwich type, with a total thickness 37.20 mm; the laminate for affixing the tubes to the panel consist of three layers of fiberglass (Mat450 – WR400 – Mat450) with a total thickness of 2.92 mm.

The loads were applied to the end of the 2-in. tube assuming three possible directions: direction Z (F1), direction X (F2), and direction XZ (F3), with a magnitude of 5 kN (Fig. 1):

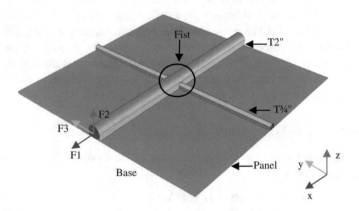

Fig. 1. Analyzed local panel

When analyzing the tubular hybrid joint, two critical zones are present: the fist (laminate at the intersection of the tubes) and the base (laminate close to the area of load application). In both zones, the magnitude of the shear stress increases from the edge of the tubes up to the crown, and finally the contact experiences "boundary bonding."

For the direction Z (F1) (parallel to the panel), the maximum shear stress is found in the fist between the vertical and horizontal tubes, as illustrated in Fig. 2. If the applied force is in direction X to the surface F2, the maximum shear stress is found at the edge of the laminate in the area where the load is applied, as illustrated in Fig. 3; this stress is considered the most critical condition for the tubular hybrid joint.

The size of the critical zone will depend on the direction of the load and its intensity. Figure 4 shows the stress produced by a direction XZ load F3; although the magnitude of the shear stress is lower than in the previous cases, the shear stress occurs at both locations.

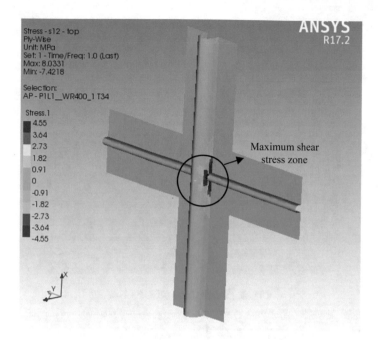

Fig. 2. Shear stress with F1 load

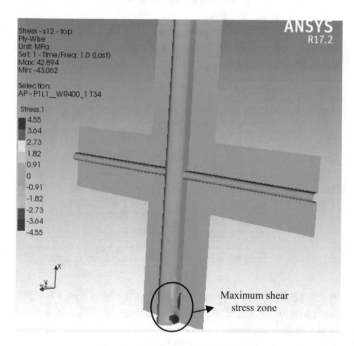

Fig. 3. Shear stress with F2 load

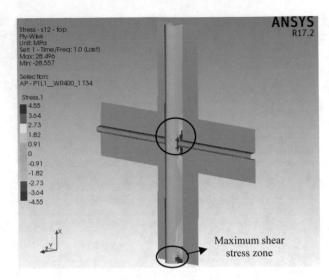

Fig. 4. Shear stress with F3 load

Diameter Variation of Welder Tubes: The results of the variation in the diameter of the welder tubes and the effect on the hybrid joint are presented below. A comparative analysis of the shear stress is presented by varying the diameter of the tube and applying a F3 direction load of 5 kN.

Table 1 shows the comparative relationship of the shear stress at the second fiberglass layer of the tubes (WR400), for the different diameters of welder tube, and the stress in the ¾-in. steel tube, λ_T.

Table 1. Relationship of shear stress to the ¾-in. steel pipe at the second layer, WR 400

Item	Shear stress relationship, $\lambda_T = \gamma_{Ti}/\gamma_{T3/4}$		
	Tube diameter	Fist	Base
1	3/4"	1	1
2	1"	0.983	0.991
3	1 1/4"	0.862	0.873
4	1 1/2"	0.871	0.865
5	2"	0.855	0.844

In Fig. 5 it can be seen that a greater reduction in shear stress is achieved with a 1 ¼-in. steel tube. There is no benefit to using welder tubes with a larger diameter, which would add weight to the structure.

The use of 1 ¼-in. steel tubes is not mandatory; however, it is an option that can be considered for the reduction of shear stress in critical areas of the superstructure.

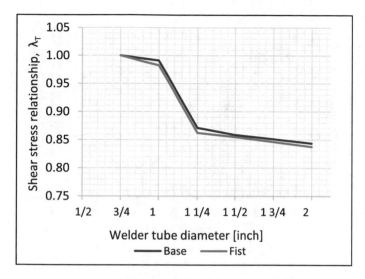

Fig. 5. Shear stress relationship λ_T vs. welder tube diameter

Influence of the Number of Layers in the Hybrid Joint: The laminate supports the entire impact of the loads on the hybrid joint; for this reason it is very important to select a suitable laminate. If this laminate is oversized, the weight will increase.

Table 2 presents several laminate combinations in order to compare the relationship between the maximum shear stress and the maximum shear stress obtained if only one layer of Mat 450 is used, λL. We can see that the greatest reduction in stress occurs in the fist, where using only three layers of fiber reduces the effort by 73% compared to the first layer, while for the base there is a reduction of 54%.

Table 2. Relationship of shear stress to first layer, MAT 450

# of Layers	Shear stress relationship, λ_L		
	Tube laminate	Fist	Base
1	Mat	1	1
2	Mat – Mat	0.442	0.767
3	Mat – WR – Mat	0.274	0.664
4	Mat – WR – Mat – Mat	0.185	0.572
5	Mat – WR – Mat –WR – Mat	0.136	0.484
6	Mat – WR – Mat –WR – Mat – Mat	0.100	0.438

Figure 6 shows the graph of the relationship λL, where it can be seen that for the laminate near the base, the trend after the second layer is almost linear, while the trend in the fist behaves linearly from layer 4 on.

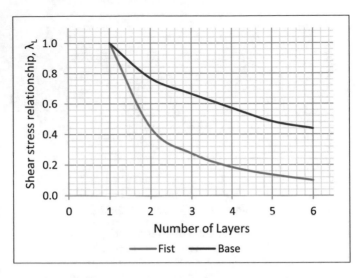

Fig. 6. Shear stress relationship λ_L vs. number of layers

2 Proposed Hybrid Joints

2.1 Study of the Proposed Hybrid Joints

For the study of a typical hybrid joint, two critical areas of the superstructure have been considered, since they are assumed to support greater pressure: the forward bulkhead and the sides of the superstructure. Based on these two zones, a hybrid connection with several geometries is considered. In all cases, the vertical tubes have a length of 60 cm and are joined by a $3 \times \frac{1}{4}$ in. flat bar at the base. The tested options are:

1. Vertical 2-in. tubes spaced 1 m apart and a ¾-in. welder tube at 30 cm from the deck. All tubes are laminated with three layers of fiberglass.
2. Vertical 2-in. tubes spaced 1 m apart and double ¾-in. welder tubes at 30 and 60 cm from the deck, respectively. All tubes are laminated with three layers of fiberglass.
3. Vertical 2-in. tubes spaced 1 m apart and double 1-in. welder tubes at 30 and 60 cm from the deck, respectively. The 2-in. tubes are laminated with four layers of fiberglass and the welder tubes have three layers of fiberglass.
4. Vertical 2-in. tubes spaced 60 cm apart and double 1-in. welder tubes at 30 and 60 cm from the deck, respectively. All tubes are laminated with three layers of fiberglass.
5. Vertical 2-in. tubes spaced 60 cm apart and double 1-in. welder tubes at 30 and 60 cm from the deck, respectively. The 2-in. tubes are laminated with five layers of fiberglass and the welder tubes have three layers of fiberglass.

Of the tests carried out, options 3 and 5 are the most convenient because less inter-laminar shear stress is produced; therefore, option 3 is recommended for side panels, while option 5 is recommended for the forward bulkhead. See Fig. 7.

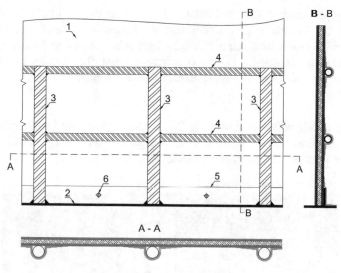

Fig. 7. Tubular hybrid joint

In Fig. 7, the numbers refer to the following:

1. Frontal bulkhead – reinforced fiberglass composite panel, laminated according to Lloyd's Register standards
2. Hull deck, thickness 6 mm – naval steel
3. 2-in. tube – steel Sch40
4. 1-in. tube – steel Sch40
5. Flat bar 75 × 6 mm – naval steel
6. Bolt 9 mm – stainless steel

Hybrid Joint Considerations: For the design of the hybrid joint, the following specifications must be used:

- The laminate of all the tubes must have an overlap in contact with the panel ("boundary bonding"), with a minimum of 25 mm between the layers, following the recommendations of Lloyd's Register [6]. A greater overlap does not improve the laminate.
- The laminate of the 75 × 6 mm flat bar should be two layers of fiberglass Mat 450.
- For the welder tubes it is recommended to use three fiberglass layers: Mat 450 – WR 400 – Mat 450. This is recommended for all zones in the superstructure.
- For the vertical tubes it is recommended to use:
    - Superstructure side: four fiberglass layers, Mat 450 – WR 400 – Mat 450 – Mat 450
    - Superstructure bulkheads: five fiberglass layers, Mat 450 – WR 400 – Mat 450 – WR 400 – Mat 450
- All tubes should be completely welded to the tubes on either side and to the flat bar; for details see [3].

- For the superstructure sides and bulkheads, there should be a vertical tube for each hull-side reinforcement or division and an intermediate reinforcement between them. For the front bulkhead, two intermediate tubes should be located between the reinforcements, depending on the structural scratch. The separation between the beams of the panels of the superstructure depends on the structural calculations made.

Structural Models of Study Areas: In Fig. 8, the loads applied to the superstructure side model are shown. The equivalent loads are generated by a heel of 30° to starboard and a design pressure on the side of 5.16 kN/mm^2.

Fig. 8. Structural model – option 3: superstructure side

Figure 9 shows the equivalent loads applied to the front bulkhead model. The loads are generated by a 15° trim and a front design pressure of 10.08 kN/mm^2.

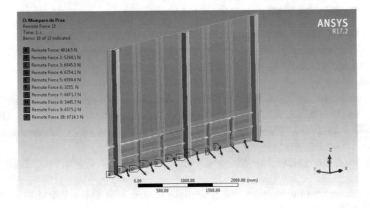

Fig. 9. Structural model – option 5: superstructure front bulkhead

3   Results

3.1   Results for the Superstructure Side Model – Option 3

The results for the shear stress obtained in the proposed model for the superstructure sides are presented in the second layer, as this layer shows greater stress. The maximum stress is generated in the intermediate cuff, with a value of 3.87 MPa, and in the base, with a value of 4.51 MPa.

In Fig. 10 it can be seen that in the fist, the stress starts from the center of the laminate of the 1-in. tube, and its magnitude is distributed approximately 45° towards the "boundary bonding." For the laminate at the base, the maximum stress starts from the corner of the laminate and is also distributed at 45°.

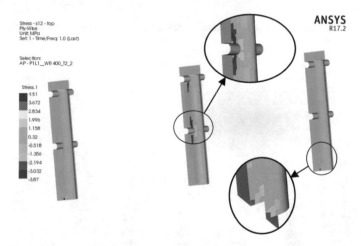

Fig. 10. Shear stress results of the 2-in. tube; second layer – WR 400, superstructure side

The lamination of the welder tubes does not present any significant shear stress. The maximum stress obtained for this case is 3.5 MPa, and occurs only in the area in contact with the fist; however, the average stresses have a value of 0.5 MPa. See Fig. 11.

The aim of the welder tubes and bolts is to contribute to the stiffening of the vertical tubes in order to avoid any possibility of failure.

3.2   Results for the Frontal Bulkhead Model – Option 5

For the proposed model of the frontal bulkhead, the stresses generated in the vertical tube laminate are shown in Fig. 12. It can be seen that the distribution of shear stress along the laminate is different from the previous case due to the forces applied. The maximum stress is generated at the base of the tubes and is distributed along the edge of the laminate to the crown. The maximum value for the stress is 4.11 MPa.

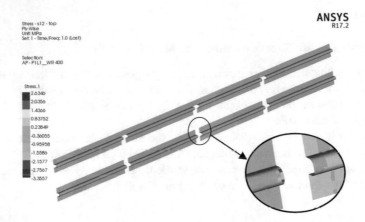

Fig. 11. Shear stress result for the welder tubes laminate; second layer – WR 400, superstructure side

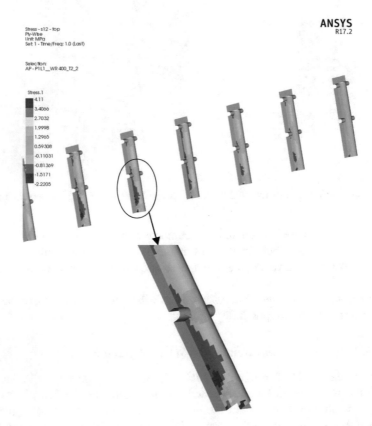

Fig. 12. Shear stress result of laminate of the vertical tubes; second layer – WR 400, bulkhead

4 Conclusion

This joining methodology can be applied to all types of vessels due to its contribution to the reduction of superstructure weight.

The hybrid joints proposed for the superstructure sides and front bulkhead meet the limiting shear stress of the critical layer of 4.55 MPa, as recommended by Lloyd's Register.

4.1 General Comments

The designer has the option to vary the diameters of the vertical tubes and the welder tubes or to increase the number of laminate layers; however, the benefit that would be obtained should be analyzed.

As has been demonstrated in this research, the greatest stress is observed in the vertical tubes. Therefore, it is sometimes convenient to add more fiberglass layers only in the critical areas.

Acknowledgements. Special thanks to ESPOL, Ecuador, for allowing the use of the Ansys software to develop this research.

References

1. Babazadeh, A., Khedmati, M.R.: Finite element investigation of performance of composite-steel double lap adhesive joint under tensile loading. Lat. Am. J. Solids Struct. **14**, 277–291 (2017)
2. Kotsidis, E., Kouloukouras, I., Tsouvalis, N.: Finite element parametric study of a composite to steel join. In: 2nd International Conference on Maritime Technology and Engineering, Lisbon, Portugal (2014)
3. Dominguez, F., Carral, L.: Superstructure design: combination of fiberglass panel and tubular structure with naval steel hull. In: COPINAVAL (2017, in press)
4. Ritter, G., Speth, D., Yang, Y.: Qualifications of adhesives for marine composite-to-steel bonded applications. J. Ship Prod. **25**(4), 198–205 (2009)
5. Rudiger, J., McGeorge, D.: Science and technology of bolt-adhesive joints. Adv. Struct. Mater. **6**, 177–199 (2011)
6. Lloyd's Register: Hull Construction in Composite. Rules and Regulations for the Classification of Special Service Craft (2016)
7. Hentinen, M., Hildebrand, M., Visuri, M.: Adhesively bonded joints between FRP sandwich and metal – Different concepts and their strength behaviour. Technical Research Centre of Finland (1997)
8. Salih, N., Patil, M.: Hybrid (bonded/bolted) composite single-lap joints and its load transfer analysis. Int. J. Adv. Eng. Technol. IJAET **III**, 213–216 (2012). E-ISSN 0976-3945

Information and Communication Technology in Education

Playability Applied in Education

Lídice Haz[1(✉)], Teresa Guarda[2], Marcelo León[2], Marjorie Chalén[3],
and León Arguello[4]

[1] Universidad de Guayaquil, Ciudadela Universitaria Salvador Allende Av. Delta y Av. Kennedy,
Guayaquil, Ecuador
victoria.haz@hotmail.com
[2] Universidad Estatal Península de Santa Elena, Avda. principal La Libertad,
Santa Elena, Ecuador
[3] Escuela Superior Politécnica del Litoral, Campus Gustavo Galindo km 30.5 Vía Perimetral,
Guayaquil, Ecuador
[4] Universidad Técnica Estatal de Quevedo, Campus Ingeniero Manuel Agustín Haz Álvarez, Av.
Quito km. 1 1/2 vía a Santo Domingo de los Tsáchilas, Quevedo, Ecuador

Abstract. In educational institutions, the use of technology to develop academic and administrative activities is increasing. Different mobile devices allow users to perform common tasks quickly and easily. The constant growth of technologies promotes new educational paradigms that integrate diverse learning styles with communication networks and information technologies. The teaching and learning process has undergone modifications and must be coupled to the new digital era. Self-learning through information networks and the use of software allows the creation of new and innovative educational pervasive environments. Playability in education is presented as a didactic resource, which promotes and elevates the students' cognitive and communicative abilities. The proposed system will provide means to aid teaching, making learning more dynamic, and the student environment automated, interactive and motivating. Due to the interoperability of the system, it can be distributed to a larger number of educational institutions, which can be used, thus enhancing the academic landscape.

Keywords: Education · Learning · Technology · Playability · Pervasive games

1 Introduction

Fast technological advancement and the development of computer tools have caused notable changes in education. Teaching strategies and didactic resources used by teachers have been modified by the use of information and communication technologies [1]. Today's society lives in a digital age, which reveals a very clear trend towards larger use of technologies [2]. The use of virtual environments aimed at the educational field, in the sense of providing the generation of knowledge in an innovative and appropriate way to the current society. By itself the existence of technology in the school environment does not guarantee that the student learns, with the teachers being responsible for the organization and planning, in a coherent way, of the resources available for the construction of knowledge. At present the teaching methods that teachers put into practice are very different compared to those used in the past. In the educational field, every day is to implement new technological tools that have in them new material for the teaching of students [3].

Educational materials and didactic resources constitute a medium between the object of knowledge and the cognitive strategies used by teachers. These materials facilitate the transfer of knowledge and its evaluation in the learning process. They promote in students the creativity, the ability to observe, classify, interact, discover or complement knowledge already acquired within their training [4].

With a context-sensitive system it is possible to aid in learning, making the classroom a pervasive experience, where computer devices are connected by wireless or Bluetooth network, providing students with classroom dynamism using the system to be proposed. We intend to analyze the impact of the use of technologies in the learning process and discuss the benefits of using a web platform of parametrizable educational games, categorized in strategies, memory and logic development. Key factors that improve students' skills and abilities, allowing optimize the teaching process.

The paper is organized as follows: the second section provides an overview of the importance of using ICTs in education; the third section discusses some of the main opportunities of playability in education; the fourth section presents the criteria applied in the design and development of the web application of educational games; section five corresponds to the preliminary results of the execution of this project, and finally in the last section the conclusions are presented.

2   Importance of Information and Communication Technologies in Education

The current society is developing in a changing world, with transformations and periods of transition to adapt to the new information and communication technologies. The internet, electronic devices and specific software occupy an important space in the citizen context, influencing behaviors and attitudes [5].

Education has received noticeable changes in its structure since the appearance of technology. ICT's have changed the teaching and learning process, teachers and students access different knowledge networks, capable of integrating collaborative works due to the use of technology. The immediate access to the information, allows acquiring updated knowledge from any place [6].

The UNESCO World Report on Education in 1998 mentions "Teachers and teaching in a changing world", specified the impact that ICTs would have on the educational context, transforming the teaching and learning process, and the way where teachers and students access knowledge and information; creating new interactive learning environments [7].

The technologies have promoted changes in the teaching profession by modifying old educational paradigms by a focus on student-centered teaching.

The current educational innovation is in the use of ICTs. These tools allow teachers and students to generate decisive changes in class sessions; using new strategies, techniques and teaching methods that facilitate the learning process. Content diversification, experimentation, innovation, creativity, the creation of learning communities and openness to dialogue contribute to the development of a holistic training of the human being, [8, 9].

The use of specific computers and software helps the teacher to generate new and innovative learning environments that facilitate the cognitive and creative development of students, obtaining significant learning.

For children, learning through educational games promotes creativity, imagination, design of strategies and cooperation among them. This statement is based on the fact that most of the students' knowledge is constructed through experimentation [10].

The importance of ICT in education not only lies in innovating learning processes; it also functions as a means or channel of communication and exchange of knowledge and experiences. In addition to providing tools to store, process and analyze technical and academic information for the administrative management of an educational entity [9]. The three reasons that justify the use of ICTs in education are the students' digital literacy; productivity and the innovation in the teaching practice. In the case of students digital literacy, allows students to acquire the basic skills in the use of technology. With regard to productivity it optimize working time by taking advantage of the use of ICTs in academic activities, information search, communication, information dissemination, library management, among others. The innovation in the teaching practice generates playful environments that involve the use of technological didactic resources.

For the above mentioned, it is necessary that the educational processes are in accordance with the technological social context that the students live. Teachers must generate new strategies or teaching techniques. Actively link the use of technological tools that help develop or improve capabilities and student skills.

3   Playability in Education

The term playability refers to the set of properties that describe the player's experience before a specific game system, whose main objective is to entertain and successfully entertain one or a group of users [11].

The use of software that implement educational games, encourage the experience and constructivist learning. The student builds his knowledge, based on previous experiences that generate discussions and collaborative debates between the game participants or classmates [12].

The human being is distinguished by the ability to manage their time and leisure/entertainment activities as part of their socio-cultural development. The activities that are carried out during a game promote human creativity and improve the communicative and strategic competences to solve a certain problem [12].

According to studies carried out, it is shown that games help to develop psychological, cognitive, physical and/or social qualities and/or abilities. For an effective application of the playful activity this must be added to the curricular content of the disciplines, and presented as complementary tasks or didactic resources that allow improving the learning process [13].

Education-oriented computer games help improve the teaching and learning process. The proper use of this type of software has a positive impact on students' academic performance. Fun and innovation in classrooms should be a teaching goal. The play activity is attractive and motivating, allows capturing the attention of students in specific subjects in different disciplines. The class is impregnated with a playful communicative environment, allowing each student to develop their own learning strategies [14].

The didactic game as a teaching strategy can be used at any educational level. The structure of a game for educational purposes, must be regulated and include moments of pre-reflective action and symbolization or logical abstract appropriation of the experiences that allow to achieve the teaching objectives established in the curriculum [14].

Several studies and educational reports [12, 14] justify and describe the benefits of using computer games as tools or educational resources, highlighting among their advantages the following: increase in reading comprehension capacity; development of cognitive skills through learning environments based on discovery and creativity; the use of computer games suppose a motivational stimulus for children, which improves the learning process; and it increases the concentration and attention of the students in the resolution of concrete problems due to its playful nature.

The use of a pervasive web platform of educational games promotes the improvement of the teaching and learning process, through the development of cognitive and collaborative skills such as strategy design, attention, communication, tolerance to errors and teamwork.

4   Pervasive Web Application Game-Logic

The use of virtual environments aimed at the educational field, in order to provide the generation of knowledge in an innovative and appropriate way to the current computerized society, is becoming increasingly popular.

Pervasive computing, systems are allied to socio-environmental information (contextual information), which results from data relevant to the system (date, time, location, text, image, video, among others) [15].

Pervasive games are games that use principles of ubiquitous computing to present a new paradigm in which computers should be part of everyday life in an "invisible" way, so that individuals/users do not realize its existence [16]. There are two approaches applied to the pervasiveness of games: a technological and a cultural one. In technological the game is analyzed according to the use of pervasive technology; and in the cultural approach takes into account the world of the game and its relation with the real world [17].

The project Game-Logic has as main objective that the child learn by playing; "Learning to play, playing to learn". Game-Logic software facilitates the acquisition and understanding of knowledge determined by the teacher, and presented implicitly in the game. Educational content is "hidden" in the game mechanisms. According to the model proposed by Vigostky, computer games should be used as a didactic resource and act as mediator in the learning process [18]. The software is focused so that the teachers use it as a didactic resource that strengthens the classes and facilitates the academic evaluation of the student.

The following functional requirements were defined in the development of the application: design oriented for children between 5 and 10 years of age; use of credentials to validate users and access the system; registration of the programmatic content for each subject associated with the teacher and the academic period; visualization of the games for all students according to their work group; the parameterization of games allows the teacher to select the information that will be displayed in the game according to the topic, establish the level of complexity, date of qualification, maximum time of solution, number of attempts allowed, and the reward to be obtained; statistical reports in real time of the solutions obtained by each game, allowing evaluate individually and in groups the results for each student or group of students. These results help determine if it is necessary to reinforce the topic addressed in a next session.

In the first version of Game-Logic have been implemented: puzzle; hanged and memory. In its second version is expected to develop other cognitive games such as alphabet soup, crossword puzzle, riddles, among others. The software design and development process combines usability and playability criteria. This helps to ensure the quality of the game as an experimental interactive system for the player, taking as a reference the ideas shown in of player-centered videogame design methodology [19]. The non-functional requirements of the software are aligned to the specification of the playability requirements. These requirements are analyzed according to the different phases of the playability. In the design phase of the software computer game, design patterns are adapted [11]. Style guides allows qualify the development of the playability based on software properties [11].

Finally, different test scenarios were executed, allowing measure the usability and functionality of the system, as well as the parameters of playability.

The functional architecture is structured in three interactive components that make up the gameplay process implemented in the Game-Logic software. This process is presented as a continuous cycle for the improvement of teaching and learning. The first component allows selecting the theme, to establish the objective to be achieved by selecting different levels of difficulty. These levels are chosen by the teacher according to the complexity of the contents and the age of the student. In case these criteria are not considered, the success of the process of playability in education can be affected. The second component, allows the selection and parameterization of the game, according to the skills and abilities to be developed. The teacher establishes the context of the game determined by the roles of the student, according to the type of group activity or individual. The parameterization consists of setting the difficulty level (high, medium, low), the time to solve the game (free or defined), the number of attempts (minimum 1, maximum 5), and the goal to achieve in the game. The third component, presents the management and follow-up that must be done by the teacher to the students once the game is over. It presents the evaluation of knowledge and skills acquired according to the type of game selected. This information is analyzed by student or group of students.

Defining clear objectives in the teaching process and determining the skills and competences that are to be developed or improved in students are important factors considered in the playful experience of the game. This creates a balance between the player's context and the game's technical design, creating a more constructivist approach to learning. The way in which teachers perceive the teaching process through playability combined with technologies facilitates their socio-technological development and generates greater interaction with students who are immersed in the world of ICTs. Students actively participated in the development of educational games, and that was evidenced in academic performance.

The development of educational game software involves integrating concepts such as interaction-centered design, user experience and pedagogical parameters for teaching children. These criteria will determine the success or failure of computer based learning.

In the initial tests of software functionality, it was observed that the motivation of the students is related to four factors described as:

- Combine criteria of usability and playability in the design of the game, in order to improve the experience and satisfaction of the player;
- Integrate pedagogical concepts in the game mechanics, whose purpose is to organize the contents presented in the game;
- Produce games based on the constructivist theory that determines actions of learning by error and correction of the player;
- Design more realistic gaming experiences by applying new technologies to create immersive or semi-immersive virtual environments;
- Software portability and compatibility, it fundamental that the game can be run on various operating systems and mobile devices.

For future versions of Game-Logic, the project migration is set to run on different operating systems and mobile devices. It also enhances user interfaces and integrates a wider selection of framed games always in the development of students' cognitive and communicative skills.

5 Conclusions

In the classroom there are activities of various areas of knowledge. Scientific efforts are aimed at reducing repetitiveness in teaching. During the learning process, students share tasks, so the challenges are resolved in a dynamic way, when there is a sum of individual efforts of each student. Education has been shrouded in a continuous process of changes since the inclusion of technology, teachers use new teaching methods for their students. Learning theories such as connectivism and constructivism play an important role in student learning, experimenting, sharing ideas and having access to information from the internet allows to acquire new knowledge from these activities.

Computer games are presented as a teaching strategy in basic education. They complement the traditional teaching, showing the advantages that contribute to the cognitive and communicative development of students.

The use of pervasive web educational games allows the teacher to strengthen the contents of the curriculum and improve the presentation of the material taught in classes. The proposed software involves the integration of various logical games that together create a playful environment and generate meaningful learning. Its architecture in web environment facilitates the distribution and easy access to this didactic resource, in addition to allow evaluating in an effective and fast form the knowledge of the students. Functionality testing in pre-production environments facilitates the evaluation of software functionality and usability based on evaluation of teachers and students scores in relation to the improvement of learning.

References

1. Gómez Gallardo, L., Macedo Buleje, J.: Importancia de las TIC en la educación básica regular. Inv. Educ. **14**(25), 209–226 (2014). ISSN 1728-5852
2. Moreira, F., Rocha, A.: A special issue on new technologies and the future of education and training. Telematics Inform. **34**(6), 811–812 (2017)
3. Domingo, M., Marqués, P.: Aulas 2.0 y Uso de las TIC en la Práctica Docente. Comunicar **19**(37), 169–174 (2011)
4. Moreira, A.: Algunos principios para el desarrollo de buenas prácticas pedagógicas con las TIC en el aula. Comun. pedagog. Nuevas Tecnol. Recur. Didáct. **222**, 42–47 (2007). ISSN 1136-7733
5. Salas, J.: Importancia de las TIC en Educación. http://www.importancia.org/tic-en-educacion.php. Accessed 2012
6. Karsenti, T., Collin, S.: TIC ct éducation: avantages, défis et perspectives futures. Éduc. Francoph. **41**(1), 1–6 (2013)
7. Unesco: Conferencia Mundial sobre la Educación Superior: la Educación Superior en el Siglo XXI: Visión y Acción, París (1998)
8. Cabero Almenara, J., Llorente Cejudo, M.C., Román, P.: La tecnología cambió los escenarios: el efecto pigmalión se hizo realidad. Comunicar **15**(28) (2007)
9. Libedinsky, M.: Diseño y Documentación de Experiencias de Aula, pp. 139–141. Docencia Universitaria (2011)
10. Carreño, L.: Constructivismo y Educación. Propues. Educ. **32**, 112–113 (2009)
11. González Sánchez, J.L., Padilla Zea, N., Gutiérrez Vela, F., Cabrera, M.: De la Usabilidad a la jugabilidad: diseño de videojuegos centrado en el jugador. In: Proceedings of Interaction, pp. 99–109 (2008)
12. Mcfarlane, A., Sparrowhawk, A., Heald, Y.: Report on the educational use of games: an exploration by TEEM of the contribution which games can make to the education process (2002)
13. Ibrahim, A., Gutiérrez Vela, F.L., González Sánchez, J.L., Padilla Zea, N.: Educational playability analyzing player experiences in educational video games. Int. J. Game Based Learn. **2**(4), 23 (2012)
14. Zea, N.P., Sánchez, J.L.G., Gutiérrez, F.L., Cabrera, M.J., Paderewski, P.: Diseño de videojuegos colaborativos y educativos centrado en la jugabilidad, pp. 191–198
15. Portela, F., Santos, M., Silva, Á., Machado, J., Abelha, A., Rua, F.: Pervasive and intelligent decision support in intensive medicine – the complete picture. In: Bursa, M., Khuri, S., Renda, M.E. (eds.) Information Technology in Bio- and Medical Informatics. Lecture Notes in Computer Science (LNCS), vol. 8649, pp. 87–102. Springer, Cham (2014)
16. Guarda, T., Vaca, O., Pinguave, M., Maldonado, E., Augusto, M., Orozco, W., Pinto, F.: Wargames applied to naval decision-making process. In: Rocha, Á., Correia, A., Adeli, H., Reis, L., Costanzo, S. (eds.) World Conference on Information Systems and Technologies, pp. 399–406. Springer, Cham (2017)

17. Nieuwdorp, E.: The pervasive discourse: an analysis. Comput. Entertain. (CIE) **5**(2), 13 (2007)
18. Payer, M.: Teoría del constructivismo social de Lev Vygotsky en comparación con la teoría Jean Piaget, Caracas, Vanezuela (2005)
19. Gonzalez Sánchez, J.L., Gutiérrez Vela, F.L., Cabrera, M.: Diseño de videojuegos colaborativos adaptados a la educación especial. In: V Taller en Sistemas Hipermedia Colaborativos y Adpatativos, vol. 1(7), pp. 17–24 (2007). ISSN 1988-3455

Teaching-Learning System Using Virtual Reality and Haptic Device to Improve Skills in Children of Initial Education

Marco Pilatásig[1(✉)], Emily Tobar[1], Liseth Paredes[1], Franklin Silva[1], Andres Acurio[1], Edwin Pruna[1], and Zulia Sanchez[2]

[1] Universidad de las Fuerzas Armadas ESPE, Sangolquí, Ecuador
{mapilatagsig,ektobar,mlparedes2,fmsilva,adacurio,
eppruna}@espe.edu.ec
[2] Unidad Educativa Mario Cobo Barona, Ambato, Ecuador
zulia.sanchez@educacion.gob.ec

Abstract. A virtual system for teaching - learning is presented using a haptic device, to improve the skills in children in initial education. Two interactive games were created using Unity 3D software. The first allows identifying the primary colors and notions of space. The second identifies the geometric figures as circle, triangle, square and rectangle. Both games has visual, auditory and strength feedback, which allows performing the tasks correctly. The outcomes allow evaluating the children learning process in initial education in a qualitative way by a educator.

Keywords: Virtual reality · Haptic device · Initial education
Learning strategies

1 Introduction

This template, modified in MS Word 2003 and saved as "Word 97-2003 & 6.0/95 – RTF" for the PC, provides authors with most of the formatting specifications needed for preparing electronic versions of their papers. All standard paper components have been specified for three reasons: (1) ease of use when formatting individual papers, (2) automatic compliance to electronic requirements that facilitate the concurrent or later production of electronic products, and (3) conformity of style throughout a conference proceeding. Margins, column widths, line spacing, and type styles are built-in; examples of the type styles are provided throughout this document and are identified in italic type, within parentheses, following the example. Some components, such as multi-leveled equations, graphics, and tables are not prescribed, although the various table text styles are provided. The formatter will need to create these components, incorporating the applicable criteria that follow.

The learning in childhood is the base of educational development. Studies show that human intelligence can be greatly increased if it is stimulated early at home and school. Nowadays, the new technologies could help a lot in education; unfortunately, early childhood does not have the enough researching [1]. A complete education is more useful when it is a commitment of all members, especially teachers, who must structure each process [2] and organize the study groups correctly [3]. This is possible when the child has the resources and the support enough from the family to develop as expected. Many factors influence learning like family, society, nutrition and cognitive abilities [4]. A little-known fact that affects the children development is the learning trouble, when child does not progress in the way expected. According to studies carried out by UNICEF in Ecuador, it is said that the main problems in children learning are "attention deficit, difficulty of compression and low level of reasoning" [5]. Nevertheless, problems related to learning are not always linked to a low intellectual capacity. These problems can differ in specific abilities like reading, writing or calculation deficits [6].

The teaching-learning process in scholar centers in the country is based on an Initial Education Curriculum, which considers interculturality in equal learning opportunities [7]. The contributions mainly of Vigotsky are taken into consideration in the curricular design issued by the Ministry of National Education in Ecuador, where learning in boys and girls is both a process and a product. "Consider that learning promotes the development and establishes that teaching always anticipates learning" [8]. Game is the methodological principle that stands out. This essential activity in children involves them in a general way, also stimulates the development and learning in all areas [7].

Games encourage children to be active, relate to society and the environment that surrounds them when comparing knowledge with objects of daily life. When children play, they explore and experiment safely. While they are learning about the environment, they solve problems and acquire new skills [9].

The existence of new alternatives for teaching must be constantly considered. Technology advances can't be ignore because they can contribute to the improvement and update of education. Several years ago, countries seek to introduce new technologies in education [10–15].

Computer is a very versatile tool in the teaching-learning process, it has advantages and disadvantages and these have been analyzed in education [16]. Computers advantages are a lot, to mention a few: computers can work with virtual reality, augmented reality, create 3D environments and work together with other devices to give sensations to users. This makes it a very useful even for education of people with special abilities.

Virtual Reality (VR) allows creating realistic and dynamic images along with diverse sensory information like hearing and touch. This fact allows the user to interact within the virtual 3D model [9]. In recent years, the field of Virtual Reality application has extended from entertainment to interactive teaching for engineering or medical training [17, 18]. In addition, it has been possible to encourage learning and convert it from a routine teaching process to a fun activity. This process showing that it is helpful when delivering knowledge in any subject [19]. A research includes virtual reality in games with a dynamic interface in a multi-touch panel, where child can make drawings and identify figures. These techniques are entertaining and they have had good outcomes in children [20–22].

When virtual reality works with haptic devices allows the child to touch, feel, manipulate and remodel virtual objects. Through haptic feedback the user feels the geometry, texture, smoothness, vibrations of objects [25]. Thus it can achieve a greater concentration of the infant and reinforce the knowledge acquired [23]. The results are satisfactory like the improving of handwriting in children of five years. Through using a haptic device combined with a virtual reality, software allows children to practice and improve calligraphy in an early age [24]. An entertaining tool is helpful in learning when leaves the typical teaching methodology. User is completely immersed in a virtual world and face situations based on real cases with the certainty that he maintains his personal security. User can apply for increase their knowledge in a shorter time [25].

In this context, a 3D virtual reality application is presented that complements the teaching-learning process in children in early childhood education. A haptic device that obtains the movement generated by child builds the system. This fact allows evaluating skills, notions and knowledge of the user.

2   Ease of Use

This chapter exposes the different stages of the virtual system; in Fig. 1 each one is indicated in a block diagram.

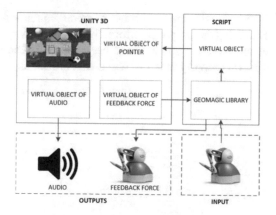

Fig. 1. Block diagram of virtual system stages.

2.1   Input Peripherals

In order to get the input data in the system, a haptic device called Geomagic Touch is used. This device has digital encoders inside, which allows getting the variables of user movement like position and angles while traveling in a virtual environment.

2.2   Scripts Development

The administration of the inputs and outputs are handled by the scripts made in Visual Studio with C# language. Scripts execute several functions in the virtual environment according to the data that is received through the input peripherals. Data is interpreted with the support of the libraries in C#. The handling of the information is illustrated in the flow diagram of Fig. 2.

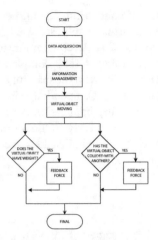

Fig. 2. Flowchart of information management in the virtual system

In general, the different environments follow the same sequence for interaction with virtual objects. Programming the scripts the information is stored for the number of times that both the correct and incorrect options have been selected. The auditory feedback allows the user to perform properly in the virtual environment, as well as the response

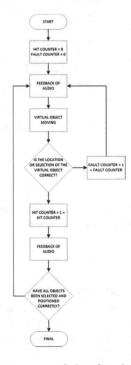

Fig. 3. Operation sequence of virtual environments flowchart.

to a hit or a failure. The forces feedback provides sensations that simulate the management of real objects (Fig. 3).

2.3   Design of Environments in Unity3D

The environments are designed with several virtual objects, which are created based on images or obtained directly from the Unity3D Asset Store. Virtual objects are assigned with a set of properties like animations, rigidity, weight and audio. These features belong to Unity libraries and provide a better experience for the user when operating the virtual environment. Features can also be managed from a script, where the programming is done according to the task that wants to be executed. This script is associated to a virtual object with the sequences of orders and tasks to be executed in the system.

3   Use of the System

In the main screen a menu is displayed which allows choosing the game according to the level of knowledge that the user is. Figure 4 presents the main menu of the system.

Fig. 4. Main menu of the teaching-learning system

Level 1 is a virtual interface that has an introduction of the primary colors. Then the user chooses the color from the elements that appear randomly and places them according to the objective organized by the notions of space. The indications are provided by an audio (primary colors and notions of space are reinforced). The system provides a force feedback when a virtual object collides with another. In addition, the audio indicates when the task was completed (Fig. 5).

Level 2 provides a introduction of the geometric figures. This interface shows a landscape with objects that can be related to the geometric figures (door-rectangle, ball-circle). User must select the element according to the indications given by audio. The elements are hidden as long as their selection is correct (Fig. 6).

Fig. 5. Game interface of primary colors and notions of space.

Fig. 6. Game interface of geometric figures task.

4   Test and Results

4.1   Test

This process must be carried out under the supervision of an educator or kindergarten teacher. This person evaluates the child's performance. The game starts by presenting the menu with the two levels. When the first level starts child can listen and see a presentation about the primary colors, once this is done, the first instruction is indicated by selecting a color (Fig. 7(a)).

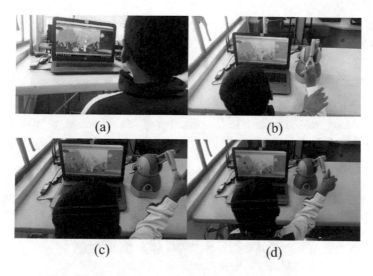

Fig. 7. First level of the virtual system being used by a child.

If the chosen color does not correspond to the order, the system sends an audio response indicating that the option is not correct, as soon as the correct color is selected, a silhouette of an object located in the interface is colored (Fig. 7(b)).

The following instructions give the place where object should be located by taking the object with the cursor. The child can feel the weight that has the object due to the forces feedback of the haptic device (Fig. 7(c), (d)).

In the second level, the main geometric figures are presented as an introduction (Fig. 8(a), (b)). Next, the child listens to the first instruction and selects the figures according to the specified form. In the previous level, the audio response indicates when

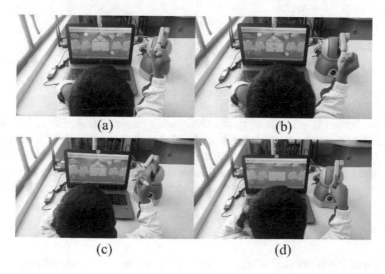

Fig. 8. Second level of the virtual system being used by a child

the selected object is incorrect or correct; the objects disappear from the environment when they are selected (Fig. 8(c), (d)).

4.2 Results

The table of results for level 1 and 2 is presented in Figs. 9 and 10, respectively. The respective successes and failures to the concepts of special notion, primary colors and geometrical figures are shown.

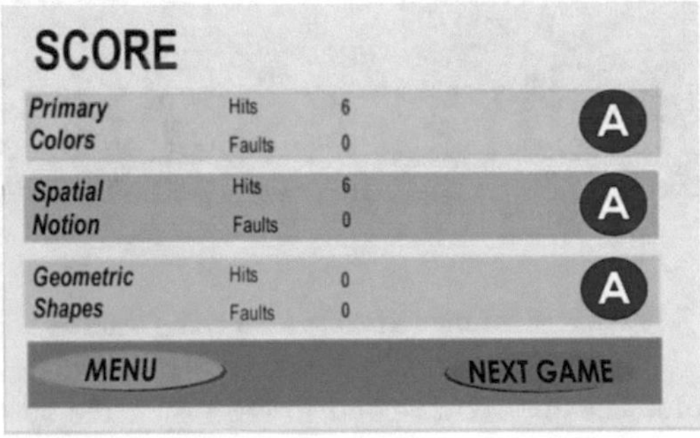

Fig. 9. Table of results in level 1

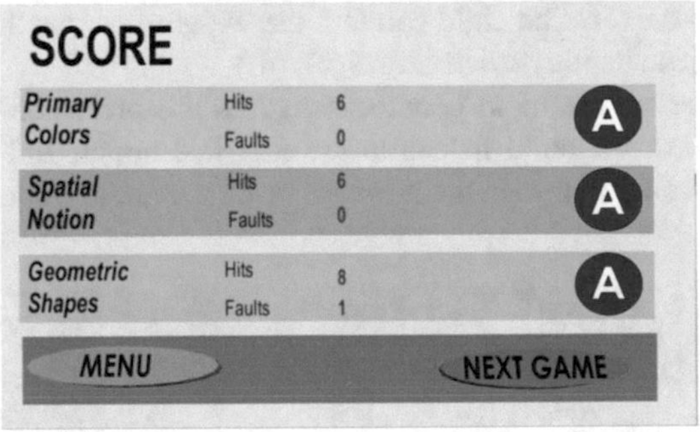

Fig. 10. Table of results of the first level and second level

The evaluation is carried out qualitatively with parameters A, EP and I. Where A represents "Acquired", which means that the child has the necessary knowledge in the subject. EP "In Process", is equivalent to a regular performance of the student. It has bases in the topics discussed, but is still learning. I corresponds to "Insufficient", and indicates that the child does not have the knowledge enough or has difficulty in learning.

5 Conclusions

The design of the environments is attractive for children in early childhood education: These allows them to focus their attention on the system. Teaching through interactive games let capturing and learning faster the topics. The use of objects that can be found in real life helps them to associate what they have learned. The use of Geomagic Touch as a haptic device provides a better experience with the environment. The use of force feedback gives the child the feeling of managing a real environment.

This system helps to complement and evaluate basic topics that are taught in the classrooms. It allows the educator to encourage the child's interest in learning and provides the evaluation of the student's performance in qualitative assessments. The development of new environments to cover the different subjects of teaching in initial education will help to complement the development of cognitive and psycho-motor skills. The system represents a support for the teachers.

Acknowledgment. We acknowledge to "Universidad de las Fuerzas Armadas ESPE" by financing fund the research project 2016-PIC-0017.

References

1. Ankshear, C., Knobel, M.: New technologies in early childhood literacy research: a review of research. J. Early Child. Lit. **3**, 59–81 (2003)
2. Skaalvik, E., Skaalvik, S.: Teachers' perceptions of the school goal structure: relations with teachers' goal orientations, work engagement, and job satisfaction. Int. J. Educ. Res. **62**, 199–209 (2013)
3. Blatchford, P., Kutnick, P., Baines, E.: Toward a social pedagogy of classroom group work. Int. J. Educ. Res. **39**, 153–172 (2003)
4. Méndez, R.: Investigación y Planificación Para El Diseño De Un Aula De Apoyo Psicopedagógico y Aporte De La Misma Al Desarrollo y Seguridad De La Educación De Niños Con Dificultades De Aprendizaje (Tesis de Maestría), Instituto de Altos Estudios Nacionales, Quito (2003)
5. Briones, M.: Guía metodológica correctiva integral neuropsicológica para dificultades específicas de lectura y escritura en niños/as de 3.er año de educación básica del Colegio Experimental El Sauce de Tumbaco (Tesis de Pregrado), Universidad Politécnica Salesiana, Quito (2013)
6. Muñoz, X.: Representaciones y actitudes del profesorado frente a la integración de Niños/as con Necesidades Educativas Especiales al aula común. Rev. Latinoam. Educ. Inclusiva **3**, 25–35 (2008)
7. Guía metodológica para la implementación del currículo de educación inicial, pp. 7–15. Ministerio de Educación del Ecuador, Quito (2015)
8. Currículo de educación inicial 2014, pp. 12–16. Ministerio de Educación del Ecuador, Quito (2014)
9. Satava, R.: Virtual reality: current uses in medical simulation and future opportunities & medical technologies that VR can exploit in education and training. Proc. University of Washington Medical Center, USA, March 2013
10. Vahtivuori-Hänninen, S., Halinen, I., Niemi, H., Lavonen, J., Lipponen, L.: A new finnish national core curriculum for basic education (2014) and technology as an integrated tool for learning. In: Niemi, H., Multisilta, J., Lipponen, L., Vivitsou, M. (eds.) Finnish Innovations and Technologies in Schools, pp. 21–32. Sense Publishers (2014)

11. Martínez, E.V., Villacorta, C.S.J.: Spanish policies on new technologies in education. In: Plomp, T., Anderson, R.E., Kontogiannopoulou-Polydorides, G. (eds.) Cross National Policies and Practices on Computers in Education. Technology-Based Education Series, vol. 1, pp. 397–412. Springer, Dordrecht (1996)
12. Wall, K., Higgins, S., Smith, H.: The visual helps me understand the complicated things': pupil views of teaching and learning with interactive whiteboards. Br. J. Educ. Technol. **36**, 851–867 (2005)
13. Mar, N.Y.: Utilizing information and communication technologies to achieve lifelong education for all: a case study of Myanmar. Educ. Res. Policy Pract. **3**, 141–166 (2004)
14. Selwyn, N., Bullon, K.: Primary school children's use of ICT. Br. J. Educ. Technol. **31**, 321–332 (2000)
15. Peltenburg, M., Van Den Heuvel-Panhuizen, M., Doig, B.: Mathematical power of special-needs pupils: an ICT-based dynamic assessment format to reveal weak pupils learning potential. Br. J. Educ. Technol. **40**, 273–284 (2009)
16. Mangen, A., Walgermo, B., Bronnick, K.: Reading linear texts on paper versus computer screen: effects on reading comprehension. Int. J. Educ. Res. **58**, 61–68 (2013)
17. Dinis, F.M., Guimarães, A.S., Carvalho, B.R., Martins, J.P.P.: Development of virtual reality game-based interfaces for civil engineering education. In: 2017 IEEE Global Engineering Education Conference (EDUCON), pp. 1195–1202. IEEE, April 2017
18. Elliman, J., Loizou, M., Loizides, F.: Virtual reality simulation training for student nurse education. In: 2016 8th International Conference on Games and Virtual Worlds for Serious Applications (VS-Games), pp. 1–2. IEEE, September 2016
19. Zhang, K., Liu, S.J.: The application of virtual reality technology in physical education teaching and training. In: 2016 IEEE International Conference on Service Operations and Logistics, and Informatics (SOLI), pp. 245–248. IEEE, July 2016
20. Yu, X., Zhang, M., Xue, Y., Zhu, Z.: An exploration of developing multi-touch virtual learning tools for young children. In: 2010 2nd International Conference on Education Technology and Computer (ICETC), vol. 3, pp. V3–V4. IEEE, June 2010
21. Chaney, C.: Language development, metalinguistic skills, and print awareness in 3-year-old children. Appl. Psycholinguist. **13**, 485–514 (1992)
22. Clements, D., Swaminathan, S., Zeitler, M., Sarama, J.: Young children's concepts of shape. J. Res. Math. Educ. **30**(2), 192–212 (1999)
23. Merwan, A., Maud, M., Adrian, G., Hiroyuqui, K.: FlexiFingers: multi-finger interaction in VR combining passive haptics and pseudo-haptics. In: IEEE Symposium on 3D User Interfaces (3DUI)/Los Angeles, CA, USA, pp. 103–106, March 2017
24. Palluel-Germain, R., Bara, F., De Boisferon, A.H., Hennion, B., Gouagout, P., Gentaz, E.: A visuo-haptic device-telemaque-increases kindergarten children's handwriting acquisition. In: Second Joint EuroHaptics Conference and Symposium on Haptic Interfaces for Virtual Environment and Teleoperator Systems, World Haptics 2007, pp. 72–77. IEEE, March 2007
25. Ahonen, T., O'Reilly, J.: Convergence of Broadband Internet, Virtual Reality and the Intelligent Home, Digital Korea, pp. 37–54 (2007)

Use of Web 2.0 Tools for Collaborative Learning in Aspiring Officers of the Ecuadorian Navy

Belén Gómez[1,2], Gloria Valencia[2,3(⊠)], Oscar Barrionuevo[2,4], and Diego Dueñas[2,5]

[1] Escuela Superior Politécnica del Litoral, Guayaquil, Ecuador
[2] Universidad de las Fuerzas Armadas-ESPE, Sangolquí Quito, Ecuador
{mbgomez4,gmvalencia,oabarrionuevo,
dxduenas}@espe.edu.ec
[3] Universidad Nacional Mayor de San Marcos, Lima, Peru
[4] Armada del Ecuador, Guayaquil, Ecuador
[5] Fuerza Aérea Ecuatoriana-FAE, Quito, Ecuador

Abstract. The present study determined the level of knowledge and use of the different Web 2.0 tools, in order to enhance the application of these in the learning process for candidate officers of the Ecuadorian Navy. The information was obtained through an online survey where Web 2.0 questions were addressed to a sample of 128 candidates from first to fourth year. The analysis of this result shows the acceptance of these tools by students, highlighting the convenience of generating academic activities through the use of web 2.0 tools and improving collaborative and autonomous performance and learning capacity.

Keywords: Web 2.0 · Collaborative learning

1 Introduction

In higher education news methods for teaching and learning arise. When preparing students for their inclusion in the work environment the universities bet on new methods of learning based on Web 2.0 and its collaborative tools for bettering apprehension of knowledge and collaboration between students allowing them to participate actively through the interchange of ideas, information, experience and knowledge [1, 2]. Due to the range of new information technologies, traditional learning has changed. Today distance, time or access to the information is no longer a limit in the learning process of the students. Without doubt it requires planning and preparatory work in reactions, consulting guides and elements of learning that use technological resources to achieve a process of appropriate learning. To verify the frequency of use and the influence of web 2.0 tools we developed a survey for 128 candidates from the first to fourth year at the Senior Naval School, Comandante Rafael Moran Valverde (ESSUNA). It had a virtual participation and group work platform, SAKAI, where the students contribute with information, videos and links. They can also sit exams created by the teacher.

© Springer International Publishing AG, part of Springer Nature 2018
Á. Rocha and T. Guarda (Eds.): MICRADS 2018, SIST 94, pp. 317–325, 2018.
https://doi.org/10.1007/978-3-319-78605-6_27

2 Literary Revision

2.1 Web 2.0 Tools

Web 2.0 tools are a combination of instruments, similar to an inclusive social web that allows the reading, creation and sharing of content with groups of users that share common interests [3, 4]. They are internet tools and services that encourage the visitors to share, contribute and edit information [5]. The appearance of these tools has created new opportunities for creating and sharing content and interact with others. They include tools that allow the individual and collective publication, exchange of images, audio, video and the creation and maintenance of social networks online [6]. The universities should act to guarantee the use of web 2.0 tools. It's necessary to think carefully and investigate the best way to take advantage of these emerging tools to push the teaching and learning process [5]. It needs a focused investigation of the use and daily learning of the students with the web 2.0 tools inside and outside of the classroom [6]. At present there exists a wide range of web 2.0 tools that can be used for free on the internet. For our analysis we will consider the following classifications of web 2.0 tools [4]: Learning Management Systems; Wikis; Social Networks, Office 2.0; File storage; Multimedia and Collaborative groups.

The Learning Management Systems (LMS) are sites that allow the student community to manage content, collaborative and communication of different assignments of the members of a course, like Joomla, Moodle, Sakai, among their key characteristics it can be accessed by a standard web browser and it doesn't need additional programmes. The basic functions that it contains, such as user administration, authenticity mechanisms, permission management, course management, content authorship, test and evaluation of the students, make this type of tool easy to use.

Wikis are content collaboration sites that allow users to create, edit and eliminate publications from the browser that has a link or website added by the author. The most well-known tools are: Wikipedia, Wikibooks, Wiki and Wetpain. A wiki is an online system [7]. The wikis can be used successfully to share knowledge in an educational context [8]. They are used for student projects that collaborate with ideas, organize documents, individual and group resources. Also, it can be used as a presentation tool, like an electronic portfolio for projects of group investigation for a specific idea, administrate school documents and the classroom, use like a co-working folder for the students, writing, books and newspapers written by students, create and maintain frequent questions, like a discussion area in the classroom, a place to add web resources, help committees, work groups and university projects [5].

The Social Networks allow individuals to interact socially, sharing multimedia content (images, videos, songs) with other users, for example LinkedIn, Periscope and Tumblr. The characteristics that associate this type of tool are: Web services where you build a profile within a limited, closed system; and users create the structure where a list shares established connections and creates more for other users inside the same system.

These social networks promote interactions and allow users to communicate and share information and personal experiences. Indeed the objective of the social networks is the interchange of knowledge and learning both at a personal and an organizational level [9].

Office 2.0 offers users opportunities to create, edit or delete documents, spreadsheets and presentations in real time. It has storage in the cloud, among these we find Office web, Google docs, Open Office Org.

In the case of File Storage users can store, synchronize, and share any type of information in applications such as Google Drive, Sky Drive, Dropbox, iCloud. Its capacity of free storage is between 2 to 25 GB.

In the Multimedia context users can see videos in a dynamic interactive way through tools such as YouTube, Vimeo and Ustream. It's considered a web 2.0 tool that can be utilized for the generation of knowledge through observation and social interaction with an academic focus [10]. This refers to the professional development of videos, students creating videos or using video-sharing sites to find videos of current themes [5]. **Sample Heading (Third Level).** Only two levels of headings should be numbered. Lower level headings remain unnumbered; they are formatted as run-in headings.

Sample Heading (Forth Level). The contribution should contain no more than four levels of headings. The following ¡**Error! No se encuentra el origen de la referencia.** gives a summary of all heading levels.

2.2 Collaborative Learning

Collaborative learning is defined as a combination of techniques that allow individuals to interact and exchange knowledge, experiences and ideas thereby educating each other and creating a process that generates, transfers and understands new knowledge [11].

Aside from the appropriate use of collaborative tools one can identify 2 main types of platforms [12]: Learning Management Systems or tools that allow the administration of virtual courses based on the web and content management systems, whose function is the administration of content for a learning programme.

2.3 Model of Collaborative Learning

There are at least three forms of applying collaborative learning which are differentiated for the performance level of the participants [2].

(1) Couple interaction: They are groups of students with different ability levels where the teacher acts as a mediator in the learning experience and the group works together in an organised way.
(2) Couple tutoring: They're groups of higher-ability students that have had some training. The coach helps to teach lower-level classmates. This increases the level of the learners but also reinforces the coach's knowledge.
(3) Collaborative groups: They're larger-sized groups that link rookies of various ability levels. This stimulates the interdependence and ensures the involvement of everyone in the learning process as Individual success depends on the group success. In this scenario the teacher should be more than just a mediator [13].

The collaborative learning that you get through the interaction of a group of students that share information produces cognitive processes and a higher rate of learning.

The benefit of varying this with individual learning is that the interaction helps answer the doubts of some students and reinforces the knowledge of others.

2.4   Benefits of Collaborative Learning

Introducing a collaborative learning scheme helps the students to develop their social skills and actively participate in the search and generation of knowledge. It improves the performance of all the students, not just the best ones. You learn not just through books and text but also with the experiences of the participants [14].

2.5   Virtual Platforms for Collaborative Learning (MOODLE)

It is a free software tool for feedback of work done by various institutions and participants that collaborate online, which allows them to freely access and incorporate various modules and courses created by other users [15, 16]. This project is developed as a form of education that's based in the philosophy of learning. More precisely in Social Constructive Learning Theory. It's not just a tool for distance learning but increasingly being used as a complement to teaching in person [15]. The principles of the platform allow autonomous learning and collaborative teaching.

3   Methodology

This work was based in a study with a quantitative focus that allowed the collection of information about the use of web 2.0 tools in ESSUNA to analyse and measure them.

To get the information a survey was created for 128 students of ESSUNA between 18 and 24 years that study between the first and fourth year of the course. The survey consisted of two parts: one demographic, to have a general idea of the profile of the students being surveyed and a second part that gained information about the use of Web 2.0 tools. This had multiple answer questions, regarding the tools' use and application in the learning process.

This survey was carried out using the Google Forms tool that can create questionnaires. This technique was used because it offered the best way to get access to the Ecuadorian Navy officials, given the military regime they follow.

4   Analysis and Results

As a consequence of the analysis of the results obtained through the surveys, the following stood out:

4.1   Demographic Information

The results indicate that the students were in an age range of 18 to 24 years, with 28% being 21-years-old (Fig. 1).

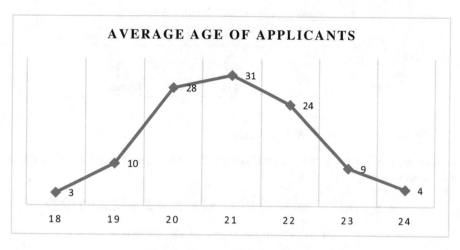

Fig. 1. Average age of applicants

The male population makes up 80%, while the female population comprises 20% of the surveyed people (Fig. 2).

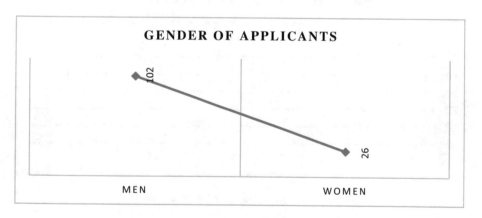

Fig. 2. Gender of applicants

4.2   Knowledge and Use of Web 2.0 Tools

In relation to the knowledge and use that the students have with the term, Web 2.0 tools, we found that 62% of the students don't understand the concept (Table 1).

Table 1. Knowledge of Web 2.0 tools

Do you know the web 2.0 tools?	Number of students	Percentage value
SI	48	38%
NO	80	62%
TOTAL	**128**	**100%**

4.3   Frequency of Use of Web 2.0 Tools

To measure the frequency that the students use the web 2.0 tools the following scheme was developed (Fig. 3):

$$A = ALWAYS - AA = ALMOST\ ALWAYS - S = SOMETIMES$$

$$VL = VERY\ LITTLE - N = NEVER$$

The results showed that 51% of the students almost always use Web 2.0 tools, and every student has used at least one of these tools.

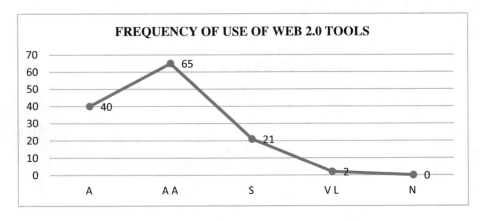

Fig. 3. Frequency of use of Web 2.0 tools

4.4   Web 2.0 Tools Most Used by the Candidates

In the study it shows that 87% of the candidates use the Course Management System (SAKAI), making it the tool they most used. In second place was Youtube an application that allows you to generate and share videos, which was used by 78%.

In third place you find wiki applications and blogs with a 70% acceptance of the candidates. These applications allow candidates to create content between different users to generate information. Another aspect to consider is that 22% of the surveyed students used other types of Web 2.0 tools (Fig. 4).

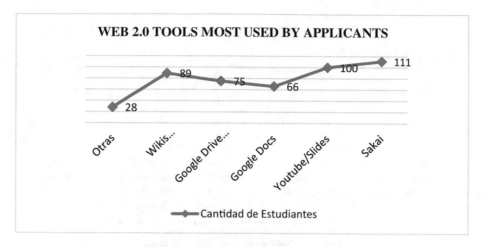

Fig. 4. Web 2.0 tools most used by applicants.

4.5 Subjects That the Candidates Use Web 2.0

Another relevant point is that this study defined how many subjects the students used the web 2.0 tools for as part of the learning process. The analysis reveals that 56 students use the tools in 3 to 5 subjects. Just 14 of them use the tools in 6 to 10 subjects. Considering that the average number of subjects that a student has per year is 11, it shows that they are using the tools in 36% of their subjects (Table 2).

Table 2. Subjects that Web 2.0 tools apply

Number of subjects	Number of students	Percentage
0 a 2	17	13%
3 a 5	56	44%
6 a 10	42	33%
10 o más	13	10%
TOTAL	**128**	**100%**

4.6 Collaboration and Web 2.0 Tools in the Learning Process

With regards to how the web 2.0 tools help the learning process we see that 55% of them agree that these tools allow the exchange of knowledge, while just 28% believe that the web 2.0 tools allow collaboration with other people (Fig. 5).

After statistical analysis the results show three aspects. One, is that the tools most commonly used were the course management system, YouTube and the wikis. Also it showed that 62% of the candidates didn't know the term 'web 2.0 tools', despite the fact that 51% of them use them almost always.

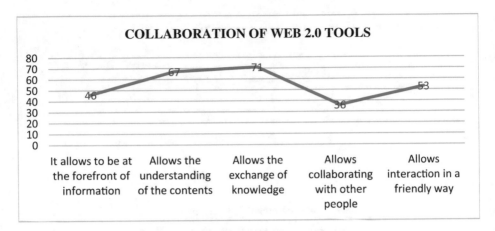

Fig. 5. Collaboration of Web 2.0 tools

On the other hand, 44% of the surveyed students say they have used web 2.0 tools in 3 to 5 of their subjects, out of an annual average of 11 subjects. So, we can conclude that just 36% of the professors use web 2.0 tools to give their classes.

5   Conclusions

It determined that the adoption of Web 2.0 tools plays an important role in the improvement of academic performance. The study showed that these tools are already present in the learning process of Ecuadorian Navy candidates but that the students should be encouraged to adopt these applications with more frequency.

The authors urge the military schools to consider the study's finding and promote the use of web 2.0 tools with more adaptability to the type of learning that navy officials students receive, with the aim of improving academic performance and autonomous learning.

References

1. Boza, C.A., Conde, V.S.: Web 2.0 en educación superior: formación, actitud, uso, impacto, dificultades y herramientas. Digit. Educ. Rev. **48**, 45–58 (2015)
2. Tudge, J.: Vigotsky: La Zona de Desarrollo Próximo y Su Colaboración en la Práctica de Aula. Universidad de Cambridge, Nueva York (1994)
3. Bennett, S., Bishop, A., Dalgarno, B., Waycott, J., Kennedy, G.: Implementing web 2.0 technologies in higher education: a collective case study. Comput. Educ. **59**, 524–534 (2012)
4. Kam, H.J., Katerattanakul, P.: Structural model of learning based on computer use web 2.0 collaboration software. Comput. Educ. **76**, 1–12 (2014)
5. Grosseck, G.: To use or not to use web 2.0 in higher education? In: World Conference on Educational Sciences 2009, Kyrenia (2009)

6. Greenhow, C., Robelia, B., Hugges, J.: Web 2.0 and classroom research: what path should we take now? Am. Educ. Res. Assoc. **38**(246), 246–259 (2009)
7. Francis, D., Boniface, S.: Consuming and creating: early-adopting science teacher, perceptions and use of a wiki support professional development. Comput. Educ. **68**, 9–20 (2013)
8. Kump, B., Moskalik, J.D., Sebastian, L.T.: Tracing knowledge co-evolution in a realistic course setting: a wiki based field experiment. Comput. Educ. **69**, 60–70 (2013)
9. Eid, M.I.M., Al Jabri, I.M.: Social networking, knowledge sharing, and student learning: the case of university students. Comput. Educ. **99**, 14–27 (2016)
10. DeWitt, D., Alias, N., Siraj, S., Yaakub, M.Y., Ayob, J., Ishak, R.: The potential of youtube for teaching and learning in the performing arts. In: 13th International Educational Technology Conference (2013)
11. Crisan, A., Enache, R.: Virtual classrooms in collaborative projects and the effectiveness of the learning process. Procedia Behav. Sci. **76**, 226–232 (2015)
12. Calzadilla, M.: Aprendizaje colaborativo y tecnologías de la información y la comunicación. OEI Rev. Iberoam. Educac., 11
13. Cavus, N., et al.: Designation of web 2.0 tools expected by the students on technology-based learning environment. Procedia Soc. Behav. Sci. **2**, 5824–5829 (2010)
14. Cela, K., Fuertes, W., Alonso, C., Sánchez, C.: Evaluación de herramientas web 2.0, estilos de aprendizaje y su aplicación en el ámbito educativo. Estilos de Aprendiz. **5**, 22 (2010)
15. Reyes, D.L.: Moodle, una plataforma formativa con gran proyección en los nuevos modelos enseñanza. DIM Rev. **19**, 1–14 (2010)
16. de Ros Martínez Lahidalga, I.: Moddle, la plataforma para la enseñanza y organización escolar. Rev. de Didáct. **2**, 1–12 (2008)
17. Hernández, S.R., Fernpandez Collado Carlos, B.L.P.: Metodología de la Investigación. Mc Graw Hill Companies, México (2014)
18. Quesada, A.: Aprendizaje colaborativo en entornos virtuales: los recursos de la web 2.0. Leng. Mod., 14 (2012)
19. Balakrishnan, V., Chin Lay, G.: An students' learning styles and their effects on the use of social media technology for learning. Int. Telematics Infor., 808–821
20. Baradaran, R.M., Khashayar, Y.: Web 2.0 embedded e-learning: a case study, Conference International IEEE Xplore Library Digital (2014)
21. Duta, N., Martinez, O.: Between theory and practice: the importance of ICT in higher education as a tool for collaborative learning. Procedia Soc. Behav. Sci. **180**, 1466–1473 (2015)
22. Traverso, H.: Herramientas de la web 2.0 aplicadas a la educación, Conference paper, p. 5 (2013)
23. Peñarroja, V., Orengo, V., Zornoza, A.: How team feedback and team trust influence information processing and learning in virtual teams: a moderated mediation model. Comput. Hum. Behav. **48**, 9–16 (2015)

Educational Technology: As a Component of the Innovative Curriculum Redesign Project for Ecuadorian Naval Officers

Marjorie Pesantes[1]([⊠]), Eduardo Pomboza[1], Marisol Gutierrez[1], Jorge Alvarez[1], and Óscar Barrionuevo[2]

[1] Universidad de las Fuerzas Armadas-ESPE, Sangolqui, Quito, Ecuador
{mepesantes, eepomboza, megutierrez2,
jmalvarez6}@espe.edu.ec
[2] Armada del Ecuador, Guayquil, Ecuador
obarrionuevo@armada.mil.ec

Abstract. The Ecuadorian Navy, in order to fulfill the duties established in the Constitution and the Nation's Security Plan, requires new generations of Officers with updated knowledge not only in academic fields but also in technological devices which influence processes for decision-making on naval units. This challenge is one of the objectives of the Naval Officers School and The Armed Forces University (ESPE) which offer majors in Naval Sciences and Naval Logistics. For this reason, the curriculum redesign was necessary to include technological components in the syllabi in order to guarantee Officers capable of facing combat threats and social challenges of this century. The technological component is the fundamental part for the development of subjects related to the Naval Officer's occupation in virtual and simulated environments. The application of Learning and Knowledge Technologies (LKT) in addition to the Empowerment and Participation Technologies (EPT) allow naval students to apply accurate decisions in the Nation's Defense area.

Keywords: Curriculum redesign · Learning cone
Learning and Knowledge Technologies · Digital competencies
Maritime interests

1 Introduction

Nowadays, Information and Communication Technologies (ICTs) play an important role in the society, its development has been part of the transformation in areas related to economy, education, politics, culture. Educational Institutions must be updated in methodological approaches that allow technology to be part of the learning resources. It is necessary a change in new generations of professionals who must apply the acquired knowledge in the responsibilities assigned in different work activities. From this point of view, the careers offered by the Officer's Naval School, which are recognized by the Ecuadorian National System of Higher Education, were redesigned in order to implement ICTs tools. These changes have taken into consideration the objectives of the Nation's Security Plan which states "Armed Forces must guarantee sovereignty and

© Springer International Publishing AG, part of Springer Nature 2018
Á. Rocha and T. Guarda (Eds.): MICRADS 2018, SIST 94, pp. 326–336, 2018.
https://doi.org/10.1007/978-3-319-78605-6_28

territorial integrity, peaceful cohabitation and contribution to national development" [1]; the Ecuadorian Constitution also stipulates that it is the Navy's duty to maintain the integrity of territorial sea and water spaces. To accomplish these missions, Naval Officers must be qualified in all aspects. This responsibility belongs to the Naval School which must prepare Officers with optimum academic and technical skills enabling them to be inserted in the Nation's Defense and Security Frame, especially in the missions of maritime sovereignty and water spaces as determined by Ecuadorian Laws [2].

2   Levels of Formation of Future Naval Officers

The curriculum redesign project divides the major in three units: Basic, Professional and Pre-degree. The integration of technology is evidenced in the ten levels of formation at the Naval School.

2.1   Basic Unit

It is conformed from the first to the third academic period of the major, in which the epistemological system considers three integrative projects that take into consideration all aspects related to educational military-naval topics. The detailed analysis of the activities that take place in the territorial sea and the description of naval platforms used for water space control. First Level: the career starts by incorporating the student, which in this case is the midshipman, to the military-naval world and its functioning as well as its organization including the digital language used in maritime operations from the perspective of a national force. The theoretical foundations of subjects like: mathematics, plane trigonometry and oral-written communication would allow the student to consolidate all the acquired knowledge to deal with scientific aspects related to navigation and naval systems. The integrative subject of the first level is Military-Naval Conditioning and the evidence of the integrative project is a theoretical essay which includes the most relevant naval data such as: maritime glossary, naval traditions, naval doctrine, among others. The objective of this project is to develop the skill of interpreting and analyzing the military-naval activities through the observation. Second level: Once the student (midshipman) knows the main characteristics of the military-naval world is able to understand the maritime environment. Subjects like Applied Mathematics for Naval aspects, Research methods, Leadership, and English (Level A1) contribute to the integrative subject Maritime Configuration and Practices. The integrative project of this level is the detailed analysis of the activities that take place in the territorial sea and its objective is to identify through research the process that these tasks demand before their implementation. Third level: Students (midshipmen) will be able to understand the basic knowledge related to naval sciences based on the principles of exact sciences as well as military-naval doctrine. Subjects like National Reality, Geopolitics, Basic Computer Programming, English (A2 level) contribute to the integrative project of the level. It consists on the "Description of naval platforms which are used to control water spaces" and its objective is to determine the fundamental characteristics, main components and systems of a naval platform.

2.2   Professional Unit

It takes place from the fourth to the eighth level of the major, in which the episte-mological system considers the development of four integrative projects: Calculations to solve problems derived from the planning stages of water spaces control programs, Roles and functions inside of a Military-Naval Ship, Use of naval resources to control water spaces and The capacity of Naval platforms. Fourth Level: The development of maritime knowledge is the focus during this level. Academic areas assigned to this period are: Vessel's theories, Internal Defense, Geometric Analysis, Advanced Military-Naval Theory and English (B1 level). The integrative subject is "Coastal Navigation and Vessel's Maneuvering". This level's project is "Calculations to solve problems derived from the planning stages of water spaces control programs" and its objective is to determine the resources needed for safe navigation. The experience of pre-professional practices onboard of a military-naval platform would also contribute to the final project of the level. Fifth Level: The development of maritime knowledge continues based on the subjects of this level. Theory of Electrical Machines, Exact Sciences, Military-Naval leadership and English (B1.2 level) contribute to the inte-grative subject "Electronic Navigation, Kinematics and Automatic Radar Plotting Aid". The project of this level is "Roles and functions inside of a Military-Naval Ship" and its objective is to use tools and follow procedures based on descriptive research to collect data in order to organize military-naval operations. Students (midshipmen) are also involved in projects oriented to Safety and Protection of Marine areas that include the participation of local population. Sixth level: The main focus of the level is to acquire knowledge and develop skills in military-naval professional fields, for this reason, subjects according to the specializations of Naval Sciences and Naval Logistics start from this level. Naval Sciences: This specialization is mainly focused on the devel-opment of skills which enable students to organize high seas tasks. Subjects like: Electricity and electronics for naval sciences, Statistics and Maritime strategies con-tribute to the integrative subject "Astronavigation, Oceanology, Hydrography and Meteorology". The integrative project of this level (according to this specialization) is "The sea and its influence on Naval systems and Platforms", its objective is to develop operational analysis based on how the environment that surrounds vessels affects navigation. This work is also product of the experience of pre-professional practices on board. Naval Logistics: This specialization is aimed at acquiring the necessary knowledge for naval operations regarding logistics. Academic fields like: Public Finances, Statistics for Naval Sciences and Maritime procedures contribute to the integrative project "Logistics Processes of Naval Platforms". Its objective is to analyze the needs of a vessel based on the experience of pre-professional practices on board of a naval unit. Seventh Level: Theoretical knowledge and professional practices are the core of the level, subjects vary according to specializations. Naval Sciences: During this level students acquire knowledge related to the use of technological tools used for naval units. The academic fields that contribute to the integrative subject "Second Instructional Cruise" are: Naval Electrical and Electronics systems, National and International Maritime Legislation, Research Methods, and Project Design and Implementation. The objective of this integrative project is to recognize how to handle naval resources during water space control. Naval Logistics: It focuses on naval logistics and the integrative subject is "Second Instructional Cruise". The academic areas established for this level are: Supplies Logistic Management, Naval Legislation for Logistics, Research Methods, and Project Design and Implementation. The integrative project is "Naval Logistics Management" and its objective is to implement processes which optimize the control of logistics. Eighth level: Theoretical knowledge and professional practices are the core of the level, subjects vary according to specializations. Naval Sciences: this level centers in the use of technological naval systems, the integrative subject is "Weapons System" (which is partially taught in English). Academic fields that are part of this level are: Naval Engineering systems and Operational Analysis which contribute to the integrative project "Naval Platforms and its different purposes". Its objective is to analyze the operational capacity of naval systems. Naval Logistics: the integrative subject of this level is "Acquisition Management" (which is partially taught in English). The academic fields that contribute to the integrative project of this level are: Engineering systems and Naval Weapons. "Acquisition processes for naval platforms in order to perform water space controls" is the integrative project, its objective is to fulfill all the acquisition requirements of national system of public procurement.

2.3   Pre-degree Unit

The ninth and tenth levels are mainly focused in the application of the acquired knowledge in specific areas. The epistemological system considers the following integrative projects: Naval systems for Sovereignty Integral Defense and Maritime Protection system. Ninth level: Theoretical knowledge and professional practices are the core of the level, subjects vary according to specializations. Naval Sciences: Naval Doctrine is the main focus of this level; the integrative subject is "Tactical Procedures and Communications". Academic fields that contribute to the integrative project are: Tactical Communications and Procedures (which will be partially taught in English), Human Rights, Humanitarian Laws (both taught 100% in English) and Operational Logistics. The integrative project is "Naval systems for Sovereignty Integral Defense" and its objective is to propose alternatives for the implementation of Naval doctrine. Naval Logistics: aspects related to naval logistics support are the core of this level. The academic fields that contribute to the integrative project are: Human Rights and International Humanitarian Rights (100% taught in English). The integrative project is "Logistics Integral Support" and its objective is to propose alternatives to implement naval logistics systems. Tenth level: During this level both specializations will be focused on acquiring knowledge for planning naval operations. The subjects that contribute to the integrative project are: Naval Management, Integrated Security Systems, Naval Military Intelligence, and Advanced Naval training. The integrative subject is "Naval Planning Foundations" and its project "The Ecuadorian Navy and the protection of maritime interests". Its objective is to plan different ways to protect Ecuador´s Naval interests. During this level, students (midshipmen) start working on their thesis and dissertation, or begin the process of a Complex Exam that includes all the subject of the major.

2.4   Integral Formation

Integral Formation is considered a transversal axis in all the levels of the major. This educational model has been denominated as "Leadership and Emotional Intelligence Requisite" considering that future Naval Officers must be trained as possible Armed Forces Leaders.

3   Curriculum Methodological Guidelines

Technology is the main component in the curriculum redesign for Naval Officers, especially in the formative process. However, this program is also focused on occupational areas which are the heart of the naval field such as: Naval Systems and Platforms, Navigation and Ship Maneuvering, Tactical procedure and Naval-Military Doctrine. Methodological guidelines, that are applied to solve problems within the naval context, establish the use of approaches, processes and protocols related to the field which make easier the explanation, comprehension and adaptation of real life scenarios that enable learning to have a pedagogical atmosphere. The Armed Forces University-ESPE is in charge of preparing future Naval Officers in all the required educational areas; and the Ecuadorian Navy provides the application of the acquired knowledge in real life working scenarios which are developed by using technological devices.

3.1   The Use of ICTs in the Naval Officer Formation

In the last decade there have been an enormous growth in the use of Internet in different areas, the educational field is among them. The evolution of Information and communications technology (ICTs) tools have made possible for the teaching-learning process to take place through a variety of devices as well as online and offline networks; which are used not only for practicing but also for storing information.

The development of teaching practices by using online or digital resources is recognized as part of an innovation process needed in all educational levels. Based on this fact, most universities are implementing virtual courses that include "just-in-time" tasks focused on individual learning; the main characteristic of these activities is the combination of face to face classes and online sessions. This is a clear example of how technology has influenced higher education and why universities are looking for new methodological approaches. The curriculum for Future Naval Officers include the implementation of B-Learning models adapted to all the subjects involved in the major, nevertheless, the "Face to Face" model will be used when students need to develop certain skills to acquire a specific level of knowledge. Online teaching and learning components will enable fast learners to advance and practice at a more rapid pace, and slow learners to take their time in the acquisition of knowledge. Remedial tasks are also considered in case of having students who fail a subject.

3.2   Simulators

Virtual environments like video games or simulators represent a potential alternative for new careers. There are video games that complement educational practices due to the fact that these simulations represent dynamic models of real life situations. When solving tasks in virtual environments, students are able to recognize the impact of decisions in determined circumstances related to their occupation. Having the possibility of analyze and evaluating the results and the contribution to solve a situation. This represents an interesting solution to involve and prepare students for a better understanding of external contexts. Knowledge built though simulations and video games challenge higher education programs to adapt their current methodology to new learning approaches. Simulation has two main uses in the redesigned educational process of Naval Officers: during the teaching-learning stage as well as the assessment process. In the former, the diverse simulators may improve the decision-making procedures involved in the naval field; contributing to the improvement of psychomotor, interactive and leadership skills. For the latter, the effectiveness of simulation results in naval issues can be measured and analyzed. This is an example of how virtual reality benefits education in comparison to other traditional teaching methods.

3.3 Software for Naval Purposes

The implementation of software in the context of naval education has extended its use worldwide. Nowadays, naval training institutions have improved their facilities in order to enable the use of educational platforms, this include the acquisition of advanced equipment and systems. Among the programs included in the formation of the Naval Officer are the following: Spectrumlab, it is a freeware for teaching subjects related to naval weapons systems; Voacap, this is an online platform to support the teaching of Propagation of Electromagnetic Waves; MATLAB student in combination with Add-on, Control Systems applications like: Simulink, Hardware, Physics modeling, among others; Hardware is also used for teaching electronics and programming of embedded systems related to the control of naval platforms.

3.4 The Flipped Classroom Technique in the Formation of Naval Officers

The modern infrastructure of the Officer's Naval School is complemented by the ICTs tools which include: online classrooms as well as virtual libraries with access to international data bases. These factors allow the implementation of the Flipped Classroom technique in subjects that require students to carry out research on their own in order to obtain the necessary information for specific topics based only on the professor's guidance as explain in Fig. 1. This is a new pedagogical model that offers an integral approach to increase the commitment of learners in the educational process; therefore, teachers will be able to monitor students individually. Flipped Classroom strategy embraces all the learning cycle stages (Cognitive Process Dimension in Bloom's taxonomy) [5, 15]: To remember: Been able to recognize information previously learnt; To understand: to build meaning from oral, written, and visual learning; To apply: to apply the skills developed in a given situation; To analyze: to divide the concept in its constituent parts and solve problems using the acquired knowledge; To evaluate: to make judgments based on criteria and standards.

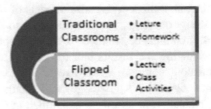

Fig. 1. Flipped classroom.

4 Technological Components of the Innovative Curriculum Redesign Project for Ecuadorian Naval Officers

The technological component is relevant in the development of Naval operations, in fact, the main characteristics are: the definition of the strategies importance as well as goals adaptation. Technological development has also influenced and marked the techniques and proceeding policies in naval operations. The Innovative Curriculum Redesign Project for Ecuadorian Naval Officers comprises learning development in real life, virtual and simulated scenarios [3]. In the model of the learning cone, Fig. 2, it is considered that real life simulations make that students acquire 90% of knowledge.

Fig. 2. Learning cone model.

Curriculum Redesign includes new professional competencies based on the application of Learning and Knowledge Technologies (LKT) which refers to the use of ICTs as formative tools, having an impact directly in the methodology and resources on educational planning. The use of Empowerment and Participation Technologies (EPT) transforms the learning environment into real life scenarios in which Naval students apply values (see Fig. 3).

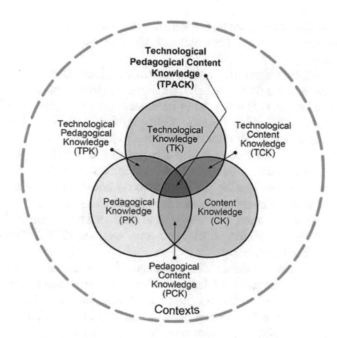

Fig. 3. Technology, pedagogy and content knowledge.

Curriculum redesign also embraces a new profile for the educators, which consists on not only been proficient in the subject matter but also been expert in teaching-learning processes and methodologies that include technology as a tool for knowledge development. Professors must also possess pedagogical techniques regarding technological aspects that make students acquire knowledge based on the use of ICTs tools. [4]. Technologies linked in the curriculum redesign for Military Naval learning will enable the student to achieve digital competencies included in the curricula. The objective is to prepare professionals in technological applications in the following areas:

(a) Electronic Chart Display and Information System (ECDIS): ECDIS y WECDIS, are navigation systems software, these must fulfill the standards defined by the International Maritime Organization and can be applied as an alternative to traditional paper maps [6]. These systems present digital maps, integrating additional information of other navigation instruments and equipment on the screen, for instance, the position, route, speed, radar data, automatic identification system (AIS), among others [7]. Digital Maps along with other modern electronic navigation systems have become an important support in the decision-making process of a ship, aimed at having a safe navigation on non-restricted and restricted waters [8]. These systems are used in addition to traditional mapping and navigation techniques. The contribution of these technologies to learning and professional practice for Naval Officers is relevant, that is why, through the use of simulators ECDIS y WECDIS, real life Naval scenarios can be created, additionally, the implementation of radar simulators, plotting generators, virtual plotting tables with environments similar to the electronic mapping system used by the Naval

Oceanographic Institute are part of the innovation in Naval careers. It is important to mention that these practices will take place on board of Military Ships and submarines.

(b) System C4ISR (Command, Control, Communications, Computers, Intelligence, Reconnaissance and Surveillance): Operational environments are full of contacts, information and data that require to be processed in real time in order to make accurate decisions before, during and after military operations. Systems C4ISR allow the compilation of information to have a clear view for the development of Naval operations which is part of the Naval Officer's duties [9]. The need of manipulating operational information is vital in Naval missions, for that reason, virtual environments similar to the existent on board scenarios are part of the officer's formation. These are based on the Naval Units or ground facilities which are used as control centers. It will be used the simulator "Poseidon" C3I of the Ecuadorian Navy, THEMI, which is an open software GCB 2.0 used by Naval Forces of other countries.

(c) Combat Integrated Systems: The exchange of information among sensors of Naval platforms, the fusion of data and the weapons designation, are essential for the decision-making process in Naval platforms; new Combat systems technologies implemented in several Naval ships require that Naval Officers have knowledge of architecture, communication protocols and use of Combat modern systems [9]. Because of the importance of these systems on Naval platforms, real Combat environments aboard Navy ships as well as simulated scenarios implemented by the Naval School are part of the Officer's training process in order to achieve practical application of tactical, navigational, kinematical, operational planning, intelligence, use of weapons procedures.

(d) Automated Control and Informatic Network of the ship: Modern Naval Units or units that have been adapted to modern systems, have been successfully introduced in the automatic system control through the application of Commercial-off-the-shell (COTS) technology. It has also been chosen the incorporation of neuronal networks in the Naval field, these modern systems are able to retrieve information on its own through autonomous devices like: Aerial Surface Vehicles (ASVs), Unmanned Aerial Vehicles (UAVs) or Unmanned Underwater Vehicles (UUVs) [10]. Based on this technological trending derived from security aspects on board Naval Units, there will be virtual platforms such as: MATLAB, LAB-WIEV, ARDUINO, that allow the future officer to have the necessary practice of control systems of Naval platforms.

(e) Managerial Information Systems: Decision-making based on reliable information is fundamental for organizations like the Ecuadorian Navy, which also have different types of missions including administrative activities. It is important that the Naval Officer knows how to work with virtual systems to optimize time and resources [11]. It will be applied the simulators for logistics information (SISLOG), Maritime and Port system (SIGMAP), and others used by the Ecuadorian Navy or Armed Forces for decision-making in Naval operations.

5   Contribution of the Project

This research contributes to the formation process of future Naval Officers. The curriculum has been developed based on the duties of the Ecuadorian Navy as well as the responsibilities that the constitution established for the Armed Forces.

The knowledge acquired and the skills developed in all the academic fields of the curriculum enable students to fulfill the objectives established for integrative subjects and projects. The combination of Military-Naval doctrine as well as Humanistic and Exact Sciences with technological components fully prepares future Naval Officers in all professional areas.

6   Desired Results of the Project

Considering the evolution of Science, Information and Communication Technology (ICT) as well as Naval in the field of innovative curriculum redesign of the Naval Officer, future generations will be able to: To provide Integrative Security and Protection of Water Spaces, that according to the last "Conference of Maritime Security" which took place in Singapore on May, 2013, the principal authorities of Naval Forces in the world included as main topic "To maintain security in the sea and reinforce the cooperation among maritime organizations to achieve it"; To secure navigational practices in which the Naval Officer knows how to make decisions through the use of established methods and techniques, based on the information provided by navigation instruments; To use Naval systems for the Nation´s Defense, related to the weapons systems applied on Naval Units, such as radars and electronic equipment; To protect and defend maritime interests which are activities that are developed by the public and private sectors in order to take advantage of their water spaces and resources; To apply naval Doctrine principles in all the missions assigned to the Ecuadorian Naval Force as well as to propose alternatives to face possible threats that Armed Forces have to deal with. This project presents the evolution of educational paradigms that is taking place around the world. The implementation of technological components would allow future Naval Officers to acquire and develop the necessary skills to face actual challenges in all the fields involved.

7   Conclusions

The implementation of Information and Communication Technology in the Curriculum Redesign of the Ecuadorian Naval Officer degree, establishes a current educational requirement for the training of future Midshipmen promotions, which will allow operation of resources and technical equipment in their professional duties for the correct decision making in the workplace. With the implementation of Information and Communication Technology, simulated environments become really important due to their access and flexibility for academic purposes, reducing the risk in real exercises and their high cost, without worsening the importance of real practice, this will ease program management of academic practice, time and resources. In a globalized world where technological advances increase rapidly, Educational Institutions must take on and meet the new technological challenges through updating their curricula so that new generations get to know and find familiar with technological advances in their professional practice.

References

1. Secretaria Nacional de Desarrollo: Plan Nacional de Desarrollo 2017–2021. Pichincha, Quito (2017). www.planificacion.gob.ec
2. Asamblea Nacional: Constitución de la República del Ecuador. Pichincha, Quito (2008). Registro Oficial 449 de 20 October 2008
3. Edgar, D.: Métodos de Enseñanza Audiovisual. Editorial Reverte, Mejico (1969)
4. INED 21: 10 ENERO 2018. https://ined21.com/tpack/
5. Bloom, B.S.: The Taxonomy of Educational Objectives: Handbook I: Cognitive Domain. David McKay, New York (1956)
6. Royal Navy: BR 45, Admiralty Manual of Navigation. ROYAL NAVY (2008)
7. Cutler, T.: Dutton's Nautical Navigation. Naval Institute Press, Annapolis (2004)
8. Bowditch, N.: The American Practical Navigator (2015)
9. Hassab, J.: El intercambio de información entre los sensores de las plataformas navales, la fusión de información y la designación de armas (2016)
10. NauticExpo: (2016). http://www.nauticexpo.es
11. Power, D., Sharda, R., Burstein, F.: Management Information Systems. Wiley, Hoboken (2015)
12. Wagenhals, L., Shin, I., Kim, D., Levis, A.: C4ISR Architectures: II. A Structured Aanalysis Approach for Architecture Design (2000)
13. Armada del Ecuador: Doctrina Básica de la Armada del Ecuador (2013)
14. Armada del Ecuador: Concepto Estratégico Marítimo (2013)
15. Gonçalves, M.J.A., Rocha, Á., Cota, M.P.: Information management model for competencies and learning outcomes in an educational context. Inf. Syst. Front. **18**(6), 1051–1061 (2016)

ICT Integration in Ecuador's Military Education: Going Beyond PowerPoint

Francisco Garay[1(✉)], Giovanna Morillo[2], and Teresa Guarda[2,3]

[1] Academia de Guerra Naval, Guayaquil, Ecuador
fgaray@armada.mil.ec
[2] Universidad de las Fuerzas Armadas-ESPE, Sangolqui, Quito, Ecuador
giovi.morillog@gmail.com, tguarda@gmail.com
[3] Algoritmi Centre, Minho University, Guimarães, Portugal

Abstract. The use of Information and Communication Technologies, ICT, as a way to generate better learning experiences by increasing student motivation in a constructivist framework, is one of the objectives to be reached by educational institutions in the current times, from which military education does not take distance, something in which the use of Information and Communication Technologies can help, being more than the use of simulators or a slide presenter. In Ecuador, the level of ICT integration in education is not high, which is reflected in military education, for which a referential model is proposed.

Keywords: Military education · Information technology and communications
Technology integration · Digital competences · Constructivism

1 Introduction

UNESCO [1] indicates that there are still people around the world who maintain that the schooling model has no future in the digital age due to the existence of e-learning, mobile learning and other digital technologies.

The importance of including computer technology in education is given by its versatility, its integrating nature and the possibilities it offers in the classroom. The didactic use of the new technologies makes them a didactic resource and/or a resource for the expression and communication, to facilitate the individualized attention to the student, the modification of the teacher's role and the access to a greater quantity of information [2]. The teacher who uses them appropriately takes on the role of accompanying the student in their learning, which has new capacities based on ICT to draw conclusions based on the available evidence [3, 55].

The amount of information available these days, requires a careful and effective planning of what will be shared with students, and then teachers become increasingly essential in this new role. New technologies facilitate access to information, and sometimes too much information, which is because teacher guidance is essential. Teachers must use ICT, along with several other aids available for teaching, as instruments of their students' learning, as elements that facilitate it, as the means and not as the end [1].

© Springer International Publishing AG, part of Springer Nature 2018
Á. Rocha and T. Guarda (Eds.): MICRADS 2018, SIST 94, pp. 337–350, 2018.
https://doi.org/10.1007/978-3-319-78605-6_29

Technology, then, must be used to remove the physical barriers that can hinder learning and in the transition between the focus on the retention of knowledge towards its real employment, that is, in the creation of competences. The ease of access provided by distance education and the availability of databases on the network, allow knowledge to be closer to students, with an appropriate guide, can develop skills based on new technologies [4].

In this context a referential model for ICT integration in Ecuadorian education is proposed, which can be used in the military education system of the country.

2   ICT Integration in Education

Records of favorable results of the use of ICT in education are in sight. Pérez, Cebrián and Blanco, cited by Fuentes and Gutiérrez [5], agree in their studies conducted between 2005 and 2006 that the use of ICT in education results in a considerable increase in students' motivation, and in a best classroom environment. A more recent study, carried out by Fuentes and Gutiérrez [5] to 15-year-old Spanish students, resulted in the technological factors employed in the houses acting positively on the students' grades.

After their experience in adult education using ICT in Andalucía et al. [6] concluded that they encourage motivation towards their use, personalized work, communication and the acquisition of knowledge. These conclusions are similar to those reached by Sánchez and Guzmán [7], when they evaluated the learning acquired by their students when they used robotic prototypes or specialized programs for pedagogical purposes.

Also, there is an evidence of unfavorable results as indicated by Calero and Escardíbul [8] when mentioning the results of Israeli, Colombian and American works that did not find an improvement in the performance of students who used ICT for learning, although they also present evidence of work in England, India and the United States in which it was evident that students of science, mathematics and English who used ICT for their learning obtained better results than those who followed a traditional method.

Experiences in Ecuador indicate that the degree of use of ICT in education is low and therefore it is still too early to establish if the results are positive. Amores [9] indicates that the students of mathematics and physics of the Central University of Ecuador in Quito do not use new technologies in their learning except for the exchange of information between them, even though their teachers consider that these tools facilitate meaningful learning. In contrast, as indicated by Ortiz [10], at Casagrande University in Guayaquil there is a greater use of new technologies, such as blogs, collaboration tools and social networks, which generates an approach to digital natives and the consequent change in the student's role, criteria that are shared by Hi Fong [11] based on studies conducted at the same University. The national experiences are varied, but the tendency is a low integration of the technology use in the teaching practice, as indicated by Rodríguez based on his experience with the Quito Educanet program of the Municipality of Quito and the Benalcázar School [12], as well as Ortiz and Chiluiza, who determined that higher education teachers linked to the Ecuadorian

Consortium for the Development of Advanced Internet (CEDIA), mainly use technology for administrative use, and a lesser extent as a support to teaching, being the lowest level is the innovative use of technology [13], despite the fact that the study group is of teachers committed to the use of technology in education.

These experiences indicate that in military educational environments, whose courses are developed in different ways depending on the objective sought, be it training, specialization or improvement, the use of ICT will be of great help, both to bring the knowledge to students to develop their competences still far from the study center, using distance education, as facilitating the development of skills such as teamwork, collaboration or professional skills, through teachers who integrate technology into their practice pedagogical.

3 Military Education in Ecuador

Now, although military education has the same characteristics as general education, it is necessary to specify some particular definitions that are used within the Ecuadorian Armed Forces, and relate them to those that are used within the pedagogical processes of civil institutions.

The Education Model of the Armed Forces in effect has as its first specific objective to implement a curricular design with a competency-based approach to military education [14], which encompasses the processes of training, preparation, specialization and improvement of Army, Navy and Air Force personnel from its entry until its retirement, processes that must be within the frame of reference that indicate the guidelines of the aforementioned model.

Training and improvement are only carried out in military institutes in the country, due to the sensitivity of the information, while training and specialization can be carried out both in national military institutes, as well as in foreign military entities or civilian educational centers.

Thereby, training and professional specialization are defined in articles 41 and 45 of the Regulation of the Armed Forces Personnel Law [15] as follows:

"The professional training is the preparation of military personnel that will be carried out through courses or seminars, which may have a maximum duration of one year, and that will be carried out without prejudice to the work activities of each military. They will be aimed at keeping the knowledge up-to-date and granting them the additional basic tools to perform in the workplace efficiently."

On the other hand, professional specialization courses group the preparation received by military personnel in a specific field of their area of higher education, after their graduation from training schools, which allows them to perfect themselves for their occupation, profession or area of expertise performance, and the positions they must assume.

Professional specialization differs from training, in that these studies help you directly for your work, such as, for example, the specialization courses with which members of the Navy choose their qualification as members of the Surface Warfare Force, Submarines, Marine Corps, Naval Aviation or Coast Guard once they graduate from training schools as officers or crew and with the title of Bachelor of Science or Naval Technologist respectively.

On the other hand, the improvement is defined by article 52 of the Armed Forces Personnel Law [16] as the educational activity through which the military, once discharged as an officer or troop, receives military and complementary knowledge to the performance in the immediate superior degree. These are courses that are carried out for promotion and develop military skills for their performance in the Armed Forces.

4 Constructivism in Military Education

The first idea of "military education" can be framed in a behavioral model by the rigidity of military forms, in which one might think that there is little space for creation, freedom and interaction, which are characteristics of a constructivist model [17].

Classes in the militia have not stopped of being mostly a transfer from the "instructor" to the "student", due to the cultural framework that, in the militia depends on traditions and inheritances [18], but for a while now it has been clear in different military institutions throughout the world, that constructivism in education can be the most effective way to reduce the gap between "knowing information" and "knowing how to use information" in complex environments, typical of an aircraft combat or an Combat and Information Center - CIC - of a ship [19].

It is for this reason that along with the Instructional model in which someone who "knows" teaches the one who can "learn" [17] each day, more experiences are generated in which a military team can use structured processes to consider alternatives. and consequences before acting, without using pre-planned answers that are not necessarily adapted in the best way to the different situations presented by reality [19]. In that sense, the pure instructional model is not adapted to the military of the present and future, defined by Szabó [20] as a person, who must be thoughtful, creative, with several competencies, great capacity for adaptation and with abilities to solve problems about the half.

Juhary [21] states that military education should be seen as a tool that teaches the student to solve problems, but from conceptions, visions, doctrine and military modalities. Military education is a balanced mix in which behaviorism and constructivism must coexist, allowing students to learn to carry out orders at first and then develop as leaders. In this regard, recognizes the need to raise significant learning using constructivist techniques, but is pragmatic to recognize that in an environment like the military, the freedoms that must be granted to achieve the construction of knowledge must coexist with the traditions and forms of the militia.

The constructivist approach makes clear the importance of the student being exposed to different perspectives on the topics under study, which go beyond sharing information or working in groups, where the objective is to guide the student to achieve learning that supports the diverse points of view and allows you to make reasoned decisions [19].

In this sense, what is stated by Juhary [21] does not depart completely from this approach, since learning is contextual and is related to what is known and believed, as stated by Annen et al. [18] when they indicate that the constructivist approach is the most appropriate for a military student, where learning becomes a social process, taking as examples cases given in the military forces of Holland, Canada and Israel. Johnson-Freese [22] reaffirms this by saying that the military can find many advantages for his intellectual development by entering a more open environment of education, which allows him to create networks of knowledge.

Then, military education adapts in the best way to the postulates of the constructivist approach, because it tries to train soldiers capable of solving problems in complex environments or situations of group pressure, because when the student works to achieve their own learning, the educator it stops being the center of that learning, passing the responsibility of the process to the student [23].

When the student is empowered by his own learning he achieves real performances, which contribute to his own motivation to continue learning, because when relating the new knowledge with real life, with those performances that are expected of him in the future in his career, he will be trained to be used in different contexts [24]. This way of learning will serve those who live in environments of rapid decision making, teamwork, high pressure situations and often away from possible additional sources of consultation that allow you to expand the knowledge you already have. The challenge in Ecuador is to overcome previous conceptions and adopt constructivism as the norm. The study carried out by Cruz and Paul [25] indicates that only 6.66% of the teachers of the Escuela Superior de Policía (Police Academy for prospective Police officers) use some constructivist pedagogical technique.

A study conducted in the Training and Specialization School of the Ecuadorian Navy (ESCAPE), demonstrated that its teachers did not agree on the importance of using e-mail and social media as means to improve their learning experience [54]. On the other hand studies conducted by Ordoñez and Rincón [53] and Ruth [28] at the Military University of Nueva Granada (UMNG) of Colombia and the Naval Postgraduate University (NPS) and the Academy Naval of the United States (USNA), respectively, coincide when affirming that the current military students would use their mobile means of learning in a good way if it were considered as valid employment by their professors. According to this research, 79% of UMNG students, 55% of distance students and 60% of NPS resident students, and more than 70% of USNA students, would use these learning ways. The Police and Navy experiences, compared to the studies conducted in Colombia and United States, tell us that our teachers may be not doing what their students expect of them to motivate their learning.

Knowledge is defined by UNESCO [1] as the information, understanding, competences, values and attitudes acquired through learning and necessarily linked to the culture, society, environment and institutions in which it is developed. In this context, Irigoyen [26], states that competency-based education is born from learning as a phenomenon of the individual who learns and from the need to train professionals capable of solving problems in real performance areas.

5 ICT Integration in Military Education

Just as new technologies have changed the lives of societies in general, this is no different for the military. The Army and the Air Force of the United States have changed their models focused on the instructor towards the new student-centered paradigm, in which relevant, interesting and tailor-made learning experiences of the person who is learning are sought, with access to knowledge through mobile devices, with advanced systems organizing expertly guided learning [27].

Studies carried out at the Postgraduate School and at the Naval Academy of the United States determined that their students, Officials and Midshipmen respectively, had similar behaviors regarding the use of technology in education. In fact, almost all the students of both institutes indicated that they used their smartphones to read their e-mail and surf the web, download or use live (streaming) audio and video files, but only 10% used them to read e-books or use podcasts of their classes. Students, in general, demand greater integration of their educational platforms - Sakai and Blackboard - towards mobile devices in such a way that they can have greater access to them [28]. This is a clear indication that the generation of "Millenials" is not only present in civil society, but also among the military, which becomes an additional reason to firmly believe in the advantages that the integration of the ICT in pedagogical practice.

Traditionally, within military educational institutions, when talking about the integration of pedagogy and technology, simulators were immediately thought of. This does not stop being true, because the use of simulators in different levels are clear examples of learning in action, collaborative work and reflection in action in the so-called "War Games", an educational resource that is used more and more every day. recognized in its validity when interpolated to the civil environment and known within gamification [29].

Recent developments have made commercial simulators more and more useful for use in professional military classes, instead of systems designed specifically for such purposes. Thus, the game of naval tactics Jane's Fleet Command is used at the United States Naval War College, USNWC, regularly, while Army training centers in that country use an improved version of the game Janus to train at the company and battalion level. A version modified by the Marine Corps of the popular Doom game is used to train four-man combat teams in concepts such as mutual fire support, automatic weapon protection, attack sequencing, munition discipline and command succession. It is known that many flight students around the world have used Microsoft Flight Simulator to practice, but it is not until recently that it is known that aspiring fighter pilots of the United States Navy who used it, have regularly obtained better scores than who did not do it [30].

NETSAFA, entity responsible for cooperation with other countries in military education in the United States Navy, has taken advantage of social networks to facilitate communication with their students after they return to their countries of origin, creating Facebook pages that Teachers use to keep in contact and exchange information and questions with their students [27].

Distance education, a subject in which the Ecuadorian Navy has some experience with the use of platforms such as Moodle or Blackboard to support distance courses for students of the Naval War College (Academia de Guerra Naval) or ESCAPE. Before entering promotion courses, officers and crew must approve distance courses that qualify them for entry, as well as pass continuing education modules at the University of Armed Forces - ESPE, which are considered as requirements to continue in the race.

Another example is the Virtual Desktop initiative of the United States Navy, which seeks to improve and facilitate the educational experience of students at a distance from that institution, by allowing students to enter from any computer, generating virtual desktops to access classified information, leaving aside the dedicated terminals, and facilitating the transmission of sensitive information that would otherwise generate serious security risks [31].

More examples of the increasing integration of ICT in military education include the employment in the United States Armed Forces of intelligent tutorial systems, or virtual worlds used for Second Life-type games for decision making, and even individualized learning models as those used by the US Marines, promoting changes in the pedagogical interaction between teacher and student [32]. These forms of using ICT in military education should only remember that their use is a means and not an end, since the integration of ICT in military education is as valid as it is in civil education.

Unfortunately, while distance education is a way in which the Ecuadorian Navy has integrated ICT in education, the experience of ESCAPE and the Police indicates that the path that "uniformed education" has to travel is long. Most ESCAPE professors, for example, believe that it is good to use technology, but at the same time, they say they do not [54].

6 ICT Integration Models

In order to integrate ICT in the teaching practice, it is necessary to have a methodology that must necessarily start with a frame of reference that allows defining the state of integration, which are defined under different models.

Majumdar [33] presented in his report to UNESCO a proposal for a model of ICT integration in education. This model consists of stages that are considered as steps that identify the development of ICT in education.

Institutions that are in the initial stage of ICT integration follow the emerging approach. These are those institutions that are just beginning to have computational infrastructure and their teachers make personal use of new technologies, such as using word processors or using the Internet to communicate with friends or family. These institutions, begin to know that they can use ICT, but are not familiar with them, and therefore use them to improve the performance of their professionals.

When institutions are in a stage of learning to use ICT, and seek to improve the ways of learning, Majumdar indicates that they are in the second stage, the application in it, teachers use ICT for professional purposes, using them to support Learning in your area of knowledge.

The third stage is known as infusion and involves integrating ICT into the curricula, using them in laboratories, classrooms and administrative sectors. It not only involves the teaching - learning processes, but also those of its management. These institutions understand when and how to use ICT and thus seek to facilitate learning.

The fourth and last stage is called the transformation stage and it happens when the educational institution renews its organization so that ICTs are an integral part of professional practice. The teachers of these institutions seek to create innovative learning environments using ICT.

Based on this model, Anderson [3] identifies the stages in which teachers could be found by comparing their ability to use ICT. Thereby, a teacher who applies tools such as word processors, programs to make presentations, databases, spreadsheets or email to support their process, will be in the first stage, while those who use software to support learning, I could already say that it is in the second.

The teacher who seeks to facilitate learning using multi-mode instruction, using a variety of multimedia tools to facilitate the learning of their students, selecting the one that is most appropriate for each task, will be in the third stage.

In contrast, the teacher who uses modeling and simulation, expert systems, interactive learning tools, will be in the fourth stage and will also be more open to supporting pedagogical innovation.

In order to determine at what stage the institution and its teachers are, there are different models that can be adapted to the reality that exists to obtain valid indicators that clearly determine the level of ICT integration in pedagogical practice. As indicated by Majumdar [33], the International Society of Technology in Education, ISTE, developed since 2000 the national standards of technology education, NETS for teachers, NETS • T, based on the national standards of technology education for students, NETS, which focus on teacher education before starting their work as a teacher, define concepts, knowledge, skills and attitudes to apply technology in education. After a few years, the ISTE changes its standards for teachers in a second edition, indicating that those who apply, design, implement and evaluate learning experiences can engage students and improve their learning using ICT. The standards are defined as follows [34]:

(1) Facilitate and inspire student learning and creativity.
(2) Design and develop learning experiences and assessments of the digital era.
(3) Modeling work and learning in the digital era.
(4) Promote and model citizenship and digital responsibility.
(5) Involvement in professional growth and leadership.

ISTE participated in the definition of the ICT Competencies Framework for UNESCO Teachers [35] together with other institutions such as Microsoft. UNESCO aims to achieve the objectives of national economic and human development through education, which means that there are differences between these two models of the integration of ICT in pedagogical practice.

Thus, the UNESCO Framework emphasizes that it is not enough for teachers to have ICT competences and be able to share them with their students, but that they should be able to help their students collaborate with each other, solve problems and learn in a creative way using ICTs to become good citizens and members of the workforce [36]. This model takes three approaches: technological literacy, the deepening of knowledge and the creation of knowledge, and in that way takes into consideration all the aspects that define within the work of the teacher, creating a matrix framework.

A third model to be taken into consideration is the Common Framework of Teaching Digital Competence of the Ministry of Education, Culture and Sport of Spain [37]. This works on 5 different domains, which take into consideration the digital competences that must be achieved and integrate teachers in their practice, referred to the domains of Information, Communication, Content Creation, Security and Resolution of Technical Problems [37, pp. 66–73].

The standards defined by UNESCO, ISTE or the Ministry of Education of Spain are clear and are accompanied by their respective indicators and as such, can be used to guide the evaluation to the pedagogical practice of teachers in military institutes, as such way that allows measuring the integration of ICT in it, and thus be able to define at what stage they are according to the model defined by Majumdar, which may provide a clear idea of the shortcomings and needs to make the best decisions that could improve those levels and take military education and its teachers to the higher stages.

For example, ISTE has indicated that a study conducted by the Richard W. Riley College of Educational and Leadership, cited by Beglau [38] found that teachers who use technology to support learning in their classrooms have frequently reported large benefits in learning, commitment and development of skills, compared to those who have not. The same study found that only 34% of 1000 teachers surveyed use technology 10% of their class or less. That is why it is necessary to train teachers according to the approach made by ISTE, which has found positive results using 3 models [38]:

(1) Cognitive training.
(2) Instructional training.
(3) Peer training focused on technology.

UNESCO [39] has carried out studies to determine the results of ICT integration using its frame of reference in the Asia-Pacific region. For example, Malaysian teachers indicated that after being trained to use ICT in their teaching practice they were able to incorporate them into their classrooms allowing them to live new forms of learning. Another important experience is that of Indonesia, a country that has committed to education using ICT with programs such as wireless networks in all schools with radio frequency links or free software licenses for donated computers. Other countries that were evaluated were the Philippines, Singapore, South Korea and Thailand, each of them with success stories referring to positive change for having applied ICT integration standards in education [40].

The Spanish case is different, since the frame of reference is more recent. However, digital literacy actions have been carried out in the community of Madrid since the nineties, which since the issuance of the frame of reference have been increased, such as courses related to the use of computer tools in curricular areas, and employment of web 2.0 applications [41]. In general, there are actions tending to integrate ICT in schools such as the provision of digital educational content platforms, tablets in classrooms, interactive digital whiteboards or even the freedom to carry their own device in some communities [42].

A large percentage of 5,000 Spanish teachers surveyed in a study conducted by Area et al. [43], indicated that the increase in the availability of resources and technological infrastructure is a positive element that has allowed integrating innovations based on ICT into pedagogical practice, through information search activities, work with word processors, on-line exercises. One of the most important findings is that 80% of teachers surveyed believe that ICT in the classroom does not cause an increase in the distraction of students or an added effort of importance to their teaching.

Now, these models are the result of the reflections of their creators and therefore define a reality that is a function of the observed. The Ecuadorian case, as can be inferred from the research carried out at universities in Guayaquil and Quito, and the Police and Navy experiences, is different. The conclusions of these studies indicate a low integration of ICT that does not even reach the lowest levels of the proposed models.

That is why, in order to establish a scale of integration of ICT in teaching practice that is better adapted to the Ecuadorian reality (including military), the models of ISTE [34], UNESCO [44] and the Ministry of Education, Culture and Sports of Spain [37], and the experiences obtained in the studies carried out by graduates of the Central University [9], Universidad Casagrande [10, 11], FLACSO [12] and the Escuela Politécnica del Litoral [13] in professors of university level are taken as reference, to define a scale of levels of integration of ICT in the teaching practice in the following way:

In the proposed model (see Fig. 1), in the 1st level: Digital Literacy, the class is standard, with a basic knowledge of technological tools and practical integration of technology on very basic topics such as generating documents or using a slide presenter. The 2nd level: Design of digital experiences, the class is standard, but using digital tools beyond PowerPoint. Design and evaluate authentic learning experiences to develop knowledge and skills using digital tools as support. In the case of 3rd level: Work modeling and digital learning, the class uses collaborative groups, applying knowledge to solve complex problems. I use complex digital tools to achieve collaborative work. In 4th level: Citizenship modeling and digital responsibility, teacher manage and guide the class. I exhibit a behavior that guides my students to integrate technology in an adequate, legal and ethical way in their practice, while creating their knowledge. In the, 5th and last level: Involvement in professional growth and leadership, we improved the professional practice, while learning for life model demonstrating the effective use of digital tools and resources and encouraging students to create knowledge by taking me as a learning model. Students are able to solve technical problems, and innovate the use of technology in a creative way.

Echegaray [45] indicates that some time ago there has been an insistence on the need to change teaching strategies in the learning process, but even so, research shows that teachers do not have enough digital skills to integrate ICT into their practice. Driskell [19], Méndez Cortes [17], Szabó [20] and Juhary [21] indicated that military education can benefit from the constructivist approach, while Bell and Reigeluth [32] showed that it can benefit from the increasing use of ICT for personalized learning.

Cabanelas and Rivas [46] in their research on postgraduate teachers in Spain show that although the Internet is an enriching resource in the educational context, it is very little exploited, with teachers being one of the cornerstones for the integration of ICT in education, so that their availability and attitude will be fundamental to make it possible.

In that way, the positive approach that military professors could have toward the use of technological resources in education will be a strong base on which to build knowledge about the possibilities offered by ICT to improve educational processes. But it still will be just the beginning of a long process.

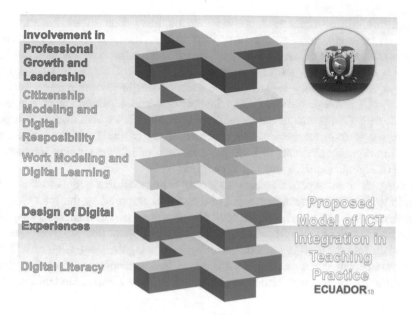

Fig. 1. Proposed model of ICT integration in teaching practice for Ecuador.

7 Conclusions

UNESCO Regional Office for Latin America and the Caribbean [47] indicates that in the reality of this region of the world, great expectations should not be created regarding rapid changes in the current situation, since ICT should not be considered a matter of specialists, but as a cross-cutting element for all educators. It is for this reason that initially it will be necessary to gradually pass digital literacy for military teachers in relation to the academic use of ICT. If based on the results obtained from the experiences in civilian educational institutions, it can be concluded that teachers in Ecuador are mostly within the first 3 levels of the ICT integration model proposed for Ecuador, it should initially aim to provide them with the necessary knowledge about the different technological resources that can be used, in such a way that they can introduce them into their teaching practice as they take more confidence and knowledge in the advantages that their use can provide for the learning experience of their students.

In order to change this situation, it is necessary to describe the training needs of teachers in the use of ICT, to improve the learning experience in the military context. If we take UNESCO's recommendation into account again, these training needs must be solved gradually and based on digital literacy. The ALEPH program in Colombia seeks the knowledge, use and incorporation of ICT in the professional training of teachers and for this, has developed a program to train them in educational services of the Internet, educational communication and audiovisual media [48], very similar to the proposal of Heinz and Lara [49], focused on the development of basic competences in ICT, development of cross-cutting competences in ICT and administrative and results management. The UNED Foundation [50] in Spain offers a training program in ICT for teachers that covers topics such as the didactic integration of the Internet in the classroom, communication with ICT applied to the classroom and the creation and management of digital content, all of these, private efforts that have counterparts led by Ministries of Education such as Argentina [51] with its Teaching Specialization in Education and ICT, which reviews topics such as teaching, learning and evaluating with ICT, ICT strategies, cross-cutting integrations and development of educational proposals with TIC; or that of the Dominican Republic [52], which intends to organize 600 courses to train teachers of the different levels of the Dominican educational system in necessary elements to implement models of ICT integration, classification and use of digital resources and ICT strategies.

Taking all these experiences and designing a proposal that adapts to the training needs in the military context, according to the model proposed by the Joint Command of the Armed Forces [14] of competency-based learning, makes it necessary first of all to take into account the need to start with digital literacy and then integrate these new digital skills into the teaching practice, starting by placing the integration of ICT in the classroom as facilitator, as a medium and not as end, technological resources that are used efficiently by properly trained teachers, will help to improve student learning.

References

1. UNESCO: Replantear la educación: Hacia un bien común mundial? (2015)
2. Rivas, M.R.: La tecnología informática al servicio de la educación (2001). http://dspace.usc.es/handle/10347/5150
3. Anderson, J.: ICT Transforming Education: A Regional Guide. UNESCO Bangkok, Bangkok (2010)
4. Courville, K.: Technology and its Use in Education: Present Roles and Future Prospects (2011)
5. Fuentes, M.D.C., Gutiérrez, J.J.T.: ¿Mejoran las TIC los resultados académicos de los estudiantes españoles? eXtoikos 9, 51–58 (2013)
6. Batanero, J.M.F., González, J.A.T: Actitudes docentes y buenas prácticas con TIC del profesorado de Educación Permanente de Adultos en Andalucía, Teacher attitudes and best practices with ICT faculty Adult Continuing Education in Andalusia (2015)
7. Sánchez, F.Á.B., Guzmán, A.F.: La robótica como un recurso para facilitar el aprendizaje y desarrollo de competencias generales. Educ. Knowl. Soc. EKS 13(2), 120–136 (2012)
8. Calero, J., Escardíbul, O.: Recursos escolares y resultados de la educación, Reflex. Sobre El Sist. Educ. Esp. 314 (2015)
9. Veloz, A.L.A.: Impacto del uso y aplicación de las TIC's en el proceso de enseñanza y aprendizaje de la Matemática de los estudiantes del primer semestre de la Carrera de Matemática y Física de la Facultad de Filosofía de la Universidad Central del Ecuador año lectivo 2010–2011 y propuesta de un software interactivo para mejorar la enseñanza y aprendizaje. Universidad Central del Ecuador, Quito (2014)
10. Ortiz Rojas, M.E.: Constructivismo y Herramientas Web 2.0 en Educación Superior. Universidad Casagrande, Guayaquil (2012)
11. Hi Fong Díaz, M.: El uso de las redes sociales en el aprendizaje. Universidad Casagrande, Guayaquil (2015)

12. Rodríguez Córdova, M.E.: Incidencia de las tecnologías de información y comunicación (TIC) en el proceso de enseñanza aprendizaje. Facultad Latinoamericana de Ciencias Sociales - Ecuador, Quito (2010)
13. Martín, O.M., Malena, C.G.: Factores y relaciones que afectan la incorporación de tecnologías de información y comunicación en la educación superíor (2009)
14. Comando Conjunto de las Fuerzas Armadas: Modelo Educativo de Fuerzas Armadas (2012)
15. Reglamento a la Ley de Personal de Fuerzas Armadas (2009)
16. Ley de Personal de Fuerzas Armadas (1991)
17. Méndez Cortes, Á.: Una mirada crítica a la educación en el ejército, June 2013
18. Annen, H., Nakkas, C., Mäkinen, J.: Thinking and Acting in Military Pedagogy. Frankfurt am Maim, Berlin (2013). Peter Lang, Wien
19. Driskell, J.E., Olsen, D.W., Hays, R.T., Mullen, B.: Training Decision-Intensive Tasks: A Constructivist Approach, November 1995
20. Szabó, J., Military pedagogy - Focusing on the fourth generation warfare. Hadtud. Szle. **6** (2013)
21. Juhary, J.: Understanding military pedagogy. Procedia - Soc. Behav. Sci. **186**, 1255–1261 (2015)
22. Johnson-Freese, J.: The reform of military education: twenty-five years later. Foreign Policy Res. Inst. **52** (2012)
23. Ordóñez, C.L.: Pensar pedagógicamente desde el constructivismo. De las concepciones a las prácticas pedagógicas. Rev. Estud. Soc. (19), 7–12 (2004)
24. Ordóñez, C.L.: Pensar pedagógicamente, de nuevo, desde el constructivismo. Rev. Cienc. Salud **4** (2010)
25. Cruz, R., Paul, R.: Propuesta de formación en Docencia Universitaria para los profesores en la Escuela Superior de Policía - ESP (2015)
26. Irigoyen, J.J., Jiménez, M.Y., Acuña, K.F.: Competencias y educación superior. Rev. Mex. Investig. Educ. **16**(48), 243–266 (2011)
27. Catanzano, K.: Enhanced training for a 21st-century military. Booz Allen Hamilton, November 2011
28. Ruth, D.M., Fricker, R., Mastre, T.M.: A Study of Mobile Learning Trends at the U.S. Naval Academy and the Naval Postgraduate School (2013)
29. Anders, F.: Gamer mode : Identifying and managing unwanted behaviour in military educational wargaming (2014)
30. Macedonia, M.: Games, Simulation, and the Military Education Dilemma. Forum Future High. Educ. (2002)
31. O'Brien, S.: Virtual Desktop Upgrade Increases Training Effectiveness, Navy Live (2013). http://navylive.dodlive.mil/2013/02/04/virtual-desktop-upgrade-increases-training-effectiveness
32. Bell, H., Reigeluth, C.: Paradigm change in military education and training. Educ. Technol. **54**(3), 52–57 (2014)
33. Majumdar, S., Anderson, J., Wai-Kong, N., Barnhart, S., Koszalka, T.A., Zhi-ting, Z.: UNESCO Office in Bangkok: Regional Guidelines on Teacher Development for Pedagogy-Technology Integration [Working Draft]. UNESCO (2005)
34. ISTE: ISTE Standards Teachers (2008)
35. Barr, D., Sykora, C.: Learning, teaching and leading. A compartive look at the ITSE Standards for Teachers and UNESCO ICT Competency Framework for Teachers. ISTE (2015)
36. UNESCO: UNESCO ICT competency framework for teachers. Version 2.0. UNESCO (2011)

37. Ministerio de Educación: Cultura y Deporte de España. Instituto Nacional de Tecnologías Educativas y de Formación del Profesorado, Marco común de competencia digital docente V 2.0 (2013)
38. Beglau, M., et al.: Technology, Coaching and Community: Power Partners for Improved Professional Development in Primary and Secondary Education. ISTE (2011)
39. UNESCO Bangkok: ICT in teacher education. Case Studies from the Asia-Pacific region. UNESCO (2008)
40. UNESCO Bangkok, UNESCO Office in Bangkok: Integrating ICT in education, lessons learned. UNESCO Asia and Pacific Regional Bureau for Education (2004)
41. Sánchez-Antolín, P., Ramos, F.J., Sánchez-Santamaría, J.: Formación Continua Y Competencia Digital Docente: El Caso De La Comunidad De Madrid (Policies for Continuous Training and the Digital Teaching Competence: The Case of Madrid). Social Science Research Network, Rochester, SSRN Scholarly Paper ID 2690387, June 2014
42. Moreira, M.A., et al.: Las políticas educativas TIC en España después del Programa Escuela 2.0: las tendencias que emergen/ICT education policies in Spain after School Program 2.0: Emerging Trends. Rev. Latinoam. Tecnol. Educ. - RELATEC 13(2), 11–33 (2014)
43. Moreira, M.A., Mesa, A.L.S., Navarro, A.M.V.: Las políticas educativas TIC (Escuela 2.0) en las Comunidades Autónomas de España desde la visión del profesorado. Campus Virtuales 2(1), 74–88 (2015)
44. UNESCO: Estándares de competencias en TIC para docentes (2008)
45. Echegaray, J.P.: ¿Y si enseñamos de otra manera? Competencias digitales para el cambio metodológico. Caracciolos 2(1) (2014)
46. Cabanelas, M.E.A., Rivas, M.R.: Los docentes de postgrado ante las nuevas tecnologías. Rev. Latinoam. Tecnol. Educ. - RELATEC 5(2), 501–512 (2006)
47. OREALC/UNESCO Santiago: Formación docente y las tecnologías de información y comunicación. Estudio de casos en Bolivia, Chile, Colombia, Chile, Ecuador, México, Panamá, Paraguay y Perú. Santiago. AMF imprenta, Chile (2005)
48. Angel, S.I.B., Moreno, G.E.J., Cuartas, I.E.M.: Capacitación de Docentes en Tecnologías de la Información y la Comunicación. en: Colombia aprende, Bogotá (2005)
49. Heinz, S., Lara, M.I.: Programa de capacitación en competencias TICs para docentes. en: Nuevas Ideas en Informática Educativa, Santiago, vol. 7 (2011)
50. Fundación UNED: Curso TIC para Profesores - Programa - Fundación UNED, Curso de competencias TIC para profesores (2016). http://www.cursoticprofesores.com/programa/
51. Ministerio de Educación y Deportes de Argentina: Especialización Docente en Educación y TIC, Especialización docente en Educación y TIC (2016). http://postitulo.educ.ar/. Accedido 09 Oct 2016
52. Ministerio de Educación de República Dominicana: Capacitación de docentes en la Integración de TIC en el Aula, Capacitación de docentes en la Integración de TIC en el Aula (2014).   http://www.educando.edu.do/articulos/docente/capacitacin-de-docentes-en-la-integracin-de-tic-en-el-aula/. Accedido 09 Oct 2016
53. Ordoñez, P.C., Rincón, J.G.C.: Las redes sociales en la Universidad Militar Nueva Granada-UMNG. Acad. y Virtualidad 6(2), 24–33 (2013). https://doi.org/10.18359/ravi.407
54. Garay, F.: La Integración de las TIC en la Práctica Pedagógica de los Docentes en la Escuela de Calificación y Perfeccionamiento de la Armada. Las TIC en ESCAPE. Más Allá del Powerpoint. Universidad Casagrande, Guayaquil (2017)
55. Sousa, M.J., Rocha, Á.: Leadership styles and skills developed through game-based learning. J. Bus. Res. (2018)

Sociometry: A University Tool to Facilitate the Cohesion of Academic Groups

Luis Miguel Mazón[1(✉)], Datzania Villao[1], Teresa Guarda[1,2],
Linda Núñez[1], María Muñoz[1], Manuel Serrano[1], and Divar Castro[1]

[1] Universidad Estatal Península de Santa Elena – UPSE, La Libertad, Ecuador
luismazon86@gmail.com, datzaniavillao@gmail.com,
tguarda@gmail.com, lnunez_ing@hotmail.com,
mardelou_80@hotmail.com, mserranoppsa@gmail.com,
d.castro.loor@gmail.com
[2] Algoritmi Centre, Minho University, Guimarães, Portugal

Abstract. The methods used by a teacher to improve the learning process, allow applying techniques that analyze the interrelations in small groups and in this way, overcome the weaknesses generated by the lack of group dynamism. In this article, the objective was to know the interrelations generated by popularity, authority, positive leadership and the complement that is the grouping of different people. For this purpose, an online sociometric test was applied to 93 students of the Business Administration Career from Peninsula de Santa Elena University-Ecuador, using a software called sociometric classroom to analyze the results. The results showed that empathy and sympathy are drivers of the searching for improvement in a group sub-structures. On the other hand, the test allowed the selection of leaders, who were students with a greater number of votes, popularity and a high academic performance.

Keywords: Sociometry · Test · Leaders · Students · Group dynamism

1 Introduction

In every working group there are situations that do not have, at first glance, a logical explanation. The interrelation that exists between members of a group determines certain reactions, stimuli, tensions, problem situations produced by an endless process of valuation and search, which affect the progress of a group (García-Martín and García-Sánchez 2016). The harmonic functioning of a group depends on the communication established between its members. Therefore, taking into account the importance of cohesion of a group, the sociometry came to search this issue, having its origin in 1936 with the work of Jacob Levy Moreno who defined sociometry as the mathematical study of psychological characteristics of social groups, the experimental technique of quantitative methods and the results obtained from their application (Moreno et al. 1972).

The definition shows that the connection of social groups is essential because people learn through group coexistence how to relate and strengthen these links. However, it is not paradoxical that different groups are built according to these links,

© Springer International Publishing AG, part of Springer Nature 2018
Á. Rocha and T. Guarda (Eds.): MICRADS 2018, SIST 94, pp. 351–359, 2018.
https://doi.org/10.1007/978-3-319-78605-6_30

which can vary with respect to the members or can be strengthened and generate durable friendships (Hellriegel et al. 2009).

In the framework of higher education, a considerable number of opportunities are created for durable friendships, as well as a numerous conflicts can be produced if there is not a team that has participatory cohesion, generates positive ideas and keep a satisfactory average of the course. In this sense, it can be pointed out that poor academic performance may be caused because fear, doubt and uncertainty do not allow students to develop social skills such as: asking for a favor, making a comment or even offering their knowledge or skills to generate stronger working groups (Yang and Tang 2003). Therefore, it can infer that sociometry is useful for teachers because it can generate an ideal climate for learning and general welfare in the classroom. Nevertheless, there are students that need help due to the lack of integration to some group or the apathy generated in various personal problems that cause a student to be backdated and maybe, shy which can affect his academic performance.

Studies of sociometry in education has showed that systematic interaction, which is formed from interpersonal relationships based on teamwork, mutual help, goals and motivation, can generate common feelings such as empathy, moral aspects and aptitudes (Evans 1964). This interaction can give integral priorities to the professionalism of higher education because it generates a preparation to the development of human resources from a methodological, scientific and academic level, as well as, the work of values to the unconditional dedication of the profession, permanent innovation, creativity and vision of the future, taking into account national cultural identity (Cañedo 2015). In this sense, Casanova, MA (1991), a scientist who promotes social facts in the field of education, points out that sociometry help to determine how are the groups formed and what are the interaction patterns among the students which lead to group cohesion in an objective manner, making the sociometric pedagogy useful for a better educational teaching performance.

Undoubtedly, sociometry, taken in its most interesting disclosed aspect, comes to be reduced to the set of experimental methods, some therapeutic or pedagogical and others of research, referred to the small group (Terry 2000). One of the most common research techniques in the framework of sociometry to discover and analyze a series of interpersonal relationships and the specific situation of a person within the group, are sociometric test, test of interaction, collective evaluation test, spontaneity test, role test, spontaneous choice test, sociometric perception test, peer comparison test and ordinal test. However, among the sociometric research techniques mentioned above, the sociometric test is, without a doubt, the most common application for knowledge of the informal structure of the class, since it presents, in addition to the advantage of its precision, the of simplicity and the speed of its use.

In particular, this paper focuses on the application of sociometry and more specifically through a sociometric test, aiming to improve social relations within the class and favor the integration of isolated and rejected students. The test was applied to the students from the first to eighth semester of the Administration Business Career from Península de Santa Elena University, Ecuador.

The article is organized as follows: Sect. 2 presents the analysis of the applications of sociometry in a classroom. Section 3 will show a case study using software called Sociometric classroom. In Sect. 4, the various results generated by the application of the sociometric test in a real case study will be shown. In Sect. 5, conclusions and future guidelines are presented.

2 Analysis of the Theoretical Foundations of Sociometry

2.1 Applications of Sociometry in a Classroom

Detection of problems related to social adaptation. Sociometry detects the existence of isolated elements in the group, not chosen by anyone, before which it must ensure that others pay attention. It also detects the existence of a student rejected by the majority of the members, before which it will be necessary to examine the causes that cause this marginalization. It is important because it helps to detect the existence of subgroups, before which it will be convenient to talk with the leaders (Casanova 1991).

Identification of methodological strategies in the classroom. Especially in the moments foreseen for the work in operative groups throughout the development of a didactic unit, the social or informal structure of a group as a whole will be a valuable piece of information. In principle, efforts will be made to ensure that a working group adjusts to a group that appears spontaneously, according to the preferences expressed by the students. They will give an optimal result, because the class will be well integrated and without special problems (García-Bacete et al. 2008). Therefore, when there is a subject with integration difficulties, the structural knowledge of the group will serve to incorporate it with students who can motivate him. It will thus be easier to break the isolation.

The impact on the academic scenarios depending on the type of subject, will increase the aspirations of educational support for the teacher lecture because teachers must know how to generate new policies of the classroom, enhancing not only the way of teaching but the multiple ways of solving group and individual doubts, that is, rethinking new teaching-learning models and the strengthening of cohesion for each classroom.

Personal and professional guidance for students. On a personal level, the integration of the individual in society will depend to a large extent on the capacity he has to relate to his classmates, friends and relatives. The result of these relationships absolutely conditions the happiness or unhappiness of the person, much more than other components of his life. At the professional level given the current model of society, it can be decisive for the individual's work development knowledge of their interpersonal relationship skills, since their choice of work will depend on them, an essential aspect for their future development (Kim 2011).

Technique for the evaluation of students. In an initial diagnosis, during the teaching-learning process and in the final reflection on the results achieved, it is essential to have data related to the sociability of a student (Ben-Yehuda et al. Ben-Yehuda et al. 2010). Thus, a formative evaluation should not leave without evaluating qualitative aspects of education, even if they are more difficult to objectify, since it would leave without evaluating what really constitutes the essence of education: attitudes, limiting itself to measuring what has always been easier: the conceptual contents acquired in teaching (Gómez and Calvo, 2005). Perhaps where it is necessary to highlight its usefulness either in the initial evaluation or when the social structure of the group is unknown at the beginning of the course. For example a good way to evaluate students in a qualitative way, would the selection of the assistant student of a lecture, through a sociometric test. This way will allow knowing the social field that is evident in the classrooms, because the group and the interpersonal relationships will be measured, analyzed and structured.

The selection of an assistant student through a sociometric test would show the importance of sociometry test because the selection would be based on the synergy among students. Thus the chosen would be the person who would be capable of generates environments of cooperation and mutual help with a social approach in order to establish personal relationships, take advantage of the skills of the group and create an atmosphere of trust. Therefore, it is necessary to mention that an assistant student that will be chosen by his classmates, will be who helps his classmates who are in situations of helplessness, confusion, personal discomfort or who have academic difficulties that are denoted by their weak academic performance.

3   Case Study Applying Sociometric Classroom Software

A sociometric test was applied to 93 students from second grade, first period of Business Administration Career. The objective of the test was to identify students that have more social skills and students that are rejected.

The research was experimental and used an online sociometric test that was applied to the students analyzed, through a Google form that was designed with 3 questions related to the way of interactions among students and was sent through a link to whatssap to their cell phones. There were 3 simple questions to answer.

1. When you have problems in the classroom, whom do you tell?
2. If you had weakness in a subject, whom would you ask for help?
3. Who would you choose as the best classmate?

The Sociometrics classroom software was use to analyzed the answers of the sociometric test. This was chosen because it is a specialized software tool applied specifically to sociometric test, which present the results in graphs, especially about the sense of the students' social status in school as well as the global climate of the classroom. The graphs distinguish between male (by means of a square) and female (by means of a circle) and present the percentage of bidirectional relations. (Kuz and Falco, 2013).

3.1   Process of the Application of Socio-Metric Test

The application of test shows the work groups of the students of Business Administration Career that goes with the regulations to enhance academic processes and facilitate mass communication, taking as an initiative the joint work between teachers, technicians and students. It has as main objective to identify students who show a higher degree of group and collective relationship. Therefore, it is essential to mention what were the steps in the application of the test:

1. It was defined the questions according to the defined objective of the research.
2. The questions were selected according to sociometry framework;
3. It was delimited the number of people that can be chosen. (It depended on the number of students, a maximum of three selections was recommended);
4. Students were encourage to make all selections according to the request;
5. The questions were understood by the members of the group. It was conducted a small induction of what was intended to be done;
6. For the application of the test it was recommended that the students must have been lived in community for at least one semester to allow a selection criterion;
7. It was selected a channel or a didactic method to apply the test. In this case, Google form was chosen because it is a fast way to send a form and get back the answers.
8. The answer were analyzed with the use of socio-metric classroom software

3.2 Selection Criteria

Once the test was applied, the work criteria were:

− Muto selection or pair: constituted by the selection between individuals of mutual form or reciprocal selection.
− Isolated: those who lack selection or to be chosen.
− Chains: one person chooses another, which in turn chooses a third and so on.
− Working islands: couples or other larger groups.
− Leaders: student with the largest number of selections.
− Grays: students chosen by the leaders, and who influence in the group

4 Results and Discussion

Taking into account the software used, the results are presented in ghaphs called socio-grams that is a data analysis technique that focuses on how social links are established within any group. The socio-gram is more than a presentation method. It makes possible the exploration of sociometric facts and to determine the appropriate place of each individual as well as all the interrelationships that exist between them.

In Fig. 1, the results show that the majority of votes are for the student Jessica, who also has a high academic performance (it was checked in her academic statement). This shows not only the grade of relationship between their workmates but also manifests a positive leadership, where the cohesion of teamwork is also appreciated. On the other hand, the other selected students are called work islands, but given the characteristics of this research, an integral student can be selected as an academic assistant.

In Fig. 1, it shows that there are islands of work since they are large groups of couples or circle of friends having Erick as the largest selection and a small minority of Trinidad votes which shows they are with a difference of 4%. However, given the elections and the objective of the research, the student with the largest number of elections is selected.

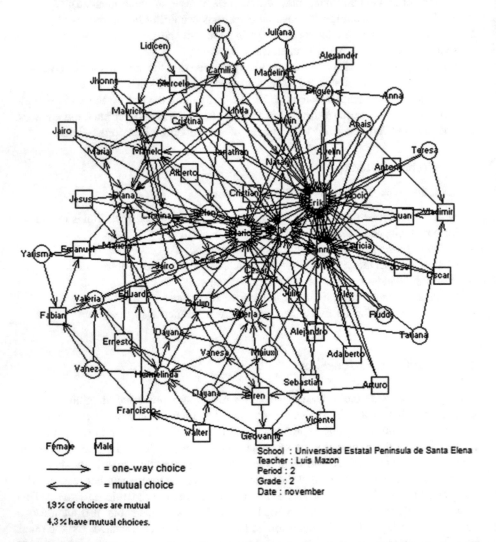

Fig. 1. Socio-gram corresponding to Question 1. If you had weakness in a subject, whom would you ask for help?

In Fig. 2, the greatest number of selection is for Stalin while the other students are chosen in minorities. It means that there are small groups as couples called islands of work, for this, a leader of a course was selected.

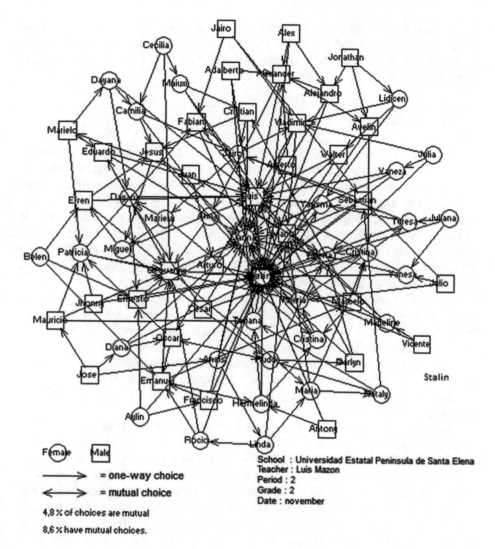

Fig. 2. Sociogram corresponding to Question 2. Who would you chooses as the best partner?

5   Limitations

Despite the sociometry test allowed getting a lot of information with few questions, the results do not represent the reasons of the elections, and why these social structures and relationships are produced. Therefore, due to its quantitative nature, qualitative aspects of high interest, escape. In addition, the results may be biased, because some students may have been influenced by a certain circumstance of a group in the moment to answer the questions.

On the other hand, not all the students answered normally to the test. On the other hand, the application of the test by the Google form facilitated the process. However, it was realized that a small group of students do not manage social networks, much less cell phones. Another aspect evidenced was that some students responded late because in their homes they do not have Internet and the areas where they live do not have internet coverage.

6 Conclusions

The classroom is a didactic space in which students take the first step in what it is known as socialization. In this area, a lot of relationships are carried out; bonds are created, links between students and even relationships of students with teachers. Therefore, sociometry and more specifically the sociometric test can be a very useful instrument to improve social relations within the class and integrate isolated and rejected students. In addition, beyond considering a group and determining the dynamics of it, the present research focused on deepening the search in a simple way in which it is plausible to help not only the teacher, but mainly the student. The application of a sociometric test was useful to determine what the feelings of students towards others.

The level of positive influence is expected to follow up to seek improvements in communication and dissemination of information, teacher-student, student-student and teacher-student as well as meet the role and function assigned by the teacher, implementing a better development of communication skills inside the classroom.

The article was involved in an educational-formative development as well as the complement of formal learning based on sociometric characteristics that reinforce teaching learning, educational values and professional ethics. It can also help to integrate a greater number of students.

Sociometric Classroom was a tool that generated many interesting data, such as students who are popular, isolated and those who are rejected. It can also determine the social status of the group. The application of the online sociometric test allowed knowing the social relations among students, as well as the level of acceptance of a person as a leader in the Business Administration Career, generating expectations that can be met efficiently. Therefore, this study becomes the basis to apply sociometric tools to the rest of the careers from Santa Elena University. Finally, due to we live in the era of social networks, an era where Facebook, Twitter and online games are everyday words, it would be important to incorporate all these social networks in the sociometric tests. More specifically in the university, in order to a teacher can get in an accurate way what is the climate of current coexistence and the level of satisfaction in which a student develops his task.

References

Ben-Yehuda, S., Leyser, Y., Last, U.: Teacher educational beliefs and sociometric status of special educational needs (SEN) students in inclusive classrooms. Int. J. Incl. Educ. **14**, 17–34 (2010)

Cañedo, I.: La diversidad en el contexto universitario: Una necesidad actual en el Ecuador. Palibrio, Bloomington (2015)

Casanova, M.: La sociometría en el aula. La Muralla, Madrid (1991)

Evans, K.: Sociometry in school – 2. applications. Educ. Res. **6**, 121–128 (1964)

García, J., García-Sánchez, J.: La Psicología Positiva en el Asesoramiento Psicopedagógico. Int. J. Dev. Educ. Psychol. 61–68 (2016)

García-Bacete, F., Sureda, I., Monjas, I.: Distribución sociométrica en las aulas de chicos y chicas a lo largo de la escolaridad. Rev. de Psicol. Soc. **23**, 63–74 (2008)

Gómez, C., Calvo, M.: Habilidades sociales en adolescencia: un programa de intervención. Rev. prof. esp. de ter. cogn. conduct. **3**, 1–27 (2005)

Hellriegel, D., Slocum, S., Hellriegel, J., Jackson, S., Slocum, J.: Administración: Un enfoque basado en competencias. Cengage, México (2009)

Ketten, D.R.: The cetacean ear: form, frequency, and evolution. In: Marine mammal sensory systems Marine mammal sensory systems, pp. 53–75 (1992)

Kim, J.: Developing an instrument to measure social presence in distance higher education. Br. J. Educ. Technol. **42**, 763–777 (2011)

Kuz, A., Falco, M.: Herramientas sociométricas aplicadas al ambiente áulico. In: Congreso Nacional de Ingeniería Informática/Sistemas de Información, pp. 2347–0372 (2013)

Moreno, J., Bouza, J., Karsz, S.: Fundamentos de la sociometría. Paidós, Buenos Aires (1972)

Terry, R.: Recent advances in measurement theory and the use of sociometric techniques. New Dir. Child Adolesc. Dev. **88**, 27–53 (2000)

Yang, H., Tang, J.: Effects of social network on students' performance: a web-based forum study in Taiwan. J. Asynchronous Learn. Netw. **7**, 93–107 (2003)

A Network Theoretical Approach to Assess Knowledge Integration in Information Studies

José María Díaz-Nafría[1,2], Teresa Guarda[3,4,5]([⊠]), and Iván Coronel[3]

[1] BITrum-Research Group, C/San Lorenzo 2, 24007 León, Spain
jdian@unileon.es
[2] Munich University of Applied Sciences,
Dachauerstr. 100a, 80636 Munich, Germany
[3] Universidad Estatal Península de Santa Elena – UPSE, La Libertad, Ecuador
tguarda@gmail.com, iacoronel@hotmail.com
[4] Algoritmi Centre, Minho University, Guimarães, Portugal
[5] Universidad de las Fuerzas Armadas-ESPE, Sangolqui, Quito, Ecuador

Abstract. The paper presents a general approach to assess knowledge integration as a basis to evaluate the performance of transdisciplinary and interdisciplinary approaches with respect to their knowledge integration capacity. The method is based on the development of Interdisciplinary-glossaries as tools for the elucidation of the conceptual networks involved in interdisciplinary studies. Such ID-glossaries are used as proxies of the corresponding knowledge integration, which is measured through the structural analysis of the co-occurrence network of terms. This approach is applied to an ID-glossary devoted to the general study of information, called glossariumBITri. The results show the capacity of the approach to detect integration achievements, challenges and barriers. Its qualitative nature is complemented by an enhanced methodology in which both the diversity of disciplines and the knowledge integration can be measured in a bi-dimensional index. To that purpose each contribution to the target ID-glossary is identified by the knowledge domains involved (using a set of knowledge domains adapted from the higher categories of the Universal Decimal Classification), while the integration is measured in terms of the small-world coefficient of the co-occurrence of terms.

Keywords: Interdisciplinarity · Network analysis · Knowledge integration
Transdisciplinarity · Information studies

1 Introduction

A scientific discipline can be characterized by its conceptual network [1, 2]. The systematic relation among the nodes of the network enables the mapping of the objects and problems that such discipline is focused on. The concepts have thus not an isolated absolute value; this is rather gained in virtue of the capacity of the whole. At the same time, each concept enables that a knowledge domain can better approach a specific part of the reality it strives to gather (or provides an operational capacity to the other concepts in such endeavor). If a node is really worth, by taking it away, the whole network would lose its ability to address its field of interest: the network separates

© Springer International Publishing AG, part of Springer Nature 2018
Á. Rocha and T. Guarda (Eds.): MICRADS 2018, SIST 94, pp. 360–371, 2018.
https://doi.org/10.1007/978-3-319-78605-6_31

(partially or globally) from the reality that it is attempting to map. These are problems of a scientific domain operating by its own.

Different problems arise when various scientific disciplines need to be gather to join their knowledge with the purpose of addressing a complex issue which none of the isolated disciplines is capable to cope with by its own capacity (for instance, the understanding of the information phenomena across the different levels of reality, from the physical to the social aspects). In the endeavor of merging a set of disciplines, we can achieve different levels of integration. UNESCO distinguishes the following levels, organized from lesser to higher integration degree: multi-, pluri-, cross-, inter- and trans-disciplinarily [3]. At the lowest level, the multidisciplinary represents a simple juxtaposition of disciplines (i.e. they solve by their own the issues they are entrusted with). Therefore, the conceptual network of the domains involved do not interact significantly. Nothing needs to be changed in their respective conceptual network to address the problems tackled. However, transdisciplinarity, at the highest integration level, "assumes conceptual unification between disciplines" [ibid, p. 9]. In other words, the conceptual network of the disciplines involved blends into a unified operative framework. In between, interdisciplinarity embraces coordination and cross-communication among participant disciplines, but "the total impact of the quantitative and qualitative elements is not strong enough to establish a [unified framework,] a new discipline". From the perspective of the conceptual network, the common concepts (for instance, 'communication', 'message' or 'data' in the general study of information) often establish different relations with the rest of the combined conceptual network because the different value of the node (term/concept) at each domain.

Because of the lack of integration provided by multidisciplinarity, the international panel of experts, convened by the UNESCO in 1985, excluded it as a level of effective knowledge integration, and agreed to consider just three interdisciplinary levels: (i) pluridisciplinarity (the disciplines are just brought together with often few contact), (ii) interdisciplinarity (there is a good knowledge of each other's concepts) and (iii) transdisciplinarity (the conceptual unification is achieved) (ibidem). Hence, while (i) and (iii) represent the extremes, (ii) occupies a broad space in between. It is our purpose to provide means to assess the interdisciplinary level applied to the general study of information, i.e., the effective distance to (i) or (iii), which can also put in terms of the effective attainment in the integration of knowledge. To this end, we rely on: (a) the development since 2009 of an interdisciplinary glossary of concepts devoted to foster the integration of knowledge in the general study of information, (named glossariumBITri) supported by an international interdisciplinary network of scientists: (b) a network approach to evaluate the effective achievement of the conceptual network deployed through the elucidation process as crystallized in the glossariumBITri edition of 2016 [4]. Building upon these results, an enhanced methodology to assess knowledge integration is presented at the end.

2   Methodology

2.1   Interdisciplinary Glossaries as Tools for the Integration of Knowledge and the Evaluation of the Integration Achieved

The concept of interdisciplinary glossary (ID-G), in which the glossariumBITri (gB) is based, differs significantly from the usual glossaries since they aim at clarifying what is meant by the terminology from the disciplinary perspective [4, 5]. The purpose of ID-G is, on the contrary, bringing together the different understandings of common terms from the summoning of various disciplinary perspectives. In the endeavour of trans-disciplinarity, the meeting of view-points targets the conceptual unification, but through the elucidation process is possible to find that there are some irreducible understandings that are worth to keep in order to preserve the consistency and integrity of the respective theorics.

According to this approach, the gB has been conceived as a tool for the conceptual and theoretical clarification in the study of information. It aims at embracing the most relevant viewpoints concerning information, relying on a board of experts coming from a wide variety of knowledge fields. From a theoretical viewpoint, gB aims to shorten the distances among the different viewpoints and increasing the linkages; while from a meta-theoretical view aims to assess the accomplishment of such integration. In other terms, gB serves as a proxy of the knowledge integration achieved by the interdisci-plinary study of information; thus assessing the interdisciplinary degree in a manner that can be generalized for other knowledge integration undertakings, as proposed in the PRIMER initiative to foster interdisciplinary capacities, which is supported by the scholar network that backs up the gB [6].

2.2   Network Approach to Assess glossariumBITri's Knowledge Integration

According to the abovementioned characterization of the interdisciplinary dialogue, the evaluation of the interdisciplinarity degree or knowledge integration is based on the scrutiny of the structural properties of glossariumBITri's semantic network. To this purpose, the semantic network structure is derived from the meaning relations estab-lished by the authors in their own writings devoted to the elucidation of the conceptual network [6, 7]. In so far as the sentence formed by the speaker implies a unit of sense, the mere syntactic co-occurrence of words (grouped in sets of derivative words) in the space of a sentence establishes a semantic linkage that can be explored in terms of the frequency of such links [8]. For instance, if we observe a high repetition in the co-occurrence of "complexity" and "algorithmic", on the one hand, o "message" and "meaning", on the other, is due to the semantic proximity of the co-occurring terms; in one case because of equivalence relation, in the other, because of consequence relation. In short, the greater or lesser occurrence of terms and links between terms have facilitated the examination of the relevance of different categories and the links between them from the perspective of the interdisciplinary research network. At the same time, the formation of semantic networks in the texts analysed with "small-world" or "scale-free" characteristic structures, whose pertinence has been studied, enables the visualization of both the categories effectively used in the generic articulation of utterances and the grouping of verbal categories circumscribed by the dealing with specific issues, for instance, "complexity" [7, 9].

According to this characterization, the semantic network analysis has been structured in the following phases:

(i) Text refinement, getting rid of those elements not corresponding to (textually) expressed utterances for which a meaningful syntactic-semantic treatment could not be performed.

(ii) Quantitative analysis of the texts by means of the application of computational linguistics "KH Coder" which enables the analysis of the semantic network in terms of the semantic links observed in the texts through the adjacency distance in sentences [10, 11].

(iii) Iterative process of relevant terms refinement according to its significance for the analysed issues which enables reviewing the aprioristic categorisation.

(iv) Co-occurrence mapping extraction of the semantic networks derived from the conceptual elucidation of the glossariumBITri.

3   Findings

3.1   Network Analysis of the *glossariumBITri - 2nd Edition*

The result shown in Fig. 1 illustrates a relevant characteristic of the glossariumBITri: the statistical degree distribution of the semantic network exhibits the properties of the free-scale networks. This means that the subsidiarity properties discussed by Díaz-Nafría [7] can be applied to glossariumBITri's semantic network. The recursive character of the corresponding structure entails a disciplinary clustering of issues that, at the same time, are well connected to the rest of the network from a semantic perspective.

The statistics of the semantic distances observed in the network and the study of the clustering offers an innovative methodological road to strengthen the interdisciplinary study of information linked to the development of the gB.

Figures 2, 3 and 4 shows the results of the gloossariumBITri's semantic network analysis. Each term/concept is represented by a node whose size is proportional to its occurrence frequency, while the thickness of links among terms is proportional to the frequency co-occurrence of the corresponding terms in the sentences of the whole text. Only the terms and links whose frequency surpasses the thresholds indicated in the figure caption are visible. At the same time, the result of the analysis of term clusters determined by intermediation distances is represented using different colours (terms with the same colour are at distances below a threshold).

As we can observe in Fig. 2 (in which the 130 most frequent terms are represented), "information" is the most dominant terms, as it could be expected. Under this nuclear

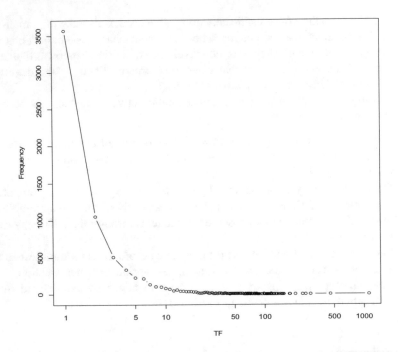

Fig. 1. Distribution of the frequency of word occurrence in the glossariumBITri 2016 edition. The statistic parameters denote that the semantic network is of the type small-world and free-scale.

term we can find other outstanding terms: 'theory', 'communication', 'knowledge', 'use', and 'concept'. They reflect, on the one hand, the general objective of the glossariumBITri (concept, theory), on the other, a significant weight of theoretical terms as communication, knowledge, use. We can also observe 4 important clusters, corresponding to domains with capacity to concentrate some specific aspects that have experience a deeper development. In addition, we only find two dominant authors, Shannon and Kolmogorov. However, while Shannon appears at a relatively central position and with a high degree of interconnectedness with the rest of the network, Kolmogorov is located at the central position of a cluster which is less connected and is more peripheral, linked to important theoretical terms as algorithm, complexity, object and other more mathematically oriented as fuzzy, set, function, etc. This cluster corresponds to one of the theoretical domains which has been incorporated in the 2016 glossariumBITri edition. Its relative disconnection with other relevant terms points to the need to devote efforts in developing missing links in order to achieve a more integrated elucidation.

Finally, it is possible to observe in the blue cluster (Fig. 2) a particularly cohesive group, composed by the terms: 'society', 'media', 'technology', 'communication', and 'critic'. Here we observe the weight of a field developed in depth, the critical theory of information focused on human, social and political aspects of information technologies.

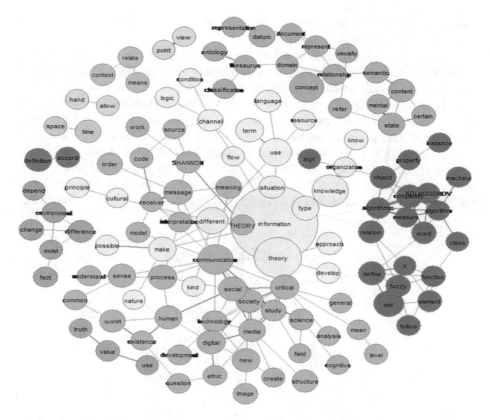

Fig. 2. glossariumBITri's co-occurrence network. Term frequency occurrence > 50; number of nodes (words-concepts): 200 most frequent ones; Colours: semantic clusters determined by intermediation measurements. Adverbial and prepositional categories are excluded.

It is worth mentioning that the article "critical theory of information" is the most read article in the interactive-glossariumBITri as determined by Internet traffic analytics.

Figure 3 corresponds to the same co-occurrence network in which only the 58 most frequent terms (nodes) are visualized (with a frequency over 75) and the 100 most frequent co-occurrences (links). According to the clustering analysis, the largest cluster is again the one aligned to the critical theory of information. At the same time the well cohesive cluster of terms related to algorithmic information theory and the General Theory of Information. At this level it can be noted that the red cluster (Fig. 2) originates from the convergence of the cluster composed by set, function, and other terms that was extensively developed in the previous edition (e.g. fuzzy logic). In both this figure and the previous one, it is interestingly possible to observe the presence of Bateson's conceptual approach to information, stated in the famous formula: "information is a difference that makes a difference" [12], which over time has gained most general support among the varied community of information studies. As we can see, this conception establishes relevant links to "environment" which reflect the concern spread along the

366   J. M. Díaz-Nafría et al.

community of information studies to go beyond the de-contextualization which is inherent to Shannon's perspective [13, 14].

Finally, Fig. 4 corresponds to a further refinement of the previous co-occurrence network in which only the 6 most frequent terms (nodes) and the 100 most frequent co-occurrence (links), which in the figure are reduced to the 12 existing among the 6 visualised terms. We observe here the 4 heaviest conceptual terms upon which the rest of the conceptual elucidation is articulated, as well as two meta-theoretical terms (theory and concept) which manifest the very goal of the gB itself. It is also worth mentioning at this level between communication and use, what shows that the gB, effectively accomplish the objective of giving account of the pragmatic aspects that was missing in the Mathematical Theory of Communication from which Shannon forged the scientific concept of information. From the inspection of the three co-occurrence

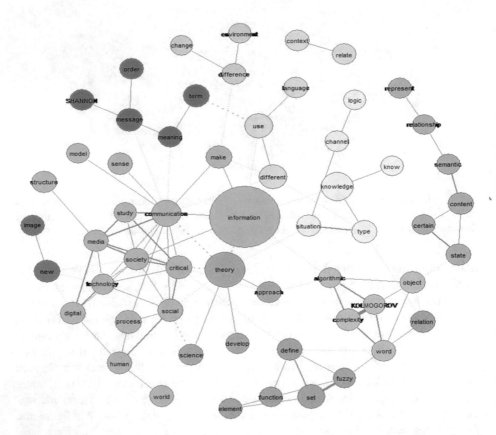

Fig. 3. glossariumBITri's co-occurrence network. Term frequency occurrence > 50; number of nodes (words-concepts): 200 most frequent ones; colors: semantic clusters determined by intermediation measurements. Adverbial and prepositional categories are excluded.

networks an absence pointing to a direction of further development and improvement of the gB: a more specific and broader consideration of metaphors. In a network structural perspective, the benefit of metaphors relies on their capacity to reduce average distances in the whole conceptual network as discussed by Díaz-Nafría (2007) [15].

Fig. 4. glossariumBITri's co-occurrence network. Term frequency occurrence > 200; number of nodes (words-concepts): 6 most frequent ones; number of links: 100 most frequent; colors: semantic clusters determined by intermediation measurements.

3.2   Enhanced Methodology to Assess Knowledge Integration

The previous results exhibit the capacity of the network approach to qualify interdisciplinarity within the broad margin left by the UNESCO's classification (referred to in Sect. 1), i.e. how far apart is from transdisciplinarity. However, this approach has not addressed how diverse the integration of knowledge is with respect to scientific knowledge in general. In addition, it provides a rather qualitative assessment that hinders the possibility of an objective evaluation. To fill the gap, building upon the referred approach, we propose–for future development–the assessment of the quality of the knowledge integration, based on two general aspects: the diversity of the disciplines involved (the more disciplines the larger the integrated knowledge); and the effective integration achieved through the meeting of different perspectives (if each discipline treats separately different aspects, the integration will be weak; if the theoretical construct gets to be merged into a general understanding of the involved phenomena, the integration will be strong).

Discipline Diversity Index. In the first place, the granularity level in the distinction of disciplines have to be determined. This can be done, in a first approximation, by fixing the number of relevant digits of the Universal Decimal Classification (UDC) used to distinguish the knowledge areas involved in a particular research [16–18]. Though the UDC offers a good and well-accepted coverage of knowledge in general, an adaptive implementation need to be introduced in the categorization of Knowledge Domains (KD): (i) some UDC categories have to be disregarded (for instance those which are not related to knowledge but to document types), (ii) other categories should be ascended from a lower granularity level in virtue of its relevance for the problems under study, and (iii) some category groups should be merged because they represent different aspects of the same knowledge, for instance, theoretical and applied.

Assuming the number of relevant KD is N, the diversity of participating disciplines can be determined through Shannon Diversity Index weighted by the maximal diversity achieved through a similar participation of the N KD, i.e. log2 N. By that means, if the N KD are homogenously distributed (i.e. they contribute equally – situation of maximal diversity) the index will be 1; and 0 in case that only one KD is contributing. Generally, the more KD are contributing in a more distributed way, the index will be closer to 1.

Calling pi the frequency of occurrence of a contribution from the ith KD (or probability that a contribution taken at random belongs to such a discipline), the diversity index will be:

$$ID = \frac{1}{\log_2 N} \sum_{i=1}^{N} p_i \log_2 (1/p_i) \qquad (1)$$

Integration of Disciplines. Nevertheless, it can be the case that despite having achieved the meeting of very diverse knowledge, its theoretical constructs do not merge at all in the explanation of the phenomena concerned, and instead each discipline devote itself to refer a different aspect of the object or problem under study. In this case the integration would be null. In the extremely opposite case, all the theoretical constructs from each involved discipline are interrelated in the explanation of the phenomena concerned.

This density of semantic relation can be measured in terms of the semantic network conformed by the theoretical terms used by the different disciplines convened in the process of interdisciplinary elucidation of concepts, metaphors, theories and problems concerning a shared problem. The development of an interdisciplinary glossary (ID-G), according to the model discussed in Sect. 2 and devoted to a specific topic, will serve to the interdisciplinary theoretical confrontation of different points of view. In such constructive process the agents establish relations between two terms whenever they occur in the same sentence, in lesser degree if they belong to subordinate ones. Thus the statistical analysis of the distance between terms in function of its grammatical function enables the analysis of the semantic distances among terms.

Quantitative assessment of knowledge integration, following the methodology depicted above. The study of the minimal average distance between any two words provides a measure of the integration achieved. In the case of natural language, taken an extended vocabulary of 66000 words, Sigman and Cecchi (2002) determined that the average minimal distance between any two words was around 7. However, when the knowledge is not well integrated, the distances increase at the same time that disconnected clusters can be identified. Thus, clustering coefficient and average minimal distance offer a characteristic of the integration achieved. Indeed, its ratio compared with the equivalent ratio for random networks, provides the small-world coefficient:

$$\sigma = \frac{C/C_{rand}}{L/L_{rand}} \tag{2}$$

from which we can evaluate, with a single parameter, whether the network satisfy or not the condition of a small-world network $\sigma > 1$ and how well integrated it is [19].

Qualitative assessment of knowledge integration. The network analysis, as the one described in Sect. 3.1, facilitates a qualitative evaluation, distinguishing specific theoretical clusters that are not well integrated, fields or concepts that are misrepresented, etc. This evaluation provides guidance for the further development of the research concerned (e.g., what disciplines need to be strengthened, what dialogue should be open up, etc.).

4 Discussion and Conclusions

In spite of the international efforts to boost interdisciplinary research, one of the barriers has been the lack of qualification criteria of interdisciplinarity itself [20, 21, 23]. This has often caused the funding assessment inefficient, disregarding promising ID research projects to which disciplinary criteria were applied. For this reason, the development of assessment criteria has been one of the objectives marked by national and international research funding agencies [21, 23]. The results discussed above shows the interest of the ID-glossaries in combination with the network analysis as a promising approach to qualify interdisciplinarity. But the benefit is not only meta-theoretic, at the level of the knowledge integration itself, it constitutes a useful tool in the advancement of knowledge integration, as it could be shown in the discussion of the co-occurrence networks with respect to the evolution of the glossariumBITri between consecutive editions. If over time the network analysis is performed to facilitate a comparative assessment of the evolution of the knowledge integration achieved, it is expected that this approach will serve to guide the theoretical work, as illustrated with a few examples in the results argued above.

The possibility to use this approach to the assessment of educational processes has been discussed by one of the authors, showing its capacity to detect the development of soft skills for which formal education is practically blind [23, 25]. Its application to the development of knowledge integration skills, as intended in the abovementioned PRIMER initiative [6], is straightforward from the methodology and results discussed herewith.

Nevertheless, the approach, on which the results presented is based, do not provide a quantitative evaluation, which prevents an objective assessment. To circumvent this limitation, the enhanced methodology discussed in Sect. 3.2 fills the gap with a bi-dimensional measure in which both the diversity of disciplines and the effective integration is measured at the same time. An ongoing international project devoted to the enhancement of the glossariumBITri and the creation of an Encyclopaedia of Systems and Cybernetics Online (ESSCO) is currently applying the described categorization of knowledge domains to deploy the described approach.

References

1. Hempel, C.G.: Fundamentals of Concept Formation in Empirical Science, vol. 2. University of Chicago, University of Chicago Press (1952)
2. Losee, J.: A Historical Introduction to the Philosophy of Science. OUP Oxford, New York (2001)
3. d'Hainault, L.: Interdisciplinarity in General Education. UNESCO, Paris (1986)
4. Díaz-Nafría, J.M., Salto, F., Pérez-Montoro, M.: Glossarium BITri 2016: Interdisciplinary Elucidation of Concepts, Metaphors, Theories and Problems Concerning Information, La Libertad (Ecuador)-León. Universidad Estatal Península de Santa Elena, Spain, Santa Elena (2016)
5. Lattuca, L.R.: Creating Interdisciplinarity: Interdisciplinary Research and Teaching Among College and University Faculty, pp. 1–20. Vanderbilt University Press, Chicago (2001)
6. PRIMER consortium: PRomoting Interdisciplinary Research and Education, BITrum-Research Group, 2018. http://primer.unileon.es. Accessed 20 Jan 2018
7. Drieger, P.: Procedia - Social and Behavioral Sciences, pp. 4–17. Elsevier Ltd. (2013)
8. Díaz-Nafría, J.M.: Cyber-subsidiarity: toward a global sustainable information society. In: Carayannis, E.G., et al. (eds.) Handbook of Cyber-Development, Cyber-Democracy, and Cyber-Defense, pp. 1–30. Springer, Berlin (2017)
9. Jackson, K.M., Trochim, W.M.: Concept mapping as an alternative approach for the analysis of open-ended survey responses. Cornell Univ. 5(4), 307–336 (2002)
10. Barabási, A.-L.: The New Science of Networks, vol. 30, p. 288. Perseus Publishing, Cambridge (2002)
11. Koichi, H.: KH Coder 3 Reference Manual. Ritsumeikan University, Kioto (2016)
12. Shinobu, A., Chie, M.: Missions of the Japanese National University corporations in the 21st century: content analysis of mission statements. Acad. J. Interdiscip. Stud. 2(3), 197–207 (2013)
13. Bateson, G.: Mind and Nature: A Necessary Unity. Hampton Press, New Jersey (2002)
14. Díaz-Nafría, J.M.: What is information? A multidimensional concern, pp. 77–108. TripleC, España (2010)
15. Díaz-Nafria, J.M., Curras, E.: Systems Science and Collaborative Information Systems: Theories, Practices and New Research, pp. 37–70. IGI Global (2011)
16. Sigman, M., Cecchi, G.: Global organization of the wordnet lexicon, vol. 99, no. 3, pp. 1742–1747 (2002)
17. McIlwaine, I.C.: The Universal Decimal Classification: A Guide to its Use, UDC Consortium (2007)
18. Slavic, A.: Innovations in Information Retrieval: Perspectives for Theory and Practice, pp. 23–48. Facet Publishing, London (2011)

19. C. UDC, UDC Fact Sheet. In UDCC website, The Hague (2017). http://www.udcc.org/index.php/site/page?view=factsheet
20. Telesford, Q.K., Joyce, K.E., Hayasa, S.: The ubiquity of small-world networks. Brain Connect. 1(5), 367–375 (2011)
21. Frodeman, R., Klein, J.T., Dos Santos Pacheco R.C.: The Oxford Handbook of Interdisciplinarity. Oxford University Press, Oxford (2017)
22. DEA-FBE, Thinking Across Disciplines – Interdisciplinarity in Research and Education, Australia: Danish Business Research Academy (DEA) & Danish Forum for Business Education (FBE) (2008)
23. European Research Advisory Board (EURAB), Interdisciplinarity in Research, European Commission - EURAB (2004)
24. Díez-Gutiérrez, E., Díaz-Nafría, J.: Ubiquitous learning ecologies for a critical cybercitizenship. Comunicar **XXVL**(54), 54, 49–58 (2018)
25. Barão, A., de Vasconcelos, J.B., Rocha, Á., Pereira, R.: A knowledge management approach to capture organizational learning networks. Int. J. Inf. Manag. **37**(6), 735–740 (2017)

Interface of Optimal Electro-Optical/Infrared for Unmanned Aerial Vehicles

Aníbal Jara-Olmedo[1], Wilson Medina-Pazmiño[1], Rafael Mesías[1],
Benjamín Araujo-Villaroel[1], Wilbert G. Aguilar[2,4(✉)],
and Jorge A. Pardo[3]

[1] CIDFAE Research Center, Fuerza Aérea Ecuatoriana, Ambato, Ecuador
[2] CICTE Research Center, Universidad de las Fuerzas Armadas ESPE,
Sangolquí, Ecuador
wgaguilar@espe.edu.ec
[3] UGT Department, Universidad de las Fuerzas Armadas ESPE,
Latacunga, Ecuador
[4] GREC Research Group, Universitat Politècnica de Catalunya,
Barcelona, Spain

Abstract. Unmanned aircraft vehicles applications are directly related to payload installed. Electro-optical/infrared payload is required for law enforcement, traffic spotting, reconnaissance surveillance and target acquisition. A commercial off-the-shelf electro-optical/infrared camera is presented as a case study for the development of interface to control the UAV payload. Based on an architecture proposed, the interface shows the information from the sensor and combines data from UAV systems. The interface is validated in UAV flight tests. The software interface enhances the original performance of the camera with a fixed-point automatic tracking feature. Results of flight tests present the possibility to adapt the interface to implement electro-optical cameras in different aircrafts.

Keywords: UAV · Payload · Interface · Electro-optical · Infrared

1 Introduction

Unmanned aircraft vehicles (UAVs) are widely used in civilian and military applications such as forest fire monitoring, law enforcement, traffic spotting, reconnaissance surveillance and target acquisition, infrastructure inspection, recognition of strategic areas and zones, filmography, surveillance, communications relay, etc. [1–3]. These applications are basically defined by the payload/sensor installed in the UAV, that must be controlled and monitored from a ground control station (GCS) [4, 5]. The GCS includes a payload station to operate the sensor. These two segments are connected by a data link that guarantees the continuous interchange of information [6]. UAV payloads include video sensors as electro-optical (EO) and infrared (IR) or thermal cameras for obtaining information to be transmitted to the GCS [7]. Both kind of lens may be combined in a gimbal as a compact platform with image, commonly simply referred to as an electro-optical system. An electro-optical system mainly consists of three parts:

© Springer International Publishing AG, part of Springer Nature 2018
Á. Rocha and T. Guarda (Eds.): MICRADS 2018, SIST 94, pp. 372–380, 2018.
https://doi.org/10.1007/978-3-319-78605-6_32

the power system, the control unit, and the vision system. The vision system contains vision lenses, stabilization systems, video outputs and inputs to accept commands from the user to perform actions [8]. The Power system provides energy to the system. Finally the control unit manage movement, focus, zoom and additional characteristics of the camera. This paper presents the development of a control interface for a commercial off-the-shelf (COTS) payload. In the first part the adaptation of the architecture of the electro-optical system is described. After that the analysis of control data is presented. Finally the control interface is developed considering the data obtained and the combination with aircraft data to present and control the payload. Finally, the results of the tests are discussed.

2   Control Interface Development

2.1   Original Architecture of the EO/IR Camera Payload

There are a number of companies dedicated to the development and commercialization of electro-optical systems. These include CONTROP, Flir Commercial Systems, MAYERS, Wescam among others. These combine electronics and optics to develop sophisticated viewing platforms to detect radiations in the optical spectrum including ultraviolet radiation, visible light, lasers, thermal imaging systems and night vision. For this paper a CONTROP TU-STAMP camera shown in Fig. 1 is used as case study. This camera is a small, lightweight, electro-optical, gyro-stabilized, airborne payload [9]. It is designed with a weight of 2.8 kg, in order to be carried on a range of UAVs, small manned aircrafts or even aerostats. The payload camera includes two vision lenses, one IR thermal imaging lens, the other lens is type Charge-Coupled Device Television (CCD-TV). Both lenses feature continuous optical zoom and ensure high quality images for day and night. These tools support the operator mission throughout the flight, as well as navigation of the aircraft.

Fig. 1. COTS EO/IR camera installed on UAV-2

The architecture of the EO system case study is shown in Fig. 2. This is the architecture for installation in a manned air vehicle such as a helicopter or small airplane, where the camera can be controlled by a operator in the aircraft. The control unit in this configuration accomplishes three functions. Firstly, it regulates the power to energize the EO/IR sensor. Secondly, the unit receives and transmits data from/to the payload. Finally, the control unit acquires the video signal of the EO/IR sensor, filters and transmits it to an external display. In this architecture, with a basic version of the COTS EO/IR camera, the operator manually detects and tracks mobile and fixed targets.

An important consideration for installation of the payload in a UAV is the limitations of weight, space and power [10]. The permissible payload weight is directly related to the autonomy of the UAV, then the UAV performance can be increased with the optimization of payload. In small UAVs the space available for the payload has to be designed in a balance with space for aircraft systems. Finally, a common configuration for the power system is the concentration of energy in a converter component that distributes different voltage values to the UAV systems equipment. Our architecture for the payload installation is developed taking these design constraints into account.

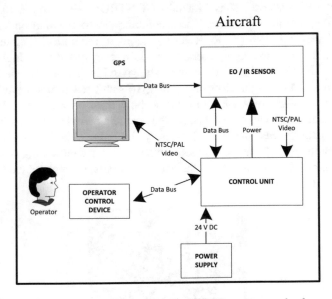

Fig. 2. Architecture for the basic version of EO/IR camera payload case study.

2.2   Proposed Architecture for UAV Control Interface

The original architecture presents a wired control system for the payload. In order to install the EO/IR in the UAV, the proposed configuration divides the payload into two segments: one in the aircraft and the other one in the ground control station. The physical control unit is converted into the GCS payload control interface. The interface software is installed in a COTS computer. The payload can be controlled and manually track a target by the operator with a USB joystick. The architecture proposed is shown in Fig. 3.

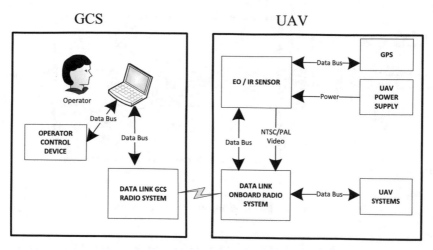

Fig. 3. Proposed EO/IR camera control architecture presents the *GCS* segment and the *UAV* segment.

The new architecture for the payload employs some changes. First, controlling and monitoring are executed remotely, with the operating range depending on that of the wireless communications system. Second, the EO/IR sensor is powered by the energy system available in the UAV. Finally, original mechanical controls are replaced by the graphical user interface GUI.

2.3   Image Analysis

The original control unit is responsible for providing the operation codes for camera horizontal and lateral movements, changes of images or vision lenses switching [11]. These codes are transmitted to the payload using the communication standard RS-232. In order to analyse these codes a software interface was implemented with Labview systems engineering software. Labview code presented in Fig. 4 is programmed to capture the control commands generated by the control unit. The communication port is configurated with a data refresh period between the control unit and payload of 50 ms. With the use of a node property tool the number of registers from the control unit can be read using the RS232 communication standard.

The control data includes 20 bytes. Table 1 shows that 16 registers are used to perform operation functions of the camera. Three registers are used for header and one register is used as error control. The error control is performed by CheckSum validation. Analysis presented in Table 1 is verified by tests of sensor operation and data captured from the control unit.

Equipment for the new configuration includes a COTS computer with a RS232 communication standard port. The values of the EO/IR movements obtained with a joystick using USB standard are interpreted by the Labview application. The data are standardized using data code established in Table 1. After that, the data is loaded into an array (computer data vector) with the respective Header and Checksum register. The logic operation algorithm is shown in Fig. 5.

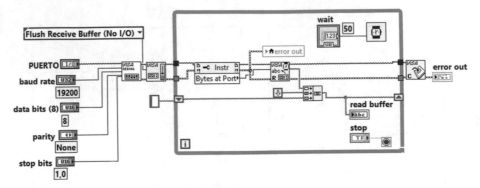

Fig. 4. LabVIEW image processing code.

Table 1. Control data for operation of the EO/IR camera

Definition	Data length	Description
Header	03 bytes	Header
Command	16 bytes	Operation functions include: Menu, Focus, Thermal video, Zoom, Video options, Vertical movement, Horizontal movement
Checksum	01 byte	Error control

2.4 Graphical Control Interface

Once the control data was acquired and coded, the graphical user interface (GUI) is developed considering the display of payload information and the automatic tracking function. In the basic version of the COTS study payload the human operator must manually follow and acquire the target. For surveillance applications the tracking of mobile or fixed points is a characteristic demanded and developed for payloads [12, 13]. This automatic tracker is essential where precise pointing is required [14]. For these reasons a fixed target detection is tested to enhance performance of the camera and to produce an automatic control mode [11].

The graphical user interface displays position of the aircraft in a two dimensional moving map, analog video from camera and the control for automatic tracking of a designed fixed target. The modules of the GUI are presented in Fig. 6. Position, movement, attitude and video data from UAV equipment is received on the GCS by the data link radio system and managed by the interface.

The fixed targets can be programmed marking the point on the map based on Google Earth or typing the coordinates into the text boxes. The method to track points automatically uses GPS data to calculate target distance from the payload. Digital altimeter data is used to calculate altitude variation. Inertial Measurement Unit (IMU) data is used to compensate vertical and horizontal movement to keep tracking the target [15]. The sensor pointing direction is based in the compensation of relative pitch, yaw and roll of the sensor with the relative pitch, yaw and roll of the aircraft. [16]. This automatic

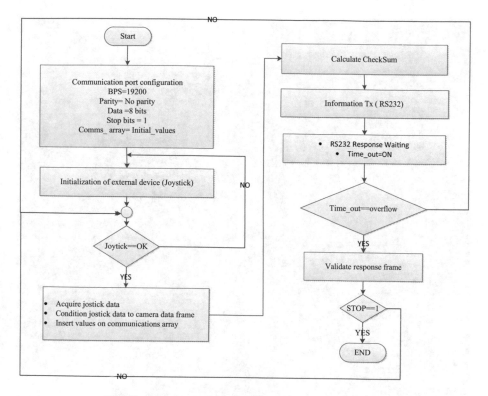

Fig. 5. Logic operation algorithm to obtain control data

control mode allows the operator to work with geo-referenced points and visualize these points defined before or during the flight. This method uses ground control points (GCPs) in order to determine the coordinates of pixel coordinates [17].

2.5   Testing and Validation of the Control Interface

In order to test the interface the EO/IR system was implemented on a UAV-2 aircraft with the architecture developed. This UAV according to UAS Classification for certification purposes [18] is a tactical short range UAV with a mass under 150 kg, range up to 30 km, flight altitude of 3.000 m and endurance between 2 and 4 h. The EO/IR camera was installed in the front part of the UAV-2 with the adaptation to the aircraft energy system. GUI was installed in the payload operator computer in the GCS. The camera will be controlled remotely by a COTS data link system in C band.

Practical flight tests were set up to verify the operation of the EO/IR camera with the new architecture and GUI. The flight test planning process begins with several terrestrial objects geo-referenced and defined as trial targets. The tests evaluate the EO/IR operating adequately and tracking automatically targets established in the trial area.

Fig. 6. Modules of the graphical control interface. On the left, the automatic target text boxes. On the right, Google earth map and the EO/IR camera video.

Ten flight tests were conducted in different geographical locations, from 200 m up to 1,000 m flight level (due to the range limit of the camera). All flights were conducted under visual meteorological conditions. The test area was limited to a maximum distance of 20 km from the trial airfield. Tests must show adequate operation of the EO/IR camera subject to changing power supply conditions and the commands implemented by software.

Before the test flight, the payload operator types latitude and longitude of five targets into text boxes of the GUI. During flight, an additional five coordinates are registered automatically when the operator clicks the target on the Google Earth interface. Manual and automatic control of the camera is achieved by buttons of the GUI and mouse in the payload operator station. In the automatic mode, the GUI registers position of targets and combines information of altitude, coordinates and UAV position to track the points.

During flight tests each of the targets were automatically tracked by the EO/IR sensor. Data from the UAV was presented along with video from the camera and recorded in GCS operator computer memory. Correct operation and precision of tracking is determined visually by the operator with the central cross tracking mark of the COTS sensor. In automatic mode, the camera keeps tracking the fixed target even with the UAV holding over test area. This feature overcomes results from manual tracking in same conditions and reduces the workload of the payload operator. These results are backed up by the airship control instrumentation installed on the airship and received by telemetry link at the Ground Control Station.

The EO/IR system with the new architecture performs as expected. Remote control of the camera responds well to commands from the GUI. Automatic tracking of previously registered targets is achieved as well as the tracking of on line targets. Following flight test parameters presented in this article, additional GUI tests were developed with another EO/IR camera installed with the same architecture on a small autogyro, and this demonstrates the possibility of a similar surveillance service with the adaptation of the GUI for an autogyro (or other aircraft) instead of the UAV.

3 Conclusion

This paper presents the process of development of an interface to control a UAV payload from the data acquisition first step to the final flight tests. The architecture of the EO/IR payload system and its control is changed from original manned aircraft (wired) configuration to a more complex architecture for installation and testing in a UAV. The new architecture adds data from other UAV systems and combines it with the control data obtained from the EO/IR camera. All information is presented in the GUI in addition to the video obtained from the sensor.

The software interface enhances the performance of the camera with automatic tracking to a fixed point on the ground. The interface can be customized to a new EO camera after the respective control data is verified and arranged. The interface and the new architecture was tested and validated on flight tests performed with the UAV and with an autogyro. As future work, new features for a basic EO/IR payload can be developed with the interface presented as a baseline.

References

1. Samad, A.M., Kamarulzaman, N., Hamdani, M.A., Mastor, T.A., Hashim, K.A.: The potential of Unmanned Aerial Vehicle (UAV) for civilian and mapping application. In: IEEE 3rd International Conference on System Engineering and Technology, pp. 313–318 (2013)
2. Yuan, C., Zhang, Y., Liu, Z.: A survey on technologies for automatic forest fire monitoring, detection, and fighting using unmanned aerial vehicles and remote sensing techniques. Can. J. For. Res. **45**(7), 783–792 (2015)

3. Ping, J.T.K., Ling, A.E., Quan, T.J., Dat, C.Y.: Generic Unmanned Aerial Vehicle (UAV) for civilian application. In: Proceedings of the IEEE Conference on Sustainable Utilization and Development in Engineering and Technology (STUDENT), pp. 289–294, October 2012

4. Gupta, S.G., Ghonge, M.M., Jawandhiya, P.M.: Review of Unmanned Aircraft System. Int. J. Adv. Res. Comput. Eng. Technol. 2(4), 1646–1658 (2013). ISSN 2278-1323

5. Wargo, C.A., Church, G.C., Glaneueski, J., Strout, M.: Unmanned Aircraft Systems (UAS) research and future analysis. In: IEEE Aerospace Conference Proceedings (2014)

6. Pastor, E., Lopez, J., Royo, P.: UAV payload and mission control hardware/software architecture. IEEE Aerosp. Electron. Syst. Mag. 22(6), 3–8 (2007)

7. Arfaoui, A.: Unmanned Aerial Vehicle: Review of onboard sensors, application fields, open problems and research issues. Int. J. Image Process. 11(1), 12–24 (2017)

8. Koretsky, G.M., Nicoll, J.F., M. Taylor, S.: A Tutorial on Electro-Optical/Infrared (EO/IR) Theory and Systems (2013)

9. CONTROP: T-STAMP, Miniature UAV Payload, Gyro Stabilized Cameras. http://www.controp.com/item/t-stamp-payload/. (Accessed: 21-Aug-2017)

10. Mian, O., Lutes, J., Lipa, G., Hutton, J.J., Gavelle, E., Borghini, S.: Direct georeferencing on small unmanned aerial platforms for improved reliability and accuracy of mapping without the need for ground control points. Int. Arch. Photogramm. Remote Sens. Spat. Inf. Sci. - ISPRS Arch. 40(1W4), 397–402 (2015)

11. Keleshis, C., Ioannou, S., Vrekoussis, M., Levin, Z., Lange, M.A.: Data Acquisition (DAQ) system dedicated for remote sensing applications on Unmanned Aerial Vehicles (UAV). In: Second International Conference on Remote Sensing and Geoinformation of the Environment, vol. 9229, p. 92290H April 2014

12. Helgesen, H.H., Leira, F.S., Johansen, T.A., Fossen, T.I.: Tracking of marine surface objects from unmanned aerial vehicles with a pan/tilt unit using a thermal camera and optical flow. In: International Conference on Unmanned Aircraft Systems, ICUAS 2016, pp. 107–117 (2016)

13. Leira, F.S., Trnka, K., Fossen, T.I., Johansen, T.A.: A light-weight thermal camera payload with georeferencing capabilities for small fixed-wing UAVs. In: International Conference on Unmanned Aircraft Systems ICUAS 2015, pp. 485–494 (2015)

14. Miller, R., Mooty, G., Hilkert, J.M.: Gimbal system configurations and line-of-sight control techniques for small UAV applications. In: Airborne Intelligence, Surveillance, Reconnaissance (ISR) Systems and Applications X, vol. 8713, pp. 1–15 (2013)

15. Eling, C., Klingbeil, L., Wieland, M., Kuhlmann, H.: A precise position and attitude determination system for lightweight Unmanned Aerial Vehicles. Int. Arch. Photogramm. Remote Sens. Spat. Inf. Sci., vol. XL-1/W2, pp. 4–6, September 2013

16. Janney, P., Booth, D.: Pose-invariant vehicle identification in aerial electro-optical imagery. Mach. Vis. Appl. 26(5), 575–591 (2015)

17. Zhou, G., Li, C., Cheng, P.: Unmanned aerial vehicle (UAV) real-time video registration for forest fire monitoring. In: Proceedings of 2005 IEEE International Geoscience Remote Sensing Symposium 2005. IGARSS 2005, vol. 3, no. 1, pp. 1803–1806 (2005)

18. Dalamagkidis, K., Valavanis, K.P., Piegl, L.A.: Current status and future perspectives for Unmanned Aircraft System operations in the US. J. Intell. Robot. Syst. 52(2), 313–329 (2008)

Generating a Descriptive Model to Identify Military Personnel Incurring in Disciplinary Actions: A Case Study in the Ecuadorean Navy

Milton V. Mendieta[✉] and Gabriel Cobeña

Armada del Ecuador, Guayquil, Ecuador
mmendieta@armada.mil.ec, ancoce@gmail.com

Abstract. The problem of navy personnel incurring in disciplinary actions has major consequences in the productivity and motivation of these individuals for getting the job done. Leaving it unsolved may negatively affect their careers, working environments, peers, families and in some cases the Navy's reputation. The implementation of data mining is widely considered as a powerful instrument for acquiring new knowledge from a pile of historical data, which is normally left unstudied. The main purpose of this paper is to use a data-driven approach to generate a descriptive model aimed at discovering knowledge, insights and interesting patterns in the personnel misconduct. The results reveal promising insights, hence the reliability of this work as a decision making and decision support tool.

Keywords: Data mining · Exploratory data analysis · Military offenses
Disciplinary actions · Personnel profile

1 Introduction and Related Work

Few working environments worldwide are as fundamentally grounded in the strict adherence to rules and regulations as the military, and the Ecuadorean Armed Forces are no exception. The Ecuadorean Navy prides itself in placing the utmost importance in the discipline of its members, and even proclaiming discipline to be one of its foremost core values, together with honor and loyalty.

The Military Discipline Statute (MDS) contains all the regulations and guidance governing the conduct of all members of the Ecuadorean Armed Forces, including Navy personnel. It is the legal framework upon which disciplinary actions are based. Any misconduct is considered an administrative action penalized by Non-Judicial Punishment (NJP) only, in the form of reprimand, arrest in quarters, demerits, and even service discharge. Military offenses are categorized as: minor, grave, and major offenses, and within each category they are further divided into subcategories as follows: (1) punctuality, (2) conduct unbecoming, (3) healthiness and hygiene, (4) morale, (5) duty and military obligations, (6) safety of military operations, (7) subordination, (8) property and infrastructure, and (9) Abuse of discretion.

Most of the research pertaining the study of military personnel behavior from a data-driven approach has been done in the fields of deployment-related injury and

© Springer International Publishing AG, part of Springer Nature 2018
Á. Rocha and T. Guarda (Eds.): MICRADS 2018, SIST 94, pp. 381–393, 2018.
https://doi.org/10.1007/978-3-319-78605-6_33

posttraumatic stress disorder [1] and alcohol misuse [2]. In the case of disciplinary actions, most of the work done lies in the field of psychology. For instance, in the 90's, the U.S. Army's Selection and Classification Project (Project A) developed several studies to measure cognitive ability and personality to predict, among other criteria, personal discipline (i.e. achievement, dependability, and adjustment) [3]. More data-driven approaches are targeted to similar problems, such as predicting minor violent crime perpetration among U.S. Army soldiers using machine learning methods (regression, random forests) [4]. The model was trained with administrative data available for all soldiers in the U.S Army during the 2004–2009 period. A military-oriented integrity test to predict future disciplinary infractions was proposed in [5], using statistics techniques.

A similar problem in the domain of Educational Data Mining (EDM), such as the student retention/student dropout problem, poses similar challenges, and we firmly believe that the knowledge available in EDM can be extrapolated to solve the problem at hand. A number of data mining methods have been developed for early detection of students at risk of dropout, hence the immediate application of assistive measure. Amery in [6] proposed a survival analysis for the early detection of student dropouts, by using a Cox proportional hazard regression model that takes into account the time of a particular event of interest.

A different approach using an ensemble of mixed-type data clustering was proposed in [7], by which the original set of features are enriched with additional variables acquired from a single data clustering, improving the accuracy of the classification problem. A similar approach is taken in [8], where the authors are not primarily concerned with predicting if a student is a graduate or dropout, especially because a predictive method fails to provide insights to understand factors and characteristics of those two student categories. They use a clustering technique, capable of revealing natural groups of objects that share similar characteristics. In [9], three data mining models: logistic regression, decision trees, and neural networks, explore important student characteristics associated with retention leading to graduation.

The data mining approach not only is used to predict student dropouts, but also student performance, such as in [10] that uses a decision tree classifier, and in [11] that implements a state-of-the-art collaborative filter in the form of Matrix Factorization, normally used in recommender systems.

The purpose of the present research is to generate a descriptive model, based on historical data and using data mining techniques in the form of exploratory data analysis, to identify a typology of Navy personnel at-risk of perpetrating military offenses. These insights will allow us to find out if there are any patterns in the data available that could be useful in future researches to predict potential individuals for misconduct.

The paper is structured as follows: Sect. 2 describes the dataset used in the research. Section 3 shows a technical discussion about the experimental results. The review is concluded in the fourth section, with a discussion of future research directions.

2 Dataset Description

The dataset used for this work was provided by the Navy's Department of Human Resources, which contains relevant personnel information since 1948 until October 2017. Among the 13 relational data tables for a total size of 300 MB, only few contribute to the problem at hand. Of particular interest are the following tables:

- *Person_table:* one register per individual, it includes: person id, year of admission into service, classification, specialty, gender, marital status, current rank, and other personal relevant information.
- *Offenses_table:* one register for each offense, which translates to multiple registers per individual, it includes: type of offense, date of the infraction, subcategory, and other relevant data.
- *Promotion_table:* one register for each promotion, which translates to multiple registers per individual, it includes: date of promotion, rank of promotion, time till next promotion, and so on.
- *Relocation_table:* one register each time an individual was relocated to a different unit, which translates to multiple registers per individual, it includes: orders, date of relocation, unit from and to relocation, and other relevant data.

The *person_table* contained 51,963 records corresponding to a similar amount of military and civilian personnel, active duty and retired from service. The military personnel were furthered classified as: midshipmen, recruits, conscripts, officers and enlisted. From this universe, we were only concerned with officers and enlisted, for a total of 18,132 records equivalent to 35% of the dataset. Similarly, the *offenses_table* contained 173,595 records, and after filtering for our target population, we were left with 141,532 records, equivalent to 82% of the total offenses.

All the queries and pre-processing techniques were done using Microsoft SQLServer. Not all the data required for the experiments was explicitly available. For instance, the rank a specific individual possessed when he or she incurred in a disciplinary action was not included in the *offenses_table*. This feature was calculated with a query between the dates of promotion from the *promotion_table* and the dates an offense was committed from the *offenses_table*. Other required features were calculated in a similar manner.

3 Experimental Results and Technical Discussion

The percentage of officers and enlisted personnel who has incurred in disciplinary actions while on active duty is very similar to each other, 62.24% and 68.38% respectively, as seen on Table 1, which demonstrates that punishment for misconduct does not favor officers over enlisted. Similarly, Tabel 2 shows the average of offenses by category committed by officers and enlisted, where it can be observed that enlisted personnel tend to incur in more disciplinary actions than officers, especially for the minor offenses category, it almost doubles the officers' offenses.

Figure 1 shows a comparison of military offenses, in absolute values, committed by officers and enlisted during the study period. There is a ratio of approximately 13:1 for

Table 1. Offenses committed by officers and enlisted.

Personnel	Total personnel	Personnel committing offenses	Percentage
Officers	2.341	1.457	62.24%
Enlisted	15.791	10.798	68.38%

Table 2. Average offenses committed by officers and enlisted.

Personnel	Offenses	# Offenses	# Personnel committing offenses	Average
Officers	Major	105	96	1.09
	Grave	1 387	693	2.00
	Minor	9 751	1 431	6.81
Enlisted	Major	2 432	1 728	1.41
	Grave	20 203	6 414	3.15
	Minor	107 110	10 556	10.15

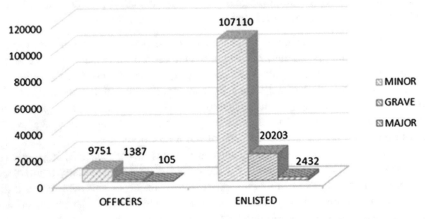

Fig. 1. Total of military offenses committed by officers and enlisted during the study period.

minor and medium offenses incurred by enlisted personnel over officers, whereas this ratio increases dramatically to 23:1 for major offenses.

As for the analysis of military offenses committed during the period of study, it is clear that the three categories of disciplinary offenses follow a similar pattern as depicted in Fig. 2. The low values from 1951 to 1967 are best explained by the fact that the Ecuadorean Navy was a young organization by then, with fewer personnel than in recent times. Moreover, it is possible that due to the lack of information systems at that time, some of the disciplinary actions were not properly registered. On the opposite side of the distribution, the three categories plateau right after 2009, which coincides with the year that the newly-amended military discipline statute came into force. This legal document set new rules for imposing punishment, appealing's, and access to a military defense attorney, mandatory for major offenses and optional for the remaining two categories. From then on, this process turned out to be more bureaucratic,

Fig. 2. Total of military offenses by year.

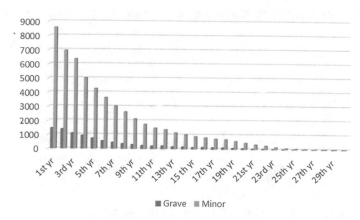

Fig. 3. Minor and grave offenses committed enlisted in 30 years of service.

which possibly explains the lack of interest in superior personnel to impose discipline to their subordinates.

For the analysis related to the year of service that a disciplinary offense took place, one would expect that junior officers and young enlisted personnel incur in minor and grave disciplinary actions during the first years in their military careers. This criteria holds true for enlisted personnel, and not for officers as depicted in Figs. 3 and 4 respectively. The findings show that enlisted personnel follow an exponential decay since the start of their career, whereas officers show this tendency after their fifth year in their careers. This evidence could be best explained by the fact that junior officers after graduating from the naval academy are assigned to naval stations or posts, where they are strictly supervised by superior officers, mainly during their first rank.

In the case of major offenses, enlisted personnel follow the same exponential decay starting from year one, officers do not follow a specific distribution, as shown in Fig. 5.

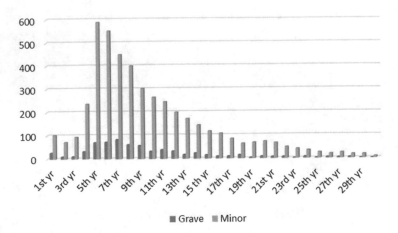

Fig. 4. Minor and grave offenses committed by officers in 30 years of service.

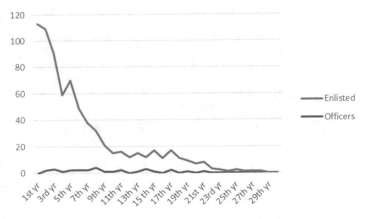

Fig. 5. Major offenses committed by officers & enlisted in 30 years of service

The Ecuadorean Navy is organized in sectors, which group several naval stations and bases that share the same functionality, as seen in Table 3.

In addition, Figs. 6 and 7 provide statistics with respect to minor and grave offenses committed by officers and enlisted respectively, quantified by navy sectors. Specific to these categories, the administrative sector, accounts for most disciplinary actions committed by officers followed by the operational. This trend is the opposite for the enlisted personnel. One would expect that most of the disciplinary actions would take place in the operational sector, where most of the navy personnel is assigned to. Moreover, this sector groups all the combat stations where discipline is more rigorous, and therefore the chances of incurring in military offenses tend to increase. The same insight holds true in major offenses for officers and enlisted, as shown in Fig. 8.

Table 3. List of sectors in the Ecuadorean Navy.

Sector	Abbreviation	Description
Personnel	SP	Including medical stations
Operational	SO	Naval Aviation, Surface Ships, Naval Infantry, Submarines
Logistics	SM	
Maritime	SIM	Units related to maritime authority
Coast Guard	SEA	
Education	SE	Training units for officers & enlisted
Administrative	SA	Land stations
Others	Others	Command posts outside of the Navy, within the Department of Defense

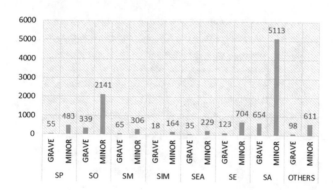

Fig. 6. Minor and grave offenses committed by officers in each sector.

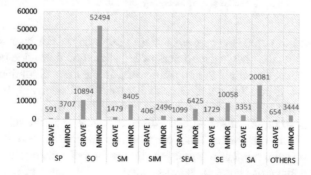

Fig. 7. Minor and grave offenses committed by enlisted in each sector.

From Fig. 9, it can be observed that the four top subcategories of military offenses are related to punctuality, conduct unbecoming, duties & obligations, and subordination, in that particular order for enlisted personnel; and punctuality, duties & obligations, conduct unbecoming, and subordination for officers. Navy officials should design

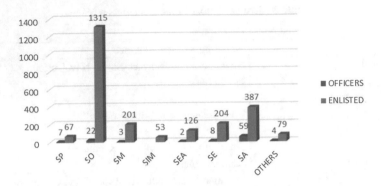

Fig. 8. Major offenses committed by officers and enlisted in each sector.

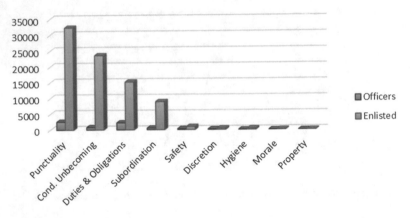

Fig. 9. Military offenses by subcategory committed by officers & enlisted.

programs targeted to reduce the causes of these misconducts. For instance, in order to improve punctuality, the navy's department of logistics should provide better means of transportation to pick up people to and from work. In the case of conduct unbecoming and subordination, a possible solution is to improve the leadership programs in the naval academy, navy's boot camp, and other training facilities.

According to article 21 of the Armed Forces Personnel Act, the military personnel is classified in three categories: Line, Service and Technical, and Specialist. Line personnel (ARM) are recruited in training institutes for enlisted and officers, and upon graduation are assigned to combat stations. Service (SRV) and Technical (TNC) personnel are also recruited in training institutes, they provide support to military operations. Specialist (ESP) personnel hold a degree from civilian universities prior to join the navy; they are assigned to non-combat units. With these definitions in mind, Figs. 10 and 11 show that line officers and enlisted personnel account for the most disciplinary actions. As previously discussed, these results make sense because line officers and enlisted personnel are mainly assigned to the operational sector where discipline is more rigorous.

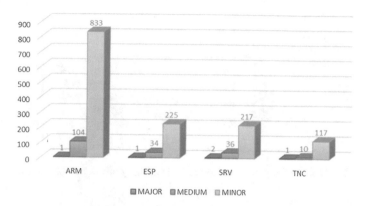

Fig. 10. Military offenses committed by category (officers).

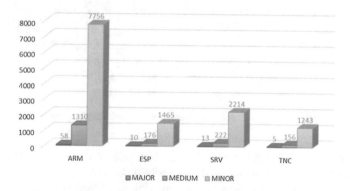

Fig. 11. Military offenses committed by category (enlisted).

Following the preceding discussion, line officers and enlisted are further classified in specialties as follows: (1) Surface Warfare (SU), (2) Submarines (SS), (3) Naval Aviation (AV), (4) Naval Infantry (IM), and (5) Coast Guard. Figure 12 shows that surface warfare officers account for the most offenses in all three categories. The Coast Guard specialty was created in 2009, and since then, they account for almost the same number of minor and grave offenses committed by the remaining specialties, except surface warfare, they even account for the only officer incurred in a major offense. With these observations, navy officials may pay a little more attention to the working environments of Coast Guard officers, especially because they are constantly involved in maritime patrols, and in some cases they deal directly with civilians, which in turn could be a cause of disciplinary actions. Similar actions should be taken for the surface warfare community.

In the case of enlisted specialties, Fig. 13 shows that naval infantry and surface warfare personnel have committed roughly the same amount of minor offenses, nonetheless, naval infantry almost doubles grave and major offenses. An appropriate assistive measure such as leadership and motivation programs may be urgently

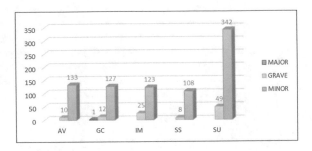

Fig. 12. Military offenses committed by line officers by specialties.

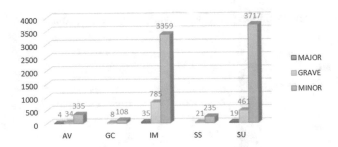

Fig. 13. Military offenses committed by line enlisted by specialties.

executed by naval authorities to reduce the amount of naval infantry and surface warfare enlisted personnel incurring in disciplinary actions.

Major offenses are a big concern for naval authorities, since they negatively impact the performance of the military personnel involved in these severe disciplinary actions, and at the same time they cause some kind of distress among their peers in the working environments. In this sense, this work also tries to find the naval stations where these severe offenses have been committed the most, for both officers and enlisted. Figures 14 and 15 show the results for the top 20 naval stations where major offenses have taken place. For the sake of security, the identity of the naval stations have been coded by sectors, only the operational sector has an extra identifier corresponding to the specialty.

In the case of officers, the operational sector accounts for 40% of naval stations, moreover, all these units belong to the surface warfare community. The trend is different for enlisted personnel. Although the operational sector accounts for over 50% of all naval stations, the naval infantry community takes the lead over surface warfare. Once again, naval authorities should analyze the working environments for officers and enlisted personnel in naval infantry and surface warfare naval units, whose causes may be the source of misconduct.

All the aforementioned results show that officers and enlisted tend to incur in disciplinary actions following different patterns, which are influenced by features such as the year of service where the offense was committed, the type of offense (minor, grave, major), the sector and naval station they are assigned to, the personnel category

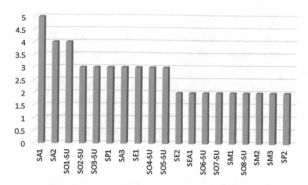

Fig. 14. Major offenses committed by officers in the top 20 naval stations.

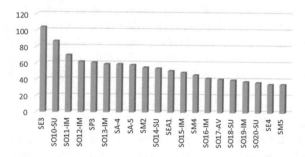

Fig. 15. Major offenses committed by enlisted in the top 20 naval stations.

or classification (ARM, SRV, TNC, ESP), the specialty for line officers and enlisted (IM, SU, SS, AV, CG), and even the offense subcategory (punctuality, conduct unbecoming, and so on).

These insights may well be useful to identify at-risk personnel with potential for misconduct. Naval authorities should encourage the implementation of leadership and motivation programs, counseling sessions, and home visits targeted to those individuals, hoping to decrease the amount of naval personnel incurring in disciplinary actions, which will turn out to be advantageous for several parties, including the naval personnel, their peers, their families and the institution.

4 Conclusions and Future Work

Although interventions in the Ecuadorean Navy exist to reduce the amount of individuals with potential for misconduct, an optimal implementation of assistive measures requires evidence-based targeting, enabling naval authorities to perform proactive interventions in a prioritized manner where limited resources are available. Data mining is the process of finding patterns in data. The proposed approach follows the basis of data mining, specifically exploratory data analysis, to collect, prepare and generate a descriptive model for the identification of at-risk naval personnel incurring in disciplinary actions. The typology of these individuals can be profiled with features such as the year of service where the offense was committed, the type of offense, the sector and naval station they are assigned to, the personnel category or classification, the specialty for line officers and enlisted, and the offense subcategory.

The results from this work are the initial steps in the development of a prediction model for the early detection of these individuals. Any prediction model must be independent for both officers and enlisted personnel, since the findings in this paper proved that the two categories follow different patterns when committing military offenses.

To strengthen this line of research, a number of important directions for future work can be highlighted. In addition to the previous insights, the model can be further improved by adding extra information such as salaries and performance grades, which may provide complementary interpretation of personnel misconduct. We can even add unstructured data, by using text mining techniques, to extract common words or phrases that superior officers normally use when evaluating their subordinates in the semester performance report. The knowledge from this descriptive model can be extrapolated to profile people involved in illegal activities, for this purpose, the dataset must have these individuals correctly identified before applying any data mining technique.

Clustering approaches can further improved the descriptive model, and even gather more insights as to what other features could potentially be included in a prediction model. A sequence classification approach in the form of Long Short Term Memory (LSTM) networks can be used to predict a class label given an input sequence. The input sequence should contain all the relevant features for each year of service until the individual retires from the Navy.

References

1. MacGregor, A.J., Tang, J.J., Dougherty, A.L., Galarneau, M.R.: Deployment-related injury and posttraumatic stress disorder in US military personnel. Injury **44**(11), 1458–1464 (2013)
2. Thandi, G., et al.: Alcohol misuse in the United Kingdom Armed Forces: a longitudinal study. Drug Alcohol Depend. **156**, 78–83 (2015)
3. Campbell, J.P.: An overview of the army selection and classification project (project A). Pers. Psychol. **43**(2), 231–239 (1990)
4. Rosellini, A.J., et al.: Using administrative data to identify US Army soldiers at high-risk of perpetrating minor violent crimes. J. Psychiatr. Res. **84**, 128–136 (2017)
5. Fine, S., Goldenberg, J., Noam, Y.: Integrity testing and the prediction of counterproductive behaviours in the military. J. Occup. Organ. Psychol. **89**(1), 198–218 (2016)
6. Ameri, S., Fard, M.J., Chinnam, R.B., Reddy, C.K.: Survival analysis based framework for early prediction of student dropouts. In: Proceedings of the 25th ACM International on Conference on Information and Knowledge Management, New York, NY, USA, pp. 903–912 (2016)
7. Iam-On, N., Boongoen, T.: Improved student dropout prediction in Thai University using ensemble of mixed-type data clusterings. Int. J. Mach. Learn. Cybern. **8**(2), 497–510 (2017)
8. Iam-On, N., Boongoen, T.: Generating descriptive model for student dropout: a review of clustering approach. Hum. Centric Comput. Inf. Sci. **7**(1), 1 (2017)

9. Raju, D., Schumacker, R.: Exploring student characteristics of retention that lead to graduation in higher education using data mining models. J. Coll. Stud. Retent. Res. Theory Pract. **16**(4), 563–591 (2015)

10. Kabakchieva, D.: Predicting student performance by using data mining methods for classification. Cybern. Inf. Technol. **13**(1), 61–72 (2013)

11. Sweeney, M., Lester, J., Rangwala, H.: Next-term student grade prediction. In: IEEE International Conference on Big Data (Big Data 2015), pp. 970–975 (2015)

Reduction of the Autoignition Risk of Propellant Based on Statistical Variables for Control in Ammunition Storage

Washington Rosero, Rolando P. Reyes Ch.[✉], and Manolo Paredes Calderón

Departamento de Seguridad y Defensa,
Centro de Investigación Científica y Tecnológica del Ejército,
Universidad de las Fuerzas Armadas ESPE, Sangolquí, Ecuador
{wnrosero1,rpreyes1,dmparedes}@espe.edu.ec

Abstract. Ammunition control systems have been in constant scientific evolution, with special emphasis on the operation and maintenance of weapons resources. However, even the studies regarding the control tests, minimization of the impact and instability of the ammunition are very limited. Determine the behavior of the ammunition when it is stored in the different ones in the different armories of the Armed Forces of Ecuador in the different regions of Ecuador. Evaluation of various functionality protocols of the powder magazines, management protocols and operation of ammunition. It was possible to better establish the probability of the state in which the facilities where the ammunition is stored based on the study variables can be found. The study allows the Armed Forces of Ecuador to take corrective and predictive measures to improve security and storage conditions immediately.

Keywords: Evaluation · Ammunition System
Ammunition storage magazines · Ammunition · Useful life
Temperature

1 Introduction

The evolution of munitions control systems has been in constant study, especially in relation to the optimization, operation and maintenance of weapons resources. In the world there is a long history of tragedies caused by ammunition storage magazines, which usually has always been caused by the instability and high probability of detonation of the ammunition due to the high amount of associated energy materials [1,2]. A representative example was in 1967 where the USS Forrestal aircraft carrier took the lives of hundreds of men and high costs in damage to aircraft carriers (approximately 700 million dollars) [2]. The events that occurred in the ammunition storage magazines are worrisome, especially in accidents that occurred due to the autoignition of the propellant. In Ecuador,

© Springer International Publishing AG, part of Springer Nature 2018
À. Rocha and T. Guarda (Eds.): MICRADS 2018, SIST 94, pp. 394–403, 2018.
https://doi.org/10.1007/978-3-319-78605-6_34

accidents have been more worrisome. In the period of twenty years have occurred at least six accidents that have occurred in different ammunition storage magazines that existed in Ecuador until before 2006 [3]. For this reason, at least in Ecuador, the monitoring of ammunition becomes essential to minimize the occurrence of accidents and improve the efficiency in the prevention to minimize the problems that occurs below the maximum temperature allowed (25 °C) [4]. In this study we intend to apply a research methodology that is based on storage criteria during the life cycle of the ammunition ranging from the acquisition of the ammunition to its end of useful life or waste. With this it will be possible to quantify and assess the permanence of the ammunition in storage magazines. We consider the historical records of the ammunition storage magazines where the accidents. Data collection (ammunition, data storage capacity, equipment lifespan, etc.) is carried out by means of a monitoring equipment. The data are the basis for its analysis and statistical evaluation of the ammunition's condition. The results are prominent, it is generated and proposes new technical alternatives that must be taken before the destruction and disposal of the ammunition at the end of its useful life, as well as a proposal of the research methodology for the ammunition monitoring in any condition in where it is stored.

The structure of the article is as follows: in Sect. 2, we mention the history of the ammunition storage magazines in Ecuador and status of literature. The illustration of the research methodology by employing specialized teams in Sect. 3. In Sect. 4 we mention the execution of our methodology where the data collection and analysis is done. Section 5 details the evaluation and results. Threats to validity is indicated in Sect. 6. Section 7 presents the discussion, conclusions and lessons learned. Finally, future work in Sect. 8.

2   Background

In the world there is a long history of tragedies caused in the ammunition storage magazines. In general, tragedies have causes related to instability and the high probability of detonation of ammunition that has high amounts of associated energetic materials [1,2]. In this regard we can say that Ecuador has not been the exception. In recent years several accidents related to the destruction of the ammunition storage magazines of its Armed Forces, due to the decomposition of the ammunition that was stored. We can mention several examples, of which we will mention the most representative: The *Balbina* in 1997 (Battalion of Engineers No. 69 "Chimborazo"), *Riobamba* in 2002 (Armored Cavalry Brigade No. 11 "Galápagos"), *Guayaquil* in 2003 (Naval Base Sur); *Sangolquí* in 2009 (Ammunition Factory "Santa Bárbara"); *Pichincha* in 2011 (Group of Intervention and Rescue of the National Police); *Latacunga* in 2016 (Weapons and ammunition warehouse of the Special Forces Brigade No. 9) [3]. Now, we have tried to find information about instability and high likelihood of detonation of ammunition due to the high amount of energy associated materials [1,2], however it has made us resort. We have found few scientific papers [1,2] studying this problem. This may be because most cases the information could be obtained from these tragedies are difficult to get to the academy, due to confidentiality. Reason for which, it is motivating to carry out a study referring to establish the statistical variables for the reduction of the risk of auto ignition of the propellant in the ammunition storage magazines. For which we propose a research methodology is detailed below.

3   Research Methodology

As mentioned above, conducting a study regarding the monitoring of ammunition, based on real historical records during the storage of ammunition in the magazines where the accidents occurred, is complex. Therefore, we propose an appropriate methodology that allows the collection of related data (e.g., ammunition, data storage capacity, equipment lifespan, among others) in order to establish statistical variables for statistical analysis and evaluation. Specifically, we talk about variables that are considered in the life cycle of the ammunition, as well as the variables that allow it to extend the usefulness of the ammunition in its storage. Then, in Fig. 1 the steps of the Research methodology for monitoring are presented.

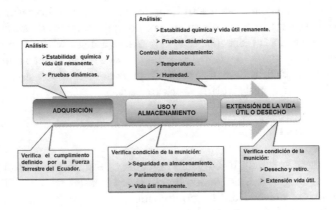

Fig. 1. Research methodology for ammunition monitoring

In this regard, our research methodology considers the extraction of data from the 3 stages of the life cycle of the ammunition: *(1) Acquisition, (2) Use and Storage* and *(3) Extension of the useful life or waste.* For the case of the *Acquisition* stage, the chemical stability and remaining life of the ammunition will be considered and dynamic tests will be carried out. For the stage of *Use and Storage* have 3 main activities are carried out: (1) analysis of the chemical stability and remaining life of the ammunition, the (2) storage control based on the temperature and humidity and the (3) verification of the condition of the ammunition with regard to storage security, performance parameters and remaining useful life. Finally, the stage of *Extension of the useful life* that is summarized in verifying the condition of the useful life referring to the parameters of: (1) waste-removal and (2) extension of the useful life. For the present study we will only consider the variables or parameters of the second stage of *Use and Storage* since in this stage it is the one that interests us for the storage of the ammunition in the ammunition storage magazines.

4   Execution

The application of our research methodology to monitor the ammunition allowed us to better understand the parameters considered regarding the treatment, storage and handling of the ammunition before destruction or disposal. We managed to have a better understanding of what happens in the second stage *Use and Storage*, obtaining a better way to obtain the degree of valuation to the found data. The details of the means used and the execution are detailed below:

4.1   Monitoring Equipment for Ammunition

The monitoring equipment for ammunition that was used during the execution, contains a temperature sensor with stainless steel metal wrap that protects it from impacts, corrosion, and humidity. An important feature of this equipment is the flexibility to measure temperatures within the range between -40 and $+125\,°C$. With this equipment it was possible to take 32,510 temperature observations, which were recorded and stored in the same equipment. Likewise, the equipment has a lithium battery of long duration, which allows the recording and storage of temperature data for up to three years, in second intervals ($1\,s$), and in a maximum of every $12\,h$. As additional information, the equipment possesses a degree of accuracy of $\pm0.2\,°C$.

To collect the temperature data, we place the monitoring equipment inside the wooden packaging where the ammunition is usually stored. In this way we obtained the measurements of the actual temperature of the ammunition inside the wooden packaging. Finally, to download the stored values, the monitoring equipment is connected directly to a USB port of a computer. Figure 2 shows the parts of the monitoring equipment in a way.

It should be noted that the monitoring equipment was adapted and built at the University of the Armed Forces ESPE.

4.2   Technical Consideration for the Storage of Ammunition

With regard to ammunition storage, special consideration was given to international regulations that are commonly used to establish the chemical stability of the propellant [5], which consists of the following parameters: simple base (SB), double base (DB), and triple base (TB), for a period of 10 years to a storage

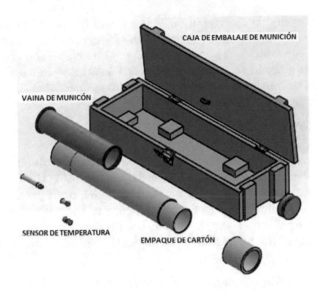

Fig. 2. Monitoring equipment for ammunition

at a temperature of 25 °C. This regulation also establishes criteria for the qualification, quality control, and extension of the ammunition's useful life for more than 10 years, as long as it is carried out with established acceptance criteria [5].

In this regard, it is also important to establish the parameters necessary to statistically quantify and assess the status and condition of the ammunition that is stored, based on a prediction with the values of temperature and storage time, according to the following Eq. (1):

$$t_{25} = t_s * e^{\frac{\left(E2*\left[\frac{1}{T_{25}} - \frac{1}{T_s}\right]\right)}{R}} \tag{1}$$

Where the calculation t_s, which refers to the safe storage time with which you can set the parameters to be evaluated. Whereas t_{25} refers to the temperature of 25 °C. It should be noted that the adequate temperature for ammunition inside the powder magazines can not be at a temperature of 25 °C since the reaction of chemical decomposition of the propellant in the ammunition can be accelerated, and as a consequence the life span of the ammunition or its safe storage. The application of the equation mentioned in the Subsect. 4.2 follows the calculation of the safe storage time, from which the data of the Table 1 is obtained.

It is important that in the application of the criteria of disposal and extension of the useful life of the ammunition, consider the analyzed data of the temperature variations to which the ammunition has been subjected while it was in storage, since such analysis will allow a better prediction of behavior [6,7].

Table 1. Predicting the safe time of ammunition in storage

Temp. storage (C)	Temp. in K grades	Secure storage time (years)
20	293.15	17.3
21	294.15	15.5
22	295.15	13.9
23	296.15	12.4
24	297.15	11.1
25	298.15	10.0
26	299.15	9.0
27	300.15	8.1
28	301.15	7.3
29	302.15	6.5
30	303.15	5.9
31	304.15	5.3
32	305.15	4.8

5   Execution and Preliminar Results

5.1   Execution

In this section we will analyze the data obtained through the monitoring that was implemented in the armament and ammunition warehouse located in the University of Armed Forces ESPE, which is located in the Province of Pichincha, in the Rumiñahui Canton. For this, a total of 16,154 data were taken belonging to a period of two years, from August 2014 to June 2016. Temperature monitoring was also carried out in the ammunition hold of the Jungle Brigade No. 19 "Napo", for which 6,080 data were taken, in a period of five months, from June to November 2016. From the samples taken, it is possible to preliminarily demonstrate the temperature variations that the ammunition undergoes in storage, just as it is described in the following subsection.

5.2   Preliminar Results

The preliminary results found are detailed with respect to the Variation of the Temperature and the Reliability of the safe storage time, as detailed below:

Temperature variation. At this point, the evaluation of the data obtained under storage conditions was carried out, according to established safety parameters. A comparison was made in the same periods from 2014 to 2016, with the temperature taken in the Province of Pichincha, Sierra region, as shown in Fig. 3a and b.

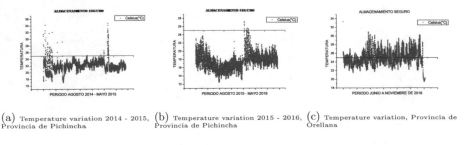

(a) Temperature variation 2014 - 2015, (b) Temperature variation 2015 - 2016, (c) Temperature variation, Provincia de
Provincia de Pichincha   Provincia de Pichincha   Orellana

Fig. 3. Temperature variation, provinces of pichincha and Orellana

As can be seen in Fig. 3a, the variation in temperature is greater, in relation to the Fig. 3b, between the same period from August to May, which has
allowed to establish that the existence of a greater probability of degradation
of the propellant exposed to temperature variations over 25 °C. It is important
to emphasize that the greater temperature variation present in an ammunition
storage warehouse, there is a higher probability of occurrence of an accident,
therefore, it also decreases safety, and increases the risk in its handling. As part
of the study carried out, the following figure is presented in Fig. 3c, where the
variation of temperature registered in the ammunition hold of the Brigade de
Selva No. 19 Napo, located in the Province of Orellana, Ecuador.

Reliability of Secure Storage Time. Under the assumption of a normal distribution, an estimate can be obtained by interval, where it is common to refer
to a confidence interval. The interval $\vartheta_L < \vartheta < \vartheta_U$, which is calculated from
the selected sample, is then called the confidence interval of $(1 - \alpha) * 100\%$, the
fraction of $1-\alpha$ is called the confidence coefficient or degree of confidence, and
the ends are called lower and upper confidence limits [6–8]. With the parameters
presented, we proceed to quantify and establish confidence criteria in the storage
of the propellant, and thereby make a projection of the maximum time that can
remain without presenting a risk as probability of occurrence of an event, and
therefore the consequences to which an accident entails, ranging from the material to the loss of human lives. As a starting point, with a normality assumption
based on measures of central tendency and measures of dispersion, the following
analyzes are presented in Fig. 4a, b and c.

Once the analysis is done, the storage reliability criteria can be determined,
key information for a continuous propellant control process, along with improving the rotation and permanence of the ammunition in a ammunition storage
magazines. What is intended to establish, is the maximum temperature at which
a powder magazine is exposed, and with it through the estimation by intervals,
it can be indicated that in propellant storage there is 95% confidence that the
interval set contains the average temperature. Thus, with the values obtained,
Table 2.

Consider that the propellant subjected to higher temperatures tends to accelerate the decomposition reaction, with the consequence of decreasing its useful
life period, as well as its safe storage. On the contrary, if said storage preserves

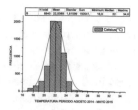 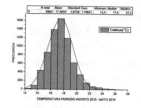 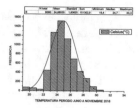

(a) Safe storage calculation, Provincia de Pichincha, Test 1 (b) Safe storage calculation, Provincia de Pichincha, Test 2 (c) Safe storage calculation, Provincia de Orellana

Fig. 4. Safe storage calculation, Province of Pichincha and Orellana, Test 1 and 2

Table 2. Prediction of safe storage at 95% y 98% confidence

	Maximum temperature	Safe storage	Maximum temperature	Safe storage
Fig. 4a	25	10	26	9
Fig. 4b	21	15	22	14
Fig. 4c	28	7	29	6

temperatures lower than 25 °C, it is understood that the decomposition rate is slower, which increases its useful life and with this its safe storage increases, thus improving the safety conditions and reducing the probability of occurrence of an accident. The Table 2 show the differences of a secure storage, in which the Fig. 4a, presents a maximum temperature reached of 25 °C, to 95% of confidence, with what his safe time of storage would be ten average years, and 98% confidence would be nine years, before being subjected to an analysis of its operation, such as static, chemical, and dynamic tests. It is also observed that Fig. 4b has a maximum temperature of 25 °C, with fifteen years of average safe storage, 95% confidence would be fourteen years, it should be noted that by regulation [2], the propellant must be subjected to ten years of technical inspection, as mentioned above. With regard to Fig. 4c of Table 2, it can be seen that the safe storage time is very low, and this is due to the direct relationship with the high temperatures recorded, which increases its probability of risk and it decreases the useful life time. It should be noted that this analysis considers the regulation [2], in isothermal conditions during a period of storage time, but what is observed according to the set of data obtained, is that the propellant is subjected to constant temperature variations that they can increase the useful life of the propellant and its stability, provided a temperature below 25 °C is maintained. The difference between values in Table 2 is a function of quantifying, through interval estimation, a safe storage with a certain degree of confidence, where the greatest amount of data obtained is within the parameters of the temperature that it is intended to analyze in order to ensure a storage without probability of risk, that is to say that security can be related to the confidence obtained from the data such as 95%, 98%, etc.

The intention to establish in this analysis is the estimation of the period and safety parameters of propellant is stored, giving rise to an alternative efficient solution in decision-making, avoiding unnecessary destruction of the material without spending large resources by the probability of occurrence of an accident.

With this application, a certain stock control is obtained that allows an effective administration and that helps in the application of consumption policies of the old ammunition, thus improving the rotation, as well as, widely comply with the guarantees given by the manufacturer in case to be new products; and, allowing to have a continuous monitoring of its history, handling, storage and transport.

6   Threat to Validity

It is important to note that in the equation mentioned in the Subsect. 4.2 sufficient care was taken that the propellant not be subject to greater temperature variations, since this could affect the stability of the same, endangering the security of storage, therefore what is estimated is the period or a security window of the propellant, where there will be evidence of reactions that put security at risk during its storage period, so that there will be an efficient tool for the administration and application of policies of rotation and consumption of the ammunition that is closer to finalizing its life cycle.

It can also be observed that the implementation of a monitoring complies with the established requirements, which are the obtaining of data and with it the determination of possible safe storage time of the propellant.

7   Discusión, Conclusions and Lessons Learned

If the risk is associated with the probability of occurrence of an accident, it is an obligation, the management that is executed with the purpose of reducing it and mitigating the collateral damages that said event may generate. It is essential to emphasize that the calculation of safe storage of ammunition is applied to comply with the life cycle process, as well as to follow safety procedures based on technical specifications. This implementation seeks, is to generate a warning signal in case of extreme temperature variations that can lead to the chemical decomposition of the propellant, affecting its condition and condition and increasing the risk in its handling and storage. Another of the great aspects pursued by this assessment, is to improve the levels of quality in order to establish percentages of reliability, which can be obtained from storage, in order to take actions to optimize the rotation of the ammunition, and improve the making decisions, avoiding unnecessary expenses for destruction due to the fact that it can present chemical and ballistic instability. It should also be taken into consideration that by improving the controls in the life cycle processes, a considerable reduction of expenses can be conceived, which substantially increases the security and efficiency in the control. This assessment could eventually make it more likely to know the state in which the facilities where the ammunition is stored can be found, so that corrective measures are taken to improve security and storage conditions immediately. It is prudent to indicate that such implementation is feasible through a centralized control of data, which allows continuous observation of its process, generating an effective consumption. The implemented system has shown that it complies with the requirements established in the regulations [2], which has proven to be an efficient indicator of storage quality, providing information that had not been previously assessed by the institution. On the other hand, the test that is executed as a complement to the process, is by the technique of micro calorimetry, which is able to predict the stability of the propellant during the storage period. Based on the regulations [2], it is possible to determine the period of time in which the propellant will remain, provided that it meets the ideal temperature conditions, without presenting decomposition reactions, avoiding causing a self-inflammation.

8   Future Works

We are currently starting the implementation of the study variables using Artificial Intelligence techniques, specifically with the prediction based on patterns. Our thesis students work together with engineers and researchers. So we have started the predictive study in the first instance with the realization of Neural Networks and later we will apply Machine Learning. We hope that our methodology based on these techniques allows a better prediction to reduce the risk of autoignition of the propellant.

Acknowledgment. Universidad de las Fuerzas Armadas ESPE, Centro de Investigación Científica y Tecnológica del Ejército y Armada del Ecuador.

References

1. Karger, B., Wissmann, F., Gerlach, D., Brinkmann, B.: Firearm fatalities and injuries from hunting accidents in germany. Int. J. Legal Med. **108**(5), 252–255 (1996)
2. Koch, E.-C.: Insensitive munitions. Propellants Explosiv. Pyrotech. **41**(3), 407 (2016)
3. Araque Salazar, M.R.: "Estándares de seguridad para la gestión de municiones en las fuerzas armadas del ecuador," Master's thesis, SANGOLQUI/ESPE/2012 (2012)
4. Eason, G., Noble, B., Sneddon, I.: On certain integrals of lipschitz-hankel type involving products of bessel functions. Philos. Trans. R. Soc. Lond. A Math. Phys. Eng. Sci. **247**(935), 529–551 (1955)
5. Avery, D., Watson, R.T.: Regulation of lead-based ammunition around the world. In: Ingestions of lead from spent ammunition: implications for wildlife and humans, pp. 161–168 (2009)
6. Jacobs, I.: Fine particles, thin films and exchange anisotropy. Magnetism **3**, 271–350 (1963)
7. Elissa, K.: Title of paper if known (2003)
8. Nicole, R.: Title of paper with only first word capitalized. J. Name Stand. Abbrev., 740–741 (1987)

Security in the Storage of Ammunitions and Explosives in Ecuador

Miguel Araque[1(✉)], Óscar Barrionuevo[2], and Teresa Guarda[3,4,5]

[1] Ejércitos Ecuatoriano, Quito, Ecuador
miguemil92@gmail.com
[2] Armada del Ecuador, Guayquil, Ecuador
oscarbarrionuevovaca@gmail.com
[3] Universidad de las Fuerzas Armadas-ESPE, Sangolqui, Quito, Ecuador
tguarda@gmail.com
[4] Universidad Estatal Península de Santa Elena – UPSE, La Libertad, Ecuador
[5] Algoritmi Centre, Minho University, Guimarães, Portugal

Abstract. Ammunition and explosives, due to their intrinsic nature, generate a potential risk of damage to the surrounding environment, and are also highly valued by illegal armed groups and criminals. In order to prevent a new accidental explosion from affecting the civilian population and at the same time preserve state assets in the best possible way, the Joint Staff Command of the Armed Forces is executing an investment project, which is being progressively complemented with technical standards of integral security (physical security, safety and occupational health); which will allow an effective and safe management of the ammunition and explosives of the three branches of the Armed Forces. After the explosion occurred in the north of Quito at the end of 2011, the need arose that this technical scientific knowledge of protection transcends the military institution and be disseminated to other public and private institutions that manufacture, handle and store ammunition and explosives, in order to protect the health and integrity of those who effectively manipulate them, protect third parties from possible harm and prevent their illegal removal.

Keywords: Storage of ammunitions · Explosives · Physical security

1 Introduction

The accidental explosion occurred on December 8, 2011, in a powder keg located in the headquarters of the Intervention and Rescue Group of the National Police north of Quito, which resulted in more than one hundred wounded, once again showed the need to carry out the relocation of storage facilities for ammunition and explosives in sites located outside the urban perimeters of cities and towns (see Fig. 1).

World statistics indicate that on average approximately every month and a half an accidental explosion occurs in a storage area for ammunition and explosives somewhere on the globe. In South America, Ecuador is the country where the greatest number of events, have happened in the last 25 years, while in the World it is Russia.

© Springer International Publishing AG, part of Springer Nature 2018
Á. Rocha and T. Guarda (Eds.): MICRADS 2018, SIST 94, pp. 404–412, 2018.
https://doi.org/10.1007/978-3-319-78605-6_35

Fig. 1. Photograph of the material damage caused in one of the urbanizations adjacent to the GIR headquarters (Source: © Diario El Comercio [6])

The analysis of world events has determined that fires and mishandling of ammunition are the main known causes of explosions.

This new accidental explosion adds to those that occurred in Ecuador in previous years. La Balbina in 1997 (Battalion of Engineers No. 69 "Chimborazo", Riobamba in 2002 (Armored Cavalry Brigade No. 11 "Galápagos", Guayaquil in 2003 (Naval Base South) and Sangolquí in 2009 (Ammunition Factory "Santa Bárbara") were the places where these tragic events arose which had negative effects on human, economic and material losses.

In April 2003 the text of the Law reforming the law of manufacture, importation, exportation, marketing and tenure of arms, municiones, explosivos and accesorios was published in the Official Registry, after the National Congress discussed and approved it. In article 24-A of the aforementioned Law, it is mainly provided that the production and storage of weapons of war, gunpowder, bombs, explosives and the like should be carried out in premises defined by the Joint Command of the Armed Forces in coordination with the Municipality and the Fire Department of the jurisdiction and authorized by the Ministry of National Defense; It also requires that they should not be located in populated centers or in community property or ancestral possession of indigenous peoples. Finally, it indicates that specialized personnel of the Armed Forces or of the authorized company will remain in said premises to perform the care and maintenance thereof and under strict security measures.

The Joint Command of the Ecuadorian Armed Forces since September 2008, is executing the project "Protection and Security of the Civilian Population in the Storage and Management of Ammunition and Explosives of Armed Forces" with the objective of implementing a protection system that allows reducing the probability of occurrence of a new accidental explosion during the activities of management of ammunition and explosives in the Armed Forces, as well as eliminating and/or minimizing the risk of affecting the civilian and military population and their assets in case of idea; through the design, construction, equipment and operation of Joint Ammunition Deposits, towards which the munitions that are currently distributed in the different units of the country will be relocated.

As part of the proposed goals of this investment project, until the present date, two Ammunition Joint Deposits have entered into operation, which has allowed the relocation of ammunition that was found in the military units located in the cantons of: Latacunga, Manta, Guayaquil, Mera and Mejía (partially), protecting 404,479 inhabitants who live or work in the vicinity of military units.

For the design and construction of the ammunition tanks and the powder magazines that conform them, international safety standards have been met, especially those set by the Department of Defense of the United States of America, which provide a certain degree of protection.

After the relocation of the ammunition, it is essential to minimize the possibility of an accidental explosion during transport, storage, handling, maintenance or final disposal, for this reason, they have also been developed and are gradually applying demanding standards of security.

So if Armed Forces have learned with the pain of the tragedies that have occurred, it is time for the work done to transcend other public and private institutions which manipulate according to their specificity ammunition and explosives.

2   Materials and Methods

Inspections carried out in the magazines of the units of the three branches of the Armed Forces and the analysis of accidental explosions in Ecuador, including the last one, have established the following vulnerabilities:

– There are storage rooms where ammunition and explosives are stored located in places near populated areas, communication routes and administrative facilities of military and police units.
– 90% of the structures of the ammunition storage warehouses and warehouses do not provide protection, nor physical and industrial security because:
  – They were not designed or built under technical safety standards to contain the effects of an accidental explosion.
  – They do not have a system of atmospheric electrical discharges (lightning rods), grounding of the structure, or personal and vehicular discharge.
– When there are two or more ammunition storages, they are not located at a safe distance calculated based on the amount of ammunition and/or explosives stored.
– The amount of facilities used to store ammunition is less than the established requirement, so:
  – In most cases, distribution by mixed storage compatibility groups is hindered or prevented.
  – The maximum stacking height is exceeded.
  – The volume of ventilation and the circulation area inside it are reduced.
– Storage of ammunition and explosives that have exceeded their useful life, or whose condition is unknown.

Lack of specialized equipment for the handling, transport and storage of ammunition and/or explosives, as well as for fighting fires (vehicles, machinery and equipment); which does not allow the tasks to be carried out completely safely.

Physical security is privileged over industrial safety.

During the storage and handling of ammunition and explosives, the knowledge and experience of tactical employment on safety standards and best practices established in the technical-scientific field prevail.

The risks that threaten ammunition storage areas and that have been identified in Table 1 based on the study of accidental explosions in the world during the last 25 years.

Table 1. Risks identified according to their source (Source: [3]).

Pure Risks		
Personal, liability and property risks		
Risks of nature	Technological risks	Antisocial risks
Lightning strike High temperature Fire (unprovoked forest fire)	Bad handling of ammunition Fire/fire Electrical fault Auto-ignition of components (spontaneous combustion) Demilitarization and final disposition of ammunition	Lack of physical security and sabotage

The project "Protection and Security of the Civilian Population in the Storage and Management of Ammunition and Explosives of Armed Forces" has the purpose of resolving the detected vulnerabilities, through the materialization of each one of its components that are:

- Possess the studies and designs of the Military Port and the Joint Munitions Depots;
- Construct engineering works for the storage and safe handling of ammunition and explosives;
- Acquire the equipment for handling, transport and safe storage of the ammunition;
- Relocate the ammunition, operate the Ammunition Joint Deposits and manage the project.

Transversal to the execution of the project, comprehensive security standards are being implemented, meaning that they cover both physical security, as well as occupational safety and health, at the international level, which guarantee the fulfillment of its objective and that place the Armed Forces Ecuadorian to the level of those of the United States of America and Europe in the safe management of ammunition.

Based on the aforementioned, the intention proposed by the Joint Command of the Armed Forces and exposed in the different meetings held in December 2011 in the Security Secretariat of the Municipality of Quito, is to transmit this knowledge and experiences to other institutions of our country, to eradicate the possibility that the disastrous effects of the accidental explosions of ammunition and stored explosives will once again affect the civilian population and its assets.

3 Evaluation of Results

When an event occurs, such as the last one in the GIR ammunition storage, the need for the relocation of installations containing munitions and explosives outside of urban areas is immediately seen, however, the simple fact of taking them to a totally remote location does not it diminishes the possibility of an accidental explosion in its new location, only transfers the effects outside the inhabited sites.

If an explosion occurred in a fuel supply facility, which would cause human and material damage, would you think about relocating all the gas stations outside the city? The answer is negative, because they are necessary within the city, but their location must correspond to a technical study and a risk analysis with the respective actions to minimize them. Before its operation, municipal authorities and firefighters must inspect them to determine the feasibility of their operation.

Technically, facilities that store ammunition and/or explosives could be located near populated areas. Normatively and prior to its operation, its location must be known and approved by the Arms Control Department of the Logistics Directorate of the Joint Command of the Armed Forces, in compliance with Art. 24-A of the Manufacturing law, importation, exportation, commercialization and tenure of arms, munitions, explosives and accessories; In addition, it is necessary that they have the facilities and equipment that provide the necessary conditions for their safe operation, which must be verified by the Fire Department and the Arms Control Department. Finally, its location will be reported to the Municipality of the Canton, to be taken into account in the planning of land use and land use of the surrounding areas.

To carry out the storage of ammunition and explosives, both the current legislation and the ordinances are clear and must be fully complied with, however, it is necessary to take into account a technical variable, which defines the degree of protection provided by an installation and the effects expected in case of an accidental explosion, this is the safety distance.

D is the safety distance in meters, K is the factor dependent on the assumed or allowed risk and Q is the quantity of explosives expressed in net weight.

$$D = K * Q1/3 \tag{1}$$

According to the K factor, different safety distances are established, thus we have the following from lower to higher:

- Distance between storages (storage distance): It is the minimum distance allowed between two powder stores that store ammunition and/or explosives. This distance is intended to prevent a simultaneous detonation, therefore, should not be omitted.
- Distance to ammunition handling facilities (interline distance): It is the distance that must be maintained between a powder keg in service and a building where munitions maintenance and surveillance tasks are performed.
- Distance to public traffic routes (public traffic route distance): It is the minimum distance allowed between a public traffic lane and an installation that contains ammunition and explosives. Where this can be a road, a railway line or a navigable river.

- Distance to inhabited buildings (inhabited building distance): It is the minimum distance allowed between an inhabited building and an installation that contains ammunition and explosives. This distance tries to avoid serious structural damage by the explosive wave, the flame and the projections.

Human safety distance is the distance at which people can be exposed, without being affected by the effects of an explosion.

The K factor also varies according to the degree of protection provided by the structure of the installation that contains explosives in front of an accidental event, therefore, it will also change the safety distance; that is why when ground-covered powder magazines are available, the safety distances are reduced, unlike when there are surface magazines. The construction of protection berms also decreases the K factor and therefore the safety distances.

Another of the factors present in the equation to determine the different safety distances is Q, which corresponds to the total amount of explosives that an installation contains or can contain at any given time. The unit used for its application in the formula will be the kilogram. This factor is essential for the location and technical design of ammunition storage facilities and/or explosives, using it as a constant that sets the maximum amount that can be stored. The net weight of military and industrial explosives will be equal to the equivalent value in TNT. In the case of ammunition, it will be the total amount of explosive substances or high explosives contained in them.

To establish the technical location of the Ammunition Joint Deposits and determine the corresponding safety distances, a design quantity of 45,359 kilograms of net weight of explosive was established, considering the most critical situation, that is, all the ammunition stored belong to the division considered to be the most dangerous, that is, the 1.1.

The United Nations (UN) to promote the safe storage and transport of hazardous materials conceived an "International Classification System" according to its danger. Within this classification, all conventional ammunition and explosives are part of Class 1.

Class 1 "Explosives" in turn is divided into six divisions that indicate, in order, the primary type of danger that is expected upon the occurrence of an accidental event (see Table 2).

Another element considered very important to guarantee the safety and prevention of accidents in an installation that contains ammunition and explosives, is that during the operation the storage is carried out according to the "mixed storage compatibility". Ammunition and explosives are considered compatible when they can be stored or transported together, without this implying a significant increase in the probability of an accident or the magnitude of the effect of such an accident. In the case of ammunition there are 13 storage compatibility groups, based on the similarity of their characteristics, properties and potential effects in the event of an accident.

On the other hand, to reduce the possibility of an accidental event in the ammunition storage facilities (powder magazines), it is necessary that these at least have the following characteristics:

- The entire structure must be properly discharged to ground (grounded), including its accessory elements such as doors and windows. This is in order to reduce the possibility of driving static electricity.

Table 2. Division for the dangerousness of class 1 "explosives" (Source: [3]).

Division	Definition
1.1	Substances and articles that have a danger of massive detonation
1.2	Substances and articles that have a danger of production and projection of fragments, but it is not expected that there will be a massive detonation
1.3	Substances and articles that are in danger of massive fire and less danger of shock wave or less danger of projection, or both; but it has no danger of massive detonation
1.4	Substances and articles that do not present a significant danger. Moderate fire without explosion
1.5	Very insensitive detonating substances with very low probability of initiation, which are in danger of massive detonation
1.6	Extremely insensitive items that have no danger of massive explosion

- The floor must be smooth, without cracks, so as to prevent the filtering of possible exudations of the explosives.
- There must be a copper plate or handle connected to the ground, which allows the discharge of static electricity from personnel entering the powder keg.
- It must have elements for the ventilation of the ammunition storage that facilitate the renewal of the air in the interior and the escape of the gases produced by the degradation of the energetic materials of the explosives and ammunition. For underground and ground-covered powder stores, air intake louvers and chimneys with extractors will be necessary for their exit.
- The lighting must have an anti-explosion specification (explosion proof), when due to costs it is not feasible installation should opt for exterior reflectors or natural lighting.
- For the shielding against atmospheric discharges, towers should be mounted outside with lightning rods, considering that their coverage should cover the entire structure.

Finally, to ensure the safety and occupational health of the operators of the magazines, it will be significant to have the following equipment:

- Forklifts for the interior of the powder magazines with the specification EE (electrical enclosed) in the case that they have an electric motor and DS (diesel safety) in the case of the combustion engine.
- CO_2 and PQS fire extinguishers, for the exterior and interior of each ammunition storage.
- Personal protective clothing such as:
  - NOMEX® III A fabric coveralls, especially for the handling of incendiary, lighting, smoke and pyrotechnic ammunition;
  - 65% polyester and 25% cotton fabric overcoat with flame retardant treatment;
  - Half-round boots with steel toe;
  - Safety gloves;
  - Safety glasses resistant to the impact of fragments;

- Half mask air purifying respirators and two disposable filter cartridges or masks with activated carbon filter;
- Head protection, by lightweight cap-type helmet;
- Mesh type vest with reflective tape;
- Lumbar support girdle.

4 Conclusions and Future Work

As part of the planning and design of ammunition storage areas and/or explosives, the technical location is essential, constituting the first step to guarantee the required protection to third parties. In order to carry it out, it is necessary to establish as a fundamental data, a quantity of design that allows determining the corresponding safety distances.

The material and economic human damages caused by the accidental explosions that occurred in Ecuador were mainly due to the fact that the ammunition storage areas and explosives where they occurred were not technically located according to the safety distances.

Undoubtedly, the accidental explosions that occurred could have been avoided, if all the safety standards required in the management of ammunition and explosives had been fully observed.

Therefore, the relocation of the ammunition to the new Joint Deposits of Ammunition will eliminate the risk of affectation to the inhabitants who live in the different cantons, where the military units are located; while the implementation of strict safety standards during its management minimizes the possibility of occurrence of a new accidental explosion. This initiative that is underway is the one that should radiate to other institutions.

It is important to point out that the standards that are being applied in the management of ammunition in the Joint Deposits are higher than those established in the technical standard INEN 2 216: 99 "Explosives". Use, storage, handling and transport, since munitions are items that have been specifically designed to create the greatest possible damage to their targets, while explosives generate mechanical effects that vary according to their type and use.

As future work, which will have a national impact will be to collaborate in the drafting of a new technical standard to replace the INEN 2 216: 99.

References

1. Araque, M., Navas, M.: Normativa para la Ubicación, Diseño y Construcción de Polvorines Militares. Proyecto de Grado previo a la obtención del título de Ingeniero Civil. Quito: Escuela Politécnica del Ejército (2004)
2. Araque, M.: Proyecto Protección y Seguridad de la Población Civil en el Almacenamiento y Manejo de Municiones y Explosivos de FF.AA. Documento SENPLADES. Quito: Comando Conjunto de las FF.AA (2010)

3. Araque, M.: Estándares de Seguridad para la Gestión de Municiones en las Fuerzas Armadas del Ecuador. Proyecto de Grado de la Maestría en Gerencia de Seguridad y Riesgos. Quito: Escuela Politécnica del Ejército (2011)
4. Comando Conjunto de las Fuerzas Armadas. Manual Técnico para el Transporte de Municiones y Explosivos. Quito, Ecuador (2011)
5. U.S. DoD (United States Department of Defense). DoD Ammunition and Explosives Safety Standars. Washington D.C., July 1998
6. Diario El Comercio. Viernes 9 de diciembre 2011. Explosión en el GIR, Fotogalerías. url: http://elcomercio.com/seguridad/Explosion-GIR-casas-aledanas-cercanas_5_605389463.html . consultado el 01 de enero del 2012

Author Index

Printed in the United States
By Bookmasters